Advances in Industrial Control

More information about this series at http://www.springer.com/series/1412

Ranjan Ganguli · Dipali Thakkar
Sathyamangalam Ramanarayanan Viswamurthy

Smart Helicopter Rotors

Optimization and Piezoelectric Vibration Control

Springer

Ranjan Ganguli
Department of Aerospace Engineering
Indian Institute of Science
Bangalore
India

Sathyamangalam Ramanarayanan Viswamurthy
Advanced Composites Division
CSIR-National Aerospace Laboratories
Bangalore
India

Dipali Thakkar
Department of Aeronautical Engineering
Sardar Vallabhbhai Patel Institute of Technology
Vasad
India

ISSN 1430-9491 ISSN 2193-1577 (electronic)
Advances in Industrial Control
ISBN 978-3-319-79675-8 ISBN 978-3-319-24768-7 (eBook)
DOI 10.1007/978-3-319-24768-7

Springer Cham Heidelberg New York Dordrecht London

Softcover re-print of the Hardcover 1st edition 2016

Printed on acid-free paper

Springer International Publishing AG Switzerland is part of Springer Science+Business Media (www.springer.com)

Series Editors' Foreword

The series *Advances in Industrial Control* aims to report and encourage technology transfer in control engineering. The rapid development of control technology has an impact on all areas of the control discipline. New theory, new controllers, actuators, sensors, new industrial processes, computer methods, new applications, new philosophies…, new challenges. Much of this development work resides in industrial reports, feasibility study papers, and the reports of advanced collaborative projects. The series offers an opportunity for researchers to present an extended exposition of such new work in all aspects of industrial control for wider and rapid dissemination.

Occasionally, the *Advances in Industrial Control* monograph series has the opportunity to publish a highly focused technical monograph on an emerging technology for a specialist domain of application. This is the case with this monograph *Smart Helicopter Rotors: Optimization and Piezoelectric Vibration Control* by R. Ganguli, D. Thakkar, and S.R. Viswamurthy. One of the many fascinations of such a monograph is to see how researchers develop and evolve a new applications paradigm. A considerable strength of this particular monograph is that the chapters have a strong narrative thread. The chapters begin from basic configurations and as the chapters proceed, build more and more realistic constructional and operational features and constraints. Essentially the chapters are studies of the rotor vibration reduction potential for different rotor and actuator configurations. The results use application and laboratory test data in determining parameters to ensure the reasonable applications validity of the results presented. These are the initial studies of possible new routes to control the vibrations that occur in helicopter rotors with the objectives of increased safety, stability, and ride comfort. The technology uses smart piezoelectric materials to give control of "better" rotor shape.

The book has a classical monograph format, with Chap. 1 presenting a detailed technical literature review of the nine topics that comprise the constituent subjects for the subsequent chapters of the monograph. This introductory chapter concludes with a short section describing the monograph's organization. Chapter 2 is about modeling that describes the context and definitions of many of the terms used in this helicopter application. The remaining chapters of the monograph is then divided into two sets. Chapters 3–6 treat rotor vibration control using active flap actuators and Chaps. 7–11 treat rotor vibration control using rotor twisting delivered by piezoelectric actuation. References are collected together in one section at the end of the monograph along with four technical appendices. The chapters of the monograph are very strongly focused on these two particular helicopter rotor developments and the chapter sequence follows an agenda of increasing applications realism. This has pedagogical value for understanding how to research and develop novel and new ideas for specific control applications. Helicopter rotors may seem to be a narrow domain but it is an important one, and the research reported here may have implications for other rotor applications, for example, the blades of wind turbine rotors.

The use of smart materials and piezoelectric materials in control applications has received some exposure in the *Advances in Industrial Control* monograph series and readers might find the following two monographs to provide insight into related fields:

- *Design, Modelling and Control of Nanopositioning Systems* by Andrew J. Fleming and Kam K. Leang (ISBN 978-3-319-06616-5, 2014); and
- *Advanced Control of Piezoelectric Micro-/Nano-Positioning Systems* by Qingsong Xu and Kok Kiong Tan (ISBN 978-3-319-21622-5, 2016).

Industrial Control Centre, Glasgow, Scotland, UK

Michael J. Grimble
Michael A. Johnson

Preface

Helicopters are susceptible to high vibration levels. The development of smart-material based piezoelectric actuators has made it possible to address this high vibration problem directly by developing a smart helicopter rotor. The basic idea of the smart rotor concept is to generate new unsteady aerodynamic forces and moments on the rotor blade which cancel the existing forces and moments which are the key sources of helicopter vibration. By appropriately actuating the piezoelectric materials using an electric field, the motion of the rotor blade can be actively controlled. The smart rotor development is multidisciplinary and requires knowledge of structural dynamics, aeroelasticity, helicopter dynamics, control theory, and piezoelectric materials. In this book, the concepts of active trailing edge flap and active twist rotor blade are investigated for the helicopter vibration reduction problem. These are the two most promising concepts for the development of smart rotor. The active trailing edge flaps are placed near the tip of the rotor blades and actuated at higher harmonics of the main rotor speed. In the active twist concept, the full blade is twisted at higher harmonics of the main rotor speed.

Fundamental issues in the trailing edge flap problem include (*a*) the optimal number of flaps, (*b*) placement of the flaps along the rotor blade, and (*c*) optimal controller design for helicopter vibration and flap deflection objective. Typically, the active flap rotor is actuated by a piezostack actuator. Such an actuator can lead to hysteresis which can cause poor performance of the controller. In this book, we present a novel hysteresis compensation algorithm to alleviate this problem. Another problem in the trailing edge flap concept is the ability to use multiple trailing edge flaps effectively. This book presents a controller algorithm to maximize the potential of multiple trailing edge flaps and a response surface-based optimization method to place such trailing edge flaps for best vibration reduction performance at least flap power. An optimization study, which shows that dual trailing edge flaps are best for vibration reduction, is also discussed. Multi-objective optimization is used and the pareto front for the flap design problem is studied.

The book also showcases the concept of using piezoceramic-induced shear actuation for the active twist rotor concept. Active twist is investigated for a rotating beam, a box-beam blade and an airfoil cross-section blade. Single-crystal piezoceramics are considered. It is also shown that nonlinearity of piezoceramic shear coefficient with respect to the applied electric field can be used to extract more actuation out of the material. A velocity feedback controller is implemented and found to be suitable for vibration control using active twist. Finally, it is shown that dynamic stall-induced vibration can be actively controlled using active twist rotor. As smart material concepts develop, the active twist concept becomes useful for reducing helicopter vibration and suppressing dynamic stall.

This book will help researchers who are engaged in the development of active vibration control methods for helicopter rotors. It is also useful for researchers and engineers in the fields of smart structures, aerospace and mechanical engineering, control theory, applied mathematics, material science, and optimization. Most of the concepts are useful in all applications of rotating systems such as wind turbines, turbomachinery, and propellers. The authors are grateful to the Rotary Wing Research and Development Centre, Hindustan Aeronautics Limited, for supporting much of this work through a sponsored research project.

Bangalore Ranjan Ganguli
Baroda Dipali Thakkar
Bangalore Sathyamangalam Ramanarayanan Viswamurthy

Contents

About the Authors

The authors are an experienced and authoritative team in this field.

Professor Ranjan Ganguli is with the Department of Aerospace Engineering at the Indian Institute of Science, Bangalore, India. He was made Professor in his Department in 2009 and has had a long involvement with helicopter research since at least 1991.

Dr. Dipali Thakkar is Professor and Departmental Head in the Department of Aeronautics at SVIT, Vasad, Gujarat, India. Her background encompasses industrial aerospace activities and material science.

Dr. Sathyamangalam Ramanarayanan Viswamurthy is a Senior Scientist at the National Aerospace Laboratory, Bangalore, India and his technical research and interests are focused on helicopter science, materials, and related issues.

Chapter 1
Introduction

Rotorcraft are widely used in civilian, military, and commercial applications due to their ability to perform remarkable maneuvers such as hover, vertical takeoff/landing, and low-speed flight. Such unique abilities set them apart from their fixed-wing counterparts. However, these advantages come at the cost of complex aeroelastic phenomena, instabilities, and severe vibration and noise issues. High noise levels greatly restrict the use of helicopters over densely populated and urban areas. High vibration levels directly lead to passenger discomfort and high maintenance costs. This chapter provides the necessary background for the smart rotor concepts. The chapter is organized as follows. Helicopter vibration problem is explained in Sect. 1.1. Section 1.2 explains the passive control of vibration. Section 1.3 explains the active control of vibration. Section 1.4 presents actuation of trailing-edge flap. Section 1.5 explains the induced shear strain actuation. Section 1.6 presents the optimal placement of actuators in active structures. Section 1.7 explains the actuator hysteresis model. Section 1.8 explains the active twist rotor. Suppression of dynamic stall-induced vibration is explained in Sect. 1.9. Section 1.10 gives the organization of the book.

1.1 Helicopter Vibration Problem

Helicopters suffer from severe vibration levels and fatigue loads compared to fixed-wing aircraft. The various sources of vibration in a helicopter are: the main rotor, the engine, tail rotor, and transmission. The primary source of vibration is the main rotor due to high degree of aeroelastic interaction between the highly flexible rotor blades and the inherently unsteady aerodynamic environment [1]. The 1/rev cyclic pitch used to control the helicopter also contributes toward harmonic airload excitations. Other sources of vibration are the asymmetry of rotor disk aerodynamics, blade–vortex interaction, high advancing blade Mach numbers, retreating blade-stall and

R. Ganguli et al., *Smart Helicopter Rotors*, Advances in Industrial Control,
DOI 10.1007/978-3-319-24768-7_1

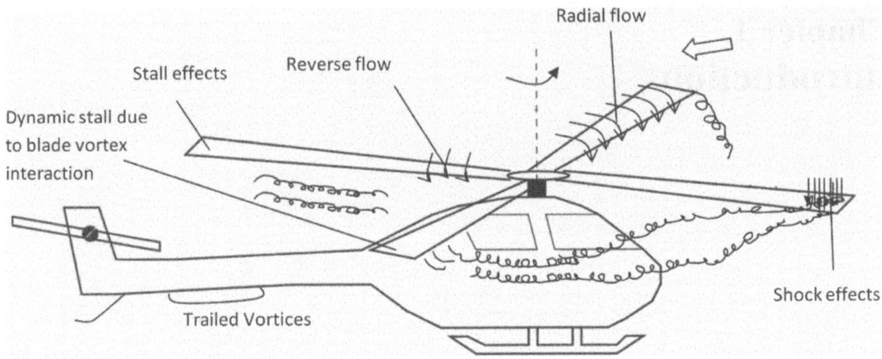

Fig. 1.1 Aerodynamic effects on a helicopter rotor

out-of-track blades (Fig. 1.1). Cabin acceleration levels in most helicopters are typically of the order of 0.15–0.2 g. However, it is desirable to reduce the vibratory loads in the cabin to less than 0.05 g (in all components) to achieve a smooth ride [2]. Severe vibration levels in helicopters lead to the fatigue damage of structural components, human discomfort, and reduced effectiveness of weapons systems. Consequently, helicopters require frequent maintenance checks and overhauls, thereby leading to high maintenance costs. Therefore, considerable effort has been devoted to study the problem of helicopter vibration prediction and reduction. The use of smart materials for helicopter vibration reduction is the main feature of this book. Many of the concepts discussed in this book are also applicable to other rotating systems, such as wind turbines, propellers, gas turbines, and steam turbines.

1.2 Passive Control of Vibration

Traditional methods to alleviate the vibration problem in helicopters are based on passive approaches [3, 4]. A direct way to reduce vibration is through structure detuning. This can be achieved in two ways: *frequency placement* method or by *modal orthogonality placement* method. In the *frequency placement* method, structural vibration is reduced by keeping the natural frequency of the structure away from the excitation frequencies (typically higher harmonics of blade rotation frequency). *Modal orthogonality placement* method (or simply *modal shaping*) involves the maximization of orthogonality of the structural mode shapes to the spatial variability of the applied loadings. A disadvantage of the modal shaping approach is that it can achieve significant reduction only in one modal component and not all of them, which may actually worsen the response. This occurs because the net response at a critical point is a vector sum of the component modal responses which often involve a lot of mutual cancellation.

The implementation of design detuning is typically done by the addition of nonstructural weight on rotor blades, since it is the only parameter that can be easily modified. This modification method leads to reduced payload capacity and higher static loads on the rotor blades. Also, the addition of nonstructural weight typically produces significant changes only in the lead-lag natural frequencies. Additional structural stiffening is required to modify of the out-of-plane bending frequencies.

1.2.1 Vibration Suppression Devices

Vibration suppression devices work on the principle that any vibration-prone structure can be relieved of dynamic loads by suitably appending an appropriately tuned spring–mass–damper system to it. These devices can be broadly classified based on their location on or in the helicopter [5]. Cabin suppression devices work on the premise that the vibratory response of the fuselage can be reduced by increasing the apparent impedance of the structure at a few select locations. They can also be employed to move the location of a node or point of minimum vibration. A dynamic vibration absorber comes under this category.

Vibration isolators attempt to decouple the dynamics of the main rotor from the fuselage dynamics thereby making the fuselage less susceptible to excitations from the main rotor. Typically, these devices replace the conventional transmission mounting and have been implemented both in passive configurations [6, 7] and as active systems [8, 9]. The drawbacks of this method are high weight penalty and high maintenance costs. Vibration absorbers like pendulum absorbers (pendab) and bifilar absorbers are passive devices used to absorb dynamic shear loads at the blade root before they propagate to the fuselage. They differ from the conventional suppression devices in that they are mounted on the main rotor itself and the centrifugal forces provide the stiffening effect [10, 11]. Typically, pendab are tuned at a frequency just below the excitation frequency and are quite effective in hub load reduction. They are easy to design, implement, and operate. However, the pendab mass required for effective cancellation of blade shear loads can be significant. Also, the friction in the pendab hinge is detrimental to its operation. The orientation of blade-mounted pendab changes due to the pitching motion of the blade leading to an undesirable coupling of blade lead–lag and out-of-plane bending motion. Pendab and bifilar absorbers also result in aerodynamic drag penalty.

1.2.2 Aeroelastic Optimization

Since the main rotor is the main source of vibration in a helicopter, an alternative and more direct approach is to design the rotor for minimum vibration. Many researchers have used optimization methods to design a low vibration rotor. Friedmann and Shanthakumaran used a structural optimization procedure to minimize vibratory hub loads of soft inplane hingeless rotor [12]. The design variables were the spanwise mass and stiffness distribution of the blades. Constraints were placed on the blade natural frequencies and blade stability in hover. A reduction of upto 40 % in vertical hub shear force was achieved. Celi and Friedmann conducted a similar study with the blade tip sweep as an additional design variable [13]. The use of tip sweep as a design variable gave an additional vibration reduction of about 10 %. Design variables in other aeroelastic optimization studies include blade stiffness, cross-sectional dimensions, taper ratio, root chord, root mass radius of gyration, twist, and nonstructural weights [14, 15].

Ganguli and Chopra used tailoring of composite structural couplings in the blade spar to reduce hub vibratory loads and blade root dynamic stresses [16, 17]. The design variables were the ply angles of the box-beam blade spars. Constraints were placed on aeroelastic stability and blade natural frequencies. They achieved a reduction of about 15–60 % in hub vibratory loads compared to the starting design. Also, the peak-to-peak flap and lag bending moments were reduced by 11 and 14 %, respectively. Though the optimal ply angles were determined at an advance ratio of $\mu = 0.30$, the optimized design was found to perform well at other forward speeds too ($\mu = 0.15$–0.40). Yuan and Friedmann carried out a similar optimization study of a composite helicopter rotor with swept tip [18]. The design variables in this study were the ply angles of the blade cross-section walls, tip sweep, and anhedral angle. They found that the tip-sweep was the most influential among the design variables considered in the study. The vibratory component of vertical hub shear force was reduced by about 30–50 % from the initial design. Many other studies have been carried out on structural and aeroelastic optimization of helicopter rotor blades. Detailed discussion of structural and design optimization studies in rotorcraft can be found in Refs. [19–27].

To summarize, passive systems are relatively simple and have no direct power supply requirements. However, they cannot adapt to change in flight conditions like forward speed, rotor rotational frequency, and/or blade dynamics due to moisture absorption or structural damage. Active methods overcome these shortcomings through the use of strategically placed actuators.

1.3 Active Control of Vibration

Active vibration control allows the vibration system to be able to adapt to several flight conditions, at a lower weight than passive devices. Also, active systems can absorb or supply power using actuators unlike passive systems which can only dissipate or store energy. Active materials like piezoelectric, electrostrictive, and magnetostrictive materials allow direct and efficient conversion of electrical energy to mechanical energy. They also reduce the need for large number of parts thereby minimizing potential failure sites. Semi-active approaches using electrorheological and magnetorheological fluid-based dampers have been studied for rotor vibration control [28–30]. Chen et al. have developed an adaptive impedance control algorithm for a smart spring whose stiffness can be actively changed to control vibrations in a helicopter rotor [31, 32]. In another study, Nitzsche and Oxley showed that the smart spring was capable of significant reduction in the vibratory loads transmitted to the fuselage through the pitch link [33]. Anusonti-Inthra and Gandhi demonstrated that optimal multicyclic variations of rotor blade root flap and lag stiffness can produce simultaneous reductions in all components of hub vibratory loads [34]. Reductions in vibration performance index of over 90 % were obtained with optimal 2/rev and 3/rev flap and lag stiffness variations. Such variable stiffness elements are possible through the use of electrorheological fluids.

Active control of structural response (ACSR) methodology employs vibration cancellation in the fuselage load paths by actively cancelling the incoming fixed-frame vibration (N_b/rev for a N_b-bladed rotor). In this approach, vibrations in the fuselage are reduced by introducing harmonically varying forces by actuators located between the rotor and the fuselage such that the sum of the response of the airframe, due to rotor loads and external excitation, is reduced to a minimum [35–37].

Among the active vibration control methods, higher harmonic control (HHC) and individual blade control (IBC) have received the most attention. For a N_b-bladed rotor with blade passing frequency Ω, the primary vibration transmitted by the rotor to the fuselage is at the $N_b\Omega$ frequency which is also known as the N_b/rev component. This N_b/rev component of fixed-frame loads is caused by $(N_b - 1)$/rev, N_b/rev, and $(N_b + 1)$/rev harmonics of the blade loads in the rotating frame. Both HHC and IBC attempt to modify the above harmonics of the blade loads, albeit through different means, such that the resultant N_b/rev hub vibratory loads are minimized.

1.3.1 Higher Harmonic Control

In higher harmonic control (HHC), the swashplate of the helicopter is excited at higher harmonics of the rotor rotational speed to generate new unsteady airloads that combine with oscillatory inertial loads to cancel the hub vibration. The excitation of the swashplate at multiples of $N_b\Omega$ frequency leads to $(N_b - 1)$/rev, N_b/rev and $(N_b + 1)$/rev excitation of the blade root pitch. The additional aerodynamic loads resulting from this excitation when properly phased leads to cancellation of the original oscillatory hub loads. Typically, higher harmonic pitch inputs of an order of $2°$ were found to be sufficient to reduce hub vibration.

Analytical studies in late 1950s and early 1960s used the concept of HHC for improvement of helicopter performance by eliminating retreating blade-stall [38, 39]. However, these studies did not explore the potential of HHC in helicopter vibration control. Molusis et al. established the applicability of HHC for vibration control in helicopters [40]. The benefits of HHC were further demonstrated in other analytical studies [41–43], wind tunnel tests [44, 45], and flight tests [46–48]. HHC implementation needs the knowledge of the frequency response relating the control input to the output measurements at multiples of the rotor angle velocity [49].

Though HHC is a mature and tested approach for helicopter vibration control, a number of issues have prevented it from actual implementation on a production helicopter. Some disadvantages of this approach are: (1) high actuation power (especially for hingeless and bearingless rotors), (2) high cost of implementation in a production helicopter, (3) large weight penalty associated with the hydraulic actuators, and (4) the limitation of HHC to integral multiples of N_b/rev excitation for a N_b-bladed rotor.

1.3.2 Individual Blade Control

The individual blade control (IBC) approach allows each blade to be controlled separately at a desired frequency. In HHC for a 4-bladed rotor, excitation of the swashplate at 4/rev allowed only 3/rev, 4/rev, and 5/rev oscillations in the rotating frame. Whereas, IBC can be used to generate 2/rev and other harmonics in the rotating frame. IBC also allows feedback loops for each blade in the rotating frame. Another advantage of IBC is that sensing in the rotating frame makes the controller less sensitive to changes in the flight conditions. Such a controller is conceptually simple and uses constant gains. The concept of IBC was pioneered by Kretz [50] and was eventually reinforced by Ham [51, 52].

The IBC approach can be implemented using three different techniques:

(1) In the conventional IBC approach, servo actuators at the blade root are used to actuate the entire rotor blade in the pitch direction [53].
(2) The entire blade is twisted by actuating the embedded piezoelectric materials. This approach is known as the active twist rotor (ATR) [54].
(3) On-blade partial span trailing-edge flaps are actuated to change the aerodynamic loading on the blade. The actively controlled flap (ACF) approach can be implemented in single- or dual-flap configurations. Each flap is individually controlled in the rotating frame [55].

In the conventional IBC approach, the pitch of each rotor blade is individually controlled in the rotating frame using servo actuators. This higher harmonic pitch input is superimposed on the 1/rev pilot control inputs. IBC was first implemented on a MBB BO-105 testbed in 1990. Whirl-tower tests and flight testing confirmed the potential of this approach. Extensive wind tunnel tests including a wide range of flight conditions were conducted at NASA Ames Research Center [56, 57]. Detailed theoretical and experimental results on the benefits of IBC can be found in Refs. [58–60].

The drawbacks of the conventional IBC approach are high actuation power and high-pitch link loads. Both HHC and conventional IBC systems actuate the blades at the root, which may not be an efficient approach in terms of actuation power and load distribution. The ACF and ATR approaches overcome the above drawback by twisting the rotor blade partially or entirely through the use of active materials (Fig. 1.2). While both ATR and ACF are IBC approaches, they differ greatly in terms of implementation.

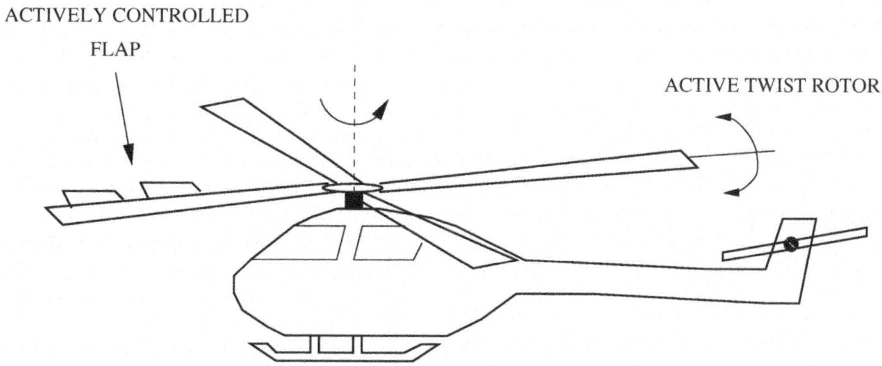

Fig. 1.2 Active twist and actively controlled flap concepts

1.3.3 Smart Rotors

More recent in the process of evolution of blade control are *active control methods* which include *trailing-edge flap control* and *active twist control* using smart materials. Fundamentally, these methods are based on the concept of IBC, but use the higher bandwidth actuators. These methods are being explored to design rotors with reduced vibration levels toward pursuit of the long-term goal of "jet-smooth-ride" for a helicopter [61]. The application of smart structures to helicopters is given by [62]. While conventional methods based on vibration absorbers and isolators have reached a point of diminishing returns, the advent of smart materials has opened up exciting research opportunities in the area. The smart-structure concept is based mainly on three elements: actuators, sensors, and microprocessors [63]. An actuator normally converts electric inputs into actuation strain/displacement that is transmitted to the host structure affecting its mechanical state. A sensor converts strain or displacement (or their time derivatives) into an electric field. The microprocessor analyzes the responses from the sensors and commands the actuators to apply localized strain/displacement to alter the system response. The range of smart materials used are piezoceramics, magnetostrictives, electrostrictives, shape memory alloys, magnetorheological fluids, electrorheological fluids, and fiber optics. These materials, when integrated in a proper manner with the host structure, lead to a change in the structural characteristics/responses.

Piezoelectric materials generate a charge in response to a mechanical deformation, and/or they develop mechanical deformation when subjected to an electric field. Hence, they are utilized either as actuators or sensors which ultimately contribute to a structure being smart. This ability of piezoceramic materials to actuate and sense simultaneously makes it a popular smart material. When a piezoelectric is used for both sensing and actuation, it is called a self-sensing layer. According to Irschik [64], the ability of structures with bonded piezoelectric patches and automatic control algorithms to react to external disturbances in a manner similar to human beings results in their being called "smart" or "intelligent" structures. Under an applied field, this material generates a small stroke and large force covering a wide range of frequency. Availability of a wider range of frequency makes it suitable for applications like active control.

Shape memory alloys are among the other promising smart materials. They are capable of achieving high force and displacements by undergoing a phase transformation at a specific temperature. They can recover their undeformed condition if their temperature is raised above the transformation temperature. But, the problem is that the response is very slow limiting the applications to 1 Hz. Electrostrictives are similar to piezoelectric, except for the fact that these materials are extremely sensitive to temperature and are monopolar. Magnetostrictive materials expand when exposed to a magnetic field. They can generate small strokes and large forces over a wide range of frequency but are monopolar and nonlinear. The setup with these materials also becomes bulky due to the elements like coil and magnetic return path. Both electrorheological and magnetorheological fluids come in the form of suspensions and respond in the presence of large electric and magnetic fields, respectively [65–68]. They change their effective damping properties in response to electric and magnetic stimuli, respectively. Comprehensive reviews on applications of smart materials are available in [69–72].

New and very promising materials known as single crystal piezoceramics are found to have induced strain actuation of an order of magnitude larger than other piezoceramics in current use [73]. Here, the major concern is to take into account of the strong dependence of material with varying orientation while building the actuator. Selection of application also depends upon the orientation with which the single crystal piezoceramic is built since the properties of these materials vary greatly with orientation. The development of smart materials has lead to considerable research in trailing-edge flap control and active twist control methods in recent years. Smart rotor for wind turbines have also been proposed. Bergrami [74] presents an aero-servo-elastic model for a wind turbine rotor equipped with trailing-edge flaps. The approach uses an aerodynamic model, a structural model, and a control model. The primary aim of this work is to reduce the fatigue loads on the turbine rotor. Model-based control algorithms are used to reduce the fatigue loads while assuming a standard wind field.

Trailing-Edge Flap Control

In the trailing-edge flap-approach, vibration control is achieved by exciting the flap at a combination of frequencies to produce incremental change in rotor lift and pitching moment which, if correctly phased, can cancel out most of the unwanted vibrations. Conceptually, the trailing-edge flap alters the maximum camber and tends to shift it toward the hinge point location. This alters the zero-lift angle and the effective angle of attack. This method has an advantage of having a high frequency bandwidth and a minimum of moving parts of compact size. Also, we need to mention here about the active passive hybrid optimization technique of rotor blades which uses a hybrid approach combining passive tailoring of the blade structure with active control using trailing-edge flap. In this technique, focus is simultaneously given on vibration reduction and control effort [75]. Basically, the trailing-edge flap controller is combined with the structural optimization routine. Apart from piezoelectric materials, vibration reduction in helicopter has also been attempted by using Terfenol-D driven flaps [76].

In the use of trailing-edge flaps, caution needs to be taken for uncoordinated motions, because these could result in local flow conditions, which enhance rather than alleviate airfoil-vortex interactions leading to an increase in vibration [77]. Also, moving parts in the structure have greater chances of wear and tear. Exposure of the trailing-edge flap to the airflow also causes a profile drag penalty. These deficiencies of the trailing edge flap point the way to using active blade twist as the long term solution to the helicopter vibration problem.

Li and Balas [78] performed aeroelastic control of a wind turbine blade using trailing-edge flap. They used a mathematical rotating blade model with periodic time-varying aerodynamic load, which was predicted by the Beddoes–Leishman dynamic stall model. Floquet theory was used to predict aeroelastic stability of the rotating blade. They showed that a trailing-edge flap-based adaptive controller can restrain flutter vibration.

Active Twist Control

Another approach to helicopter vibration reduction is to actively twist the entire rotor blade. This leads to actively controlling the blade angle of attack distribution around the rotor disk by creating a time-varying and controllable elastic twist which in turn reduces helicopter vibrations to a great extent by influencing the aerodynamic disturbances at its source. In this approach, there is an advantage of avoiding the profile drag penalty associated with trailing-edge flaps. Also, as the active material is an integrated part of the structure, exposure to airflow, and the resulting profile drag is minimum. Nevertheless, the drawback in this approach is that it requires increased actuation for twisting the entire blade, compared to the relatively small flaps. This drawback could be addressed by optimally placing the actuators to improve upon the actuation efficiency and by using novel ways of utilizing piezoceramics for high induced strain. Furthermore, active twist requires a blade redesign and is therefore suited for the next generation of helicopters and is a long-term solution of the vibration problem. Piezoceramics have found an increasing use in the active control solutions and have now become an integral part of the smart structures systems. In piezoelectric materials, electrical energy is converted to mechanical energy and vice versa. When an electric field is applied to a free layer of piezoelectric material, particles of the layer exhibit displacements from their original position in a manner similar to thermal expansion of a freely heated elastic strip. If the piezoelectric layer or patch is bonded to some non-piezoelectric structure, such as a helicopter rotor blade, without completely constraining the free displacement, the converse piezoelectric effect will actuate the structure. On the other hand, piezoelectric materials can also be used to sense deformations because an electric field is actuated due to strains because of the direct piezoelectric effect. Piezoelectric materials have been found to be the main candidates for the active twist rotor because they can operate over a wide bandwidth of frequency and can be placed in a manner such that they are removed from the primary flight control path.

Malgaca et al. [79] considered vibration control of rotating blade using root embedded piezoelectric materials. They point out that rotating blade vibration leads to failure in turbine blade of turbomachinery in the petroleum and airline industries. They use a piezoelectric sensor to capture vibration of a rotating blade at its fixed end. This sensed vibration is then fed to a piezoelectric actuator at the root of the blade via velocity feedback and proportional control. The potential of piezoelectric materials control rotating blade vibrations at different shaft speeds is shown.

1.4 Actuation of Trailing-Edge Flaps

The renewed interest in the area of helicopter rotors with trailing-edge flaps can be attributed to the recent developments in the area of smart material-based actuation technology. Smart material-based actuators are light-weight, compact, have high energy density and high bandwidth. They are ideally suited for actuation of on-blade trailing-edge flaps as they can be designed to fit inside the geometric confines of a helicopter rotor blade. Piezo-bimorph actuators, piezoelectric stack actuators, and induced shear strain actuation appear most suitable for actuation of rotor trailing-edge flaps.

1.4.1 Piezo-Bimorph Actuators

A piezo-bimorph is a bending actuator consisting of piezoceramic sheets that are bonded on both sides of a very thin brass shim with an electrically conducting adhesive coating. The bender is cantilevered, and equal but opposite fields are applied to the piezoceramic sheets on opposite sides of the shim. This causes pure bending of the actuator and the resulting tip displacement is amplified using a mechanical leverage or flexure system to provide the actuation stroke for the flap. Spangler and Hall showed the feasibility of this concept by designing and testing a nonrotating 2D-wing model with a piezo-bender actuated trailing-edge flap [80]. A 1/5th scale model with 10 % chord flap was wind tunnel tested at various airspeeds between 0 and 78 ft/sec, and at frequencies upto 100 Hz. Significant flap deflection, lift, and pitching moments were recorded, but the values were well below theoretical predictions. One reason for this performance degradation was found to be the backlash in flap hinge. Subsequently, Hall and Prechtl improved the design of the actuator by replacing the mechanical hinges with flexural amplification mechanisms [81]. They also improved the efficiency of the actuator by tapering its thickness with length. Flap deflections of 11.5° were demonstrated while operating under no-load conditions at frequencies upto 100 Hz.

In another study, Samak and Chopra used commercially available two-layered bimorph (G-1195) to actuate a model flap [82]. The flap was actuated by three piezo bender elements each measuring 1.5 in. long, 1.0 in. wide, and 0.021 in. thick. These piezo-bender elements were rigidly anchored to the blade spar to form 1.0 in. long cantilever beams. The bender elements were connected to the flap through a common hinge. A two-bladed rotor was tested on a hover stand. Flap deflection of upto ±7° was achieved under nonrotating condition but the amplitude decreased sharply with increase in rpm. At the scaled operating speed of 900 rpm, maximum flap amplitude of ±1° was achieved. Walz and Chopra built four-layered bimorph actuators in order to increase their actuation force [83]. The modified actuators provided about three times as much block force at the expense of reduction of one-third of free displacement. Also, modifications to the mechanical leverage arrangement between actuators and flap were incorporated using a rod-cusp arrangement. Nonrotating tests showed flap deflection of ±7° at 5 Hz and ±17° at 60 Hz. However, hover testing of the actuator mechanism did not yield the desired flap performance. To investigate the causes of performance degradation, Ben-zeev and Chopra built new rotor blades with several modifications to the actuation system and tested in vacuo to isolate the effects of centrifugal loading on the actuator-flap system [84]. A beam model of the piezo-bimorph including propeller moment effect as formulated. The cause of reduced flap deflections was traced to frictional forces at the flap hinge. The introduction of a thrust bearing was found to dramatically improve flap performance at higher rotational speeds (±5° at 800 rpm in the vacuum chamber and ±2° at 750 rpm on the hover stand). Again, on the hover stand, there was severe degradation of flap performance at the operating speed of 900 rpm.

Koratkar and Chopra conducted detailed testing of a 2-bladed rotor model in vacuum chamber [85]. They concluded that the performance degradation at higher rpm was due to the slippage of the piezo-bimorph under high centrifugal loads. An improved clamping of the actuator resulted in considerable increase in flap deflections. In hover, flap deflections of ±4 to ± 8° were achieved for 1/rev to 5/rev excitation at 900 rpm. This study also showed that flap deflections of ±6° at 4/rev excitation could be obtained at the Mach scaled operating speed of 2150 rpm using an 8-layer tapered piezo-bimorph in conjunction with a bias voltage. Subsequently, they built a 4-blade Mach scaled rotor model with a piezo-bimorph actuated flap located at 75 % blade span location [86]. The goal of the actuation system was to deflect the flap (8 % span, 20 % chord) by ±4°. This model was tested on the hover stand and flap deflections of ±6 to ±10° at 1850 rpm was achieved. Wind tunnel testing of the Mach scaled rotor model using Bell-412 hub was conducted at the Glenn L. Martin wind tunnel. In open-loop tests, flap deflections of ±4 to ±5° were achieved at 1800 rpm over a wide range of advance ratios and collective pitch settings. Preliminary closed-loop tests conducted using a neural network-based controller confirmed the control authority of the actuator-flap system by minimizing the 4/rev vertical hub shear force by 40 %. Fulton and Ormiston built a 7.5 ft diameter 2-bladed Froude-scaled rotor model with a plain trailing-edge flap actuated by piezo-bimorph [87]. The model was tested in a hover stand and maximum flap deflections of ±5° was achieved at

760 rpm. Open-loop wind tunnel tests successfully demonstrated the effectiveness of the actuator-flap mechanism [88].

The piezo-bimorph actuator concept appears suited for scaled rotor models and serves as a tool to study the performance of various adaptive control strategies to minimize rotor vibration using trailing-edge flaps [89]. Studies indicate that full-scale application of this concept could lead to considerable weight penalty.

1.4.2 Piezoelectric Stack Actuators

A piezoelectric stack actuator consists of a large number of thin piezoelectric sheets stacked in a series arrangement separated by electrodes. The actuator uses the induced strain of the piezoelectric material in the thickness direction (along the polarization axis). Piezostack are characterized by small free displacements but much larger block force than piezoceramic sheet actuators. The overall displacement of the piezostack can be expressed as:

$$\Delta L = n d_{33} V, \tag{1.1}$$

where ΔL is the axial displacement, n is the number of piezoceramic layers, d_{33} is the piezoelectric constant, and V is the operating voltage. For a typical piezoceramic material, the absolute value of d_{33} is about 2–3 times the absolute value of d_{31}. Also, the stroke of the actuator can be increased by using a large number of piezoceramic layers. Samak and Chopra studied the possibility of using piezostack and electrostrictive stack actuators with mechanical leverage amplification to actuate leading edge droop and flaperon of a wing model. The free displacement of the chosen piezostack was 0.72 mil at 150 V. A mechanical amplification of 25 was used. However, the measured free displacement was about 5 mils as compared to the analytical value of 18 mils. This prediction error was attributed to the slippage at knife edges and other mechanical losses.

Chandra and Chopra designed and developed an improved piezostack actuation mechanism to deflect the trailing-edge flap in a wing model [90]. A composite wing model was built around this high-performance stack actuator with integrated mechanical amplification. The free displacement and block force of this actuator at 1000 V excitation were 28 mil and 16 lb, respectively. The linear motion of the stack actuator was converted to rotary motion for flap actuation using a hinge offset mechanism. Maximum flap deflection of 8.5° was achieved. A 30 % decrease in flap deflection at free-stream velocity of 95 ft/sec in an open-jet wind tunnel was noticed. The flap deflection remained fairly constant even at an angle of attack of 10°. A study on the performance of several commercially available piezostack actuators revealed the feasibility of a full-scale trailing-edge flap actuated with specially designed piezostack actuators.

Prechtl and Hall designed a discrete trailing-edge servo-flap actuator called the X-frame actuator and conducted prototype bench-top tests [91]. The actuator consists of two active piezostack and two criss-cross frames and amplification is achieved via a shallow angle arrangement. The X-frame actuator achieves higher mass efficiency but requires larger frame dimensions than that of lever amplification mechanism. The active material is a lead magnesium niobate-lead titanate (PMN-PT) composition. They also built and hover tested a 1/6th Mach scaled rotor model of CH-47D (Chinook) with a trailing-edge flap actuated by X-frame actuators [92]. At the operating speed (tip Mach of 0.63) and collective pitch of 8°, oscillatory flap deflections of ±2.4° were achieved with an excitation of ±1000 V. A full-scale MD-900 rotor system with trailing-edge flap actuated by X-frame actuators is currently being used in wind tunnel testing and flight tests.

Lee and Chopra carried out extensive testing of various commercially available piezostack actuators under high preload and DC/AC excitations [93]. Based on these tests, they chose Physik Instrumente P-804.10 piezostack for the actuation of a full-scale trailing-edge flap. Five of these stack actuators were used to build a double-lever (L-L) amplification device [94]. The design goal was to achieve ±7° flap deflection for a full-scale Boeing MD-900 Explorer helicopter in hover condition. A pushrod and flap hinge mechanism utilizing four-bar linkage was designed to enhance the performance. Vacuum spin tests were conducted upto 600 g of centrifugal loading environment and the actuation mechanism performed satisfactorily without any apparent performance degradation.

Straub and his associates carried out the design and development of a biaxial piezoelectric stack actuator for trailing-edge flap control on a 34 ft (10.4 m) diameter, bearingless rotor [95]. The 5-bladed rotor had plain flaps with 18 % blade span (74–92 % radial location), 25 % chord and an aerodynamic balance of 40 %. The biaxial actuator performance and robustness were demonstrated through a series of bench, shake, and spin tests, subjecting the actuator to the full range of dynamic conditions inside the rotor blade, including 814 g steady and 29 g vibratory loading. Janker et al. developed a piezostack based actuator for trailing-edge flap [96]. The amplification device consists of a shallow angle flexure mechanism that is wrapped around the stacks. The performance of this actuator was demonstrated through testing of wing-section in a wind tunnel. The goal was to achieve a flap deflection of ±8° for an EC-135 helicopter rotor. They also examined a piezostack-based actuator for leading-edge droop control to delay dynamic stall. Shaner and Chopra also studied the actuation of leading-edge flaps using piezostack actuators [97]. A single lever amplification mechanism was used and bench-top tests were conducted.

It is evident that the concept of trailing-edge flap actuation using piezostack actuators has been studied extensively and thoroughly. Studies mentioned above indicate that piezostack driven actuators in conjunction with a mechanical amplification device are capable of actuating full-scale trailing-edge flaps for helicopter vibration and noise suppression.

1.5 Induced Shear Strain Actuation

Piezoelectric materials have three strain coefficients d_{31}, d_{33} and d_{15} which define their induced strain actuation behavior. For a given electric field, a higher value of the strain coefficient results in a higher strain actuation. Mechanical motion generation by most ceramic actuators is related to either the longitudinal mode or the transverse mode. It is a well accepted fact that among all piezoelectric coefficients, the one in the shear mode (d_{15}) is highest in terms of its absolute value [98, 99]. The important reasons for using the shear actuator mechanism are [100–102]: (1) d_{15} based torsional actuators produce large angular displacement and torque; (2) the performance of shear actuators is less dependent on structure's stiffness; (3) less problems of actuator debonding at its extremities; (4) d_{15} has a higher value than d_{31} and d_{33}.

This book navigates through the initial smart blade/beam formulation and basic demonstration of smart formulation by investigating dynamic response of rotating smart beams to successively investigate the performance of active twist control methodology using induced shear actuation for vibration reduction of a real rotor in forward flight. Piezoceramic is selected to be the material of choice. A comprehensive aeroelastic analysis is performed to assess the effect of active twist on blade response, hub loads, trim control, and stability. Relevant validation studies are performed to confirm the model rotor follows in the ranges of real rotor.

1.6 Optimal Placement of Actuators in Active Structures

Recent studies indicate that multiple flaps are capable of achieving higher vibration reduction compared to a single flap configuration within available actuator authority [103]. Multiple flaps provide more flexibility in altering the aerodynamic loads acting on the rotor blade as compared to a single trailing-edge flap. Most researchers use simple parametric studies to determine the ideal location of the trailing-edge flap(s) on the rotor blade, though the problem is well-suited for optimization applications. Most comprehensive helicopter aeroelastic analyses are huge computer programs involving several disciplines and it is often cumbersome to integrate them with other optimization programs. The vibration reduction results are obtained using a control algorithm which calls the helicopter aeroelastic analysis many times, making the computation even more expensive.

Several researchers have looked into the issue of optimal placement of sensors and actuators in flexible structures based on some optimality criteria. In an early study, Lim presented a new method of finding optimal locations of sensors and actuators for control of a flexible structure [104]. This method was based on the orthogonal projection of structural modes into the intersection subspace of the controllable and observable subspaces corresponding to an actuator/sensor pair. The controllability and observability grammians are then used to weight the projections to obtain a measure of controllability and observability. This method produces a three-dimensional design space wherein sets of optimal actuators and sensors may be selected. Dhingra and Lee used a combination of artificial genetic search and gradient-based search method to determine the optimal location of actuators while designing an actively controlled structure [105]. The performance criteria in this study was the damping augmentation provided by the control action. Varadan et al. used Powell's direct search method to determine the optimum location of lead-zirconate-titanate (PZT) actuators to minimize the total sound power radiated from an acoustically excited clamped isotropic plate [106]. They found that the optimal locations of the piezoelectric actuators tend to be near the corners of the plate rather than the center of the plate.

Several studies have looked into the optimal placement of actuators on a beam for active vibration control [107–110]. In Refs. [107, 111], the authors defined and used a novel performance index known as the actuator power factor. It was defined as the ratio of the total dissipative mechanical power of an actuator to the total supplied electrical power to the actuator. The concept of actuator power factor was based on the ability of an integrated induced strain actuator such as a PZT to transfer supplied electrical energy into structural mechanical energy. For a given structure, the location of an actuator directly influences the authority of the actuator toward driving the structure. A system optimized based on the actuator power factor guarantees the highest energy efficiency for single frequency and broad-band applications. Kang et al. investigated the optimal location of a collocated sensor/actuator for vibration control of laminated composite beams. Modal damping was chosen as the performance index, because of its direct effect on the settling time of vibrations. A structural damping index (SDI) is defined from the modal damping. Weights for each vibrational mode were taken into account in the SDI calculation. The optimum location of the sensor/actuator was determined as the point where the SDI is maximum. Wang et al. obtained the optimal location and size of a collocated pair of piezoelectric patch actuators surface bonded onto beams. The objective function was referred to as the controllability index, which is the singular value of a control matrix. The index measures the input energy required to achieve a desired structural control by the piezoelectric actuators. The controllability index is directly dependent on the placement and size of the piezoelectric patches. Thus maximizing the index leads to optimal design of the piezoelectric actuator. Several other studies have presented innovative design methodologies for optimal placement of actuators in flexible structures [112–114].

In general, the problem of finding optimal actuator locations in a structure can be posed as "Given a set of N possible locations, find a subset of M locations which provides the best possible performance." Simple enumeration technique can be used when M is comparable to N. However, a need for an appropriate optimization method arises when $M \ll N$. Heuristic search methods such as Genetic algorithms (GA), Simulated annealing (SA), and Tabu search (TS) have gained popularity in solving such problems. Firefly and artificial bee colony algorithms are also useful for sensor/actuator placement [115]. These methods do not guarantee convergence to a global optima, but can yield useful and innovative design solutions. However, they are often computationally expensive.

Recently, GA's have been used to determine optimal locations of piezoelectric sensors and actuators in isotropic and laminated composite plates [116–118]. The criteria in these studies was to maximize controllability and observability while preventing spillover. GA's have also been effectively used in optimization of actuator locations in non-aerospace applications such as design of earthquake resistant buildings and shape control of telescope mirrors [119–122]. GA's are a particular class of evolutionary algorithms that use techniques inspired by evolutionary biology, such as inheritance, mutation, selection, and crossover [123]. GA's are implemented as a computer simulation in which a population of abstract representations (called chromosomes) of candidate solutions (called individuals) to an optimization problem evolves toward better solutions. Traditionally, solutions are represented in binary as strings of 0s and 1s, but other encodings are also possible. The evolution usually starts from a population of randomly generated individuals and happens in generations. In each generation, the fitness of every individual in the population is evaluated, multiple individuals are stochastically selected from the current population (based on their fitness), and modified (recombined and possibly mutated) to form a new population. The new population is then used in the next iteration of the algorithm. This population-to-population approach prevents the algorithm from falling into the trap of local optima [124].

Heverly et al. formulated a hybrid optimization methodology to simultaneously determine optimal actuator locations and control actions to minimize helicopter airframe vibrations [125]. The optimization process couples an optimal control formulation with a Simulated Annealing optimization routine. It was shown that actuators in the airframe can more effectively control some of the dominant airframe modes, rather than actuators centralized near the main rotor support assembly. When compared to a representative state-of-the-art centralized control configuration, the distributed optimal actuator configuration achieved significantly greater vibration reduction with less control effort. Results showed that while a centralized actuator configuration reduced vibration levels by 49 %, the optimally distributed actuator configuration led to 90 % reduction in vibrations and required 50 % less control effort.

Padula and Kincaid give an excellent survey of sensor and actuator placement problems in various engineering disciplines and the application of optimization techniques to improve their performance [126]. This study concluded that it is important to develop approximations (surrogate models) of the performance index and constraints and employ these computationally efficient approximations to determine the optimal sensor/actuator locations. Surrogate models are computationally efficient and are therefore attractive for aeroelastic problems.

1.7 Actuator Hysteresis Models

Piezoceramic actuators are rapidly gaining popularity in vibration control applications due to their high bandwidth, compact size, and minimum moving parts. However, a major limitation of piezoceramic actuators is their lack of accuracy due to nonlinearity and hysteresis. One source of nonlinearity is due to the change in electromechanical coupling coefficients at high electrical fields where piezo actuators are operated to exploit their full authority. Hysteresis, on the other hand, is a form of nonlinearity with memory. Hysteresis causes the piezoceramic expansion to depend not only on the current voltage excitation but also on the history of excitation. Unmodeled hysteresis in the actuator can generate amplitude-dependent phase shifts in the output leading to inaccuracy in open-loop control and probably even loss of stability in closed-loop control [127, 128]. Kurdila et al. [129] derived a control methodology that accounts for overall hysteresis in PZT actuated on-blade elevons. They used a simple collocated output velocity feedback to study the stability of the aeroelastic system. They found that the introduction of actuator hysteresis led to a sharp degradation in the performance of collocated feedback control, sometimes even leading to loss of asymptotic stability. Therefore, it is necessary to address hysteresis in vibration control methods that use piezoceramic actuators. However, modeling hysteresis is a challenging problem. Hysteresis models can be further classified as rate-independent (static) or rate-dependent (dynamic). The term 'static' means that the model output is determined purely by the sequence of input values, while the speed of input variation has no influence on the its output. A 'dynamic' hysteresis model, on the other hand, requires the history of both input and input-rate to determine the model output [130, 131].

1.7.1 Static Hysteresis Models

The classical Preisach model (CPM) is a popular static model that has been extensively used in literature. It was originally proposed by the German physicist F. Preisach as a model for magnetic hysteresis and was based on some hypotheses concerning the physical mechanisms of magnetization. Krasnoselskii separated this model from its physical meaning and represented it in a pure mathematical form which is similar to a spectral resolution of operators [132]. As a result, this model can now be used for the mathematical description of hysteresis of any physical nature. The Preisach model has several appealing features including its ability to model complex hysteresis types, a well-defined identification algorithm, and a convenient numerical simulation form [133, 134].

There exists considerable similarity between the primary hysteresis mechanism of ferromagnetic and ferroelectric materials. Several researchers have successfully applied the classical Preisach model to represent hysteresis in ferroelectric material systems. Sreeram and Naganathan were the first to apply the classical Preisach model to piezoceramic material systems [135]. Ge and Jouaneh used a modified Preisach model to represent hysteresis in a stacked piezoceramic actuator and validated the model through experiments. Hughes and Wen discussed and verified the applicability of Preisach model to piezoceramic and shape memory alloy systems. The parameter identification based on the input/output data and hysteresis compensation via direct inversion of the identified model was also demonstrated. Yu et al. used a modified

Preisach model to predict the output of a piezoceramic actuator system and experimentally verified the results [136]. Hu and Mrad developed a discrete-time Preisach model to capture the hysteresis of a piezoceramic actuator and used it to develop an inversion/compensation algorithm [137]. The Preisach model has also been used to describe hysteresis in magnetostrictive (Terfenol-D) actuators [138, 139].

1.7.2 Dynamic Hysteresis Models

The Preisach model, though widely used, has several limitations. One such shortcoming is its applicability to model only static hysteresis phenomena. Mayergoyz was the first to extend the CPM to model dynamic hysteresis phenomena [140]. The 'static property' of the CPM was relaxed by introducing the dependence of Preisach distribution function (or Preisach measure) on the rate of output-variations. The study also deals with identification problems of fitting this model to experimental data and its numerical implementation. This model is described in time-domain and is therefore capable of predicting the output for any arbitrary input wave form.

Bertotti proposed a dynamic generalization of the CPM, in which the rate of change of flux associated with each elementary Preisach operator is not infinite, as in CPM, but proportional to the difference between the external field and the loop threshold fields [141]. With this assumption, the shape of the elementary Preisach loops, as well as the macroscopic hysteresis loops, become dependent on the magnetizing frequency. However, this approach has two disadvantages. First, this approach requires an analytical expression of the Preisach distribution function which is not trivial to obtain for any given piezoceramic actuator. Second, this approach is applicable only for harmonic inputs. Song and Li also developed a frequency dependent hybrid hysteresis model based on CPM and neural networks [142]. Again, this model is limited to sinusoidal input signals. Also, its validity for high frequency excitations (>50 Hz) is not clear. Recently, Dang and Tan proposed a neural-networks dynamic hysteresis model by combining the Preisach model with diagonal recurrent neural networks [143]. They replaced the rate-independent Preisach relay operators by rate-dependent hysteresis operators of first-order differential equation. The similarity in structure between this modified Preisach model and modified diagonal recurrent neural networks (MDRNN) allowed the use of MDRNN to describe the dynamic hysteresis behavior of the actuator. The effectiveness of this neural-networks dynamic hysteresis model was demonstrated through experiments.

Recent advances in smart materials technology have led to the practical realization of ACF concept. Piezoceramic materials such as PZT and PMN-PT are ideal candidates for this application due to their high-bandwidth and stiffness. However, piezoceramic materials are inherently nonlinear, particularly under high-field strengths which are likely to be encountered during normal operating conditions. Piezoceramic stack actuators are leading contenders for actuation of full-scale trailing-edge flaps on the helicopter rotors. Stack actuators are high-force and low-stroke devices and hence require a stroke amplification mechanism to move a full-scale trailing-edge flap. The amplification mechanism gain directly magnifies the material hysteresis in piezoceramics. In addition, hysteresis in the amplification mechanism due to friction and freeplay, can couple strongly with the piezoceramic material hysteresis to yield significant hysteresis at the structural level. The piezo actuator and link mechanism are therefore sources of considerable hysteresis which must be addressed in helicopter vibration control. If unmodeled, hysteresis nonlinearity can lead to amplitude-dependent phase shifts in actuator output leading to inaccuracy in open-loop control and probably even loss of stability in closed-loop control.

1.8 Active Twist Rotors

The prime objective of the works discussed below is to reduce vibration in helicopters using active blade twist. This approach is futuristic but is gaining popularity. There is substantial material available for the proof of this concept in the literature. The existing literature is classified into the following sections: (1) Early studies, (2) Smart active blade tips, (3) Active fiber composites, (4) NASA/MIT/Army active twist rotor, (5) Active composite beams, and (6) Induced shear strain mechanisms.

1.8.1 Early Studies

Earlier studies were mostly experimental in nature with simple modeling and were undertaken to prove the concept of active twist control using piezoceramic materials. The key idea was to check if the required 1–2° of twist needed for suppressing vibration could be obtained with a minimum consumption of power. An early study on the use of adaptive structures to attenuate rotary wing aeroelastic response was done by Nitzsche and Breitbach [144]. The concept in this study was to demonstrate the feasibility of using adaptive materials for sensors and actuators to attenuate the higher harmonic loads. A linear mathematical model of a single rotating blade including only flatwise bending and torsion degrees of freedom was developed. They mention that second lag mode and the first torsion mode of the blade are most important for vibration control. IBC was used to act specifically on the bending and torsion modes in order to reduce the overall dynamic response of the blade. Closed-loop performance using both sensors and actuators was evaluated. Contribution of aeroelastic modes with variation in azimuth at various advance ratios was verified. Closed-loop performance was found to be satisfactory and better than open-loop performance.

Chen and Chopra [145] built a Froude-scale model rotor and concurrently developed an analytical model to predict the behavior of the rotor blades with embedded piezoceramic actuators. For this, a uniform strain theory was formulated to predict the static torsion and bending response of rectangular section beams and rotor blades, followed by experimental validation.

Experiments were conducted on a 1.83 m diameter, 1/8th Froude-scale model of a two-bladed bearingless rotor. The blade cross-section was NACA 0012. Total blade length was 67.51 cm from tip to root and chord was 7.62 cm. Piezoceramic elements that were 11.5 mil thick were embedded under the fiberglass skin in banks of discrete actuators at angles of ±45° on the top and bottom. The first configuration had 55 pairs of actuators spaced 0.254 cm apart and the second configuration had 12 pairs of actuators spaced 3.81 cm apart. Both these rotors were assessed in order to find out the optimum configuration in terms of piezoceramic actuators. An analytical model was developed for a beam and then applied to a rotor and the results were compared with experimental data.

The first set had 80 % more actuators than the second set. It was observed that an increase in number of torsion actuators resulted in an increase in torsion stiffness of the rotor blade and the 'interference effect' eventually reduced the twist.

However, it was found that on reducing the number of actuators, stiffness reduces, and twist increases. Rotating response of the rotor blade was evaluated at various speeds with a Froude-scale operating speed of 900 rpm. The rotor collective was set at 4, 6 and 8° and the piezoceramic actuators were actuated at 100 V rms to manipulate blade twist motion at frequencies from 10 to 100 Hz. In general, the response was observed to decrease in amplitude with increasing rpm for all collectives.

Uniform strain theory could correctly predict the trends of torsion and bending responses of the rectangular beam specimens, except that it shows discrepancy at very high voltage. This theory was able to predict static torsion response with an accuracy of approximately 80 %. Blade testing showed that structural stiffness of the blade decreased as the actuator spacing was increased.

In [146, 147], the work done in [145] was extended and it was found that geometric parameters of the actuator have a significant effect over structural properties, and hence, twist performance of the blades. Blade twist performances in static and rotating modes differ due to the presence of large external loads at high rpm. By generating new unsteady airloads, blade twist is capable of altering the aerodynamic characteristics of the rotor. In this study, (experiment) a blade twist of 0.5° was attained at 900 rpm with dual layer actuators. But for viable vibration reduction, a blade tip twist of 1–2° is required at 900 rpm. The twist achieved here being small, only partial reduction was possible. The maximum tip twist values at resonance frequencies (50 and 90 Hz) were 0.350 and 1.10 respectively. At nonresonance frequencies, the response was less than 0.50° at 4/rev excitation.

Derham and Hagood used smart materials to achieve main spar twist [148]. This method is called Inter-Digitated Electrode Piezoelectric in Fiber Composite (IDEPFC), and was jointly applied by MIT and Boeing to rotor blades. It uses active materials embedded in the composite plies of rotor blades actuated through applied electric fields. Active twist for a 1/6th Mach scale CH-47 rotor blade was demonstrated and a preliminary design for a 1/6 scale active twist blade was presented. Blades were made of composite materials and had a built in pretwist of 12°. The rotor was three bladed and fully articulated. Feasibility of applying this concept to rotorcraft was discussed in terms of performance including economic merit, weight, and power consumption.

IDEPFC is a variation on traditional PFC (piezo-fiber composites). These incorporate active material fibers in a composite ply lay-up. The PFC's are actuated by digitated electrodes and hence are known as IDEPFC. A structural analysis was performed to assess the structural integrity and strength parameters of the active blade configuration. Design loads consistent with full-scale blade requirements and practices, including the effects of added actuator mass and inertias were incorporated using Boeing's TECH-01 rotor aeroelastic analysis. TECH-01 is a comprehensive rotor analysis considering unsteady aerodynamic forcing and associated elastic blade response caused by the complex flow and wake patterns around a rotor in edgewise flight.

A 1/16 Froude-scale CH-47 helicopter blade was manufactured with embedded IDEPFC actuators. The blade was bench tested for twist actuation performance to verify both an MIT developed mathematical model for predicting actuator performance and the manufacturing process for active blades. Predictions for the actuated torsional capabilities of the 1/6 Mach scale blade were made using the model that was correlated to the 1/16 Froude-scale experiment. It was observed that 3/rev and 4/rev induced twist had the greatest impact on 3/rev vertical shear. A 70 % reduction in vibratory vertical shear was achieved over the baseline (no actuation) case with a penalty of 2.5 % in required power. A 40 % reduction was obtained with a 2/rev actuation without a penalty in power.

The main conclusion from [148] was that substantial vibration reduction could be obtained using the active twist concept. But, the studies by Chen and Chopra found that the amount of twist generated was not sufficiently high with current day actuators (1996–1997) to completely suppress vibration. However, their results showed that partial suppression of helicopter vibration is possible using the active twist concept.

1.8.2 Smart Active Blade Tips (SABT)

Bernhard and Chopra worked on the development of an active on-blade vibration reduction system using smart active blade tips (SABT) [149]. SABT's are driven by piezo-induced, bending-torsion coupled actuator. The beam was designed specifically to fit within the rotor blade profile and was essentially used as a pure torsional actuator. The actuator was located lengthwise in the box-beam cavity and was supported only at the root and at the tip of the main blade. The inboard end is directly mounted on the blade root and the outboard end is connected to the SABT via a shaft, supported in a radial bearing. The bearing is capable of carrying lift, drag, flapwis, and chord-wise bending moments of the blade tip whereas the centrifugal and torsional loads were transferred to the actuator beam. Beam layup was selected as $[-45, 0, 45]$. The actuator beam applied a torque at the tip, which thereby induced the blade to twist. By actively pitching the tips, new unsteady airloads were generated. Correctly phased, they could be used to reduce the dynamic rotor hub shears and moments. The advantage here is that the large centrifugal force imposed by the moving tip on the beam is eliminated, improving the twist performance. The spanwise segments were differentially energized such that, the induced strain bending curvatures cancelled out, while the induced strain twist curvatures added up to a net tip rotation.

These SABT have distinct advantages at the expense of a few drawbacks. First, the entire blade can be used for actuation. The required twist can be achieved without any amplification mechanisms. Second, the need for the tip thrust beam is eliminated via the actuator beam. Third, the moving tip is hinged at the aerodynamic center to minimize the pitching moments. For analysis, an extension of the induced strain composite-beam model presented by Chandra and Chopra [150] was used. This was based on Vlasov-theory [151], in which the structure is modeled in two phases. First, at the local cross-sectional level, a two-dimensional (thin walled) plate-analysis was used. Second, the plate analysis was reduced to a global one-dimensional bar analysis, using the principle of virtual work. The fabrication of these SABT blades was carried out at the Maryland rotorcraft center for a 1/8th Froude-scale bearingless rotor hub. The nominal diameter was 1.83 m with a root cutout of 31 % and a nominal operating speed of 817 rpm. The main rotor blade was a rectangular, untwisted NACA 0012 blade with a 76.2 mm chord. Structural integrity tests, nonrotating tests, and hover tests were performed on this rotor.

The rotating twist amplitudes with SABT were similar to tip twist amplitudes reported by Chen and Chopra. However, here the blade twist was achieved with half the power. The reduced power requirement was because the actuator beam used a surface area of 206.45 cm^2 with 0.0254 cm thick PZT-5H elements as compared to 412.90 cm^2 with 0.0254 cm used by Chen and Chopra. It is to be noted that the SABT blades were 25 % heavier than the blades used in Ref. [146]. Hover tests at 875 rpm produced blade twist results from 0.30 at 1/rev through 0.50 at 5/rev excitation frequencies.

In [146], a Vlasov based bar analysis was used to derive a specialized, computationally efficient, one dimensional finite element, to model the nonrotating actuator beam. This was a two-node, 12-degree-of-freedom element with bending, torsion, and in-and out-of-plane warping degrees of freedom and induced strain capability. Results obtained by FEM (Finite Element Method) were validated by experiments for two cases. Good correlation was obtained with experimental data. In one case, a reverse bending region was observed between the plate cantilever root and the first bank of piezo actuators. This was captured in the analysis with the help of specialized continuous in-plane warping mode. The predicted dynamic blade tip pitch deflection was approximately 10–20 % below the measured values. Later, a Mach scale active twist rotor was fabricated [150] and tested with the above-mentioned SABT. This suggested that the concept was capable of achieving twist distribution throughout the blade.

While the SABT concept was demonstrated by wind tunnel tests and Mach scale models, tip twist amplitudes were similar to those obtained by Chen and Chopra and below the 2° needed for full vibration suppression. The primary disadvantage of SABT is that this is a localized active system, consulting a single-input single-output control devise.

1.8.3 Active Fiber Composites (AFC)

Active fiber composites (AFC) were integrated within a composite rotor blade to induce a twisting moment. AFC are active plies oriented at a 45° angle to the blade span in order to induce shear stresses and a distributed twisting moment along the blade. They consist of a laminated structure of fiberglass plies and PZT-fiber plies. The PZT-fiber plies have continuous, aligned, PZT-fibers in an epoxy layer, and polymid/copper electrode films. The electrode films are etched into an inter-digitated pattern that affects the electric field along the fiber direction, thus, activating the primary d_{33} piezoelectric effect. The AFC has a number of advantages; it is a conformable actuator, which could be integrated with a passive structure. The actuators are distributed throughout the structure providing redundancy in operation. The active blade requires no articulating components, thus, eliminating the need of amplification mechanisms and does not increase the profile drag of the blade like servo-flap actuators.

Design, manufacture, and testing of this rotor were investigated thoroughly in [152]. A 1/6th Mach scale CH-47D blade was designed, fabricated, and tested in a hover test stand facility. The span of the blade was 153.97 cm and the chord was 13.68 cm. The blade was used on a fully articulated hub. The blade had a built-in linear twist of 12° and taper of a VR7 airfoil at 0.85R to a VR8 airfoil at the tip. A configuration with three diagonally placed active plies in the spar laminate was selected for the integral blade. These active plies were uniformly distributed between 0.27R and 0.95R and were divided into seven span-wise AFC segments, where R is the radius of the blade. Activation of the diagonally placed fibers induced shear in the spar skin, which generated blade twist. Bench tests performed at frequencies up to 67.5 Hz demonstrated that a maximum twist of 1 to 1.5° peak-to-peak (±0.5 to ±0.75 amplitude) could be obtained. However, full span sections resulted in a twist performance relatively below design targets. But half-span sections successfully demonstrated twist at full authority. Moreover, excellent correlation with model predictions was achieved. Even with the reduced twist, the hub load generation at all blade loading conditions and rotor speeds was consistent and near predicted levels, suggesting that the integral twist actuation concept is feasible and worthy of further investigation.

The studies proved the feasibility of the active twist rotor concept when active fiber composites were used. However, the AFC's are more difficult to manufacture and much more costly than equivalent monoliths. Considerable information about AFC-based smart rotor is given by [74].

1.8.4 NASA/Army/MIT Active Twist Rotor

The NASA/Army/MIT Active Twist Rotor (ATR) is ambitious project in this area. This work uses the active fiber composites for actively twisting the blades using both analytical simulations and wind tunnel tests. Wilkie et al. [153] describe aeroelastic-modeling procedures and use them in the design of a piezoelectric controllable twist helicopter model rotor for wind tunnel testing. Two numerical active twist rotor approaches were developed. First, a code named PETRA (Piezoelectric Twist Rotor Analysis) was developed and was specifically intended for fundamental studies of active twist rotor designs with embedded actuators. The code PETRA was used for control law and design optimization studies. The blade equations were simplified to a linear out-of-plane bending-torsion model by using an ordering scheme. A sectional analysis was used to determine the blade structural properties and piezoelectric terms. The resulting piezoelectric terms were applied to the right-hand side of the final equation of motion as a force term along with all aerodynamic forcing terms. The aerodynamic loads were derived using strip theory and a finite-state unsteady aerodynamics formulation, which includes the ONERA model of dynamic stall. A uniform inflow model, with a linear variation across the rotor disk in forward flight was used.

The second approach is based on the CAMRAD II comprehensive rotor analysis package, which can be used for more detailed analytical studies [154]. The commercially available CAMRAD II software package has more extensive rotorcraft modeling capabilities. Forward flight simulations using CAMRAD II indicate that large reductions in 4P (4/rev, meaning 4 times the rotation speed) vertical hub shear may be achievable with significantly less than the maximum twist actuation authority of the active twist rotor. However, it should be noted that this study modeled the smart structure using applied loads and the piezo-structure was not included in the aeroelastic equations through constitutive relations.

In [54], a four-bladed, aeroelastically scaled, active twist rotor (ATR) model was designed and fabricated to be tested in the heavy gas medium of a transonic wind tunnel. Active fiber composites (AFC) were utilized in fabricating the rotor blades. Each ATR blade consisted of 24 active fiber composite (AFC) actuators to implement the active twist control. The AFC's were placed in four layers through the thickness of the blades and were oriented such that the active strain is applied at $\pm 45°$ relative to the blade span-wise axis to permit maximum torsional control on the blades. They were actuated using separate high-voltage, low-current power channels for each blade. The pretwist was linear with a twist of $-10°$ from the center of rotation to the blade tip.

NACA-0012 airfoil section was used. Hover testing was conducted to determine the basic frequency response characteristics of the ATR blades under active twist control. Tests were carried out for a wide range of frequencies. In forward flight tests, both low ($\mu = 0.14$) and high ($\mu = 0.333$) advance ratio cases were considered. Here μ is a ratio of the forward speed of the helicopter to the rotor tip speed. Results show that for both low and high-speed forward flight, active control was capable of generating significant variations in both rotating and fixed system loads, and that large reductions in fixed-system vibratory loads were achievable. Both rotating and fixed-system loads could be dramatically affected by changes in active twist control phase and a reduction in fixed-system loads of 60–95 % was obtained. An active twist control frequency of 3P was found to be most effective in reducing fixed-system loads for both the low-speed and high-speed conditions. Active twist amplitudes of 1.10–1.40° were obtained for 3P to 5P twist actuation, with 1000 V. Although the 4P blade loads were reduced using 4P active twist control, there was a significant increase in the 4P pitch link loads, tending to counter the effect in the fixed system. The power necessary to operate the four ATR blades is, at most, less than 0.9 % of the maximum rotor power required during testing.

Shin et al. [155] describes system identification and closed-loop control of the ATR system. They found that the ATR rotor system could be treated as linear and time invariant and the periodic effects can be neglected. For the ATR rotor, it was also found that the transfer functions did not vary much for level flight conditions. Therefore, a single control law could be used to effectively control the rotor over a wide range of advance ratios. The closed-loop control tests were performed using the T-matrix (Transfer-matrix) approach recast as a continuous time formulation. Transfer matrix relates the input to the vibratory hub loads. The objective was to reduce the 4/rev normal shear. The collective mode of blade actuation was found to be useful for controlling 1/rev vibrations due to rotor tracking and was less effective for hub normal shear, whereas effect of cyclic actuation was found to be much more. For closed-loop analysis, vibration at each time step was measured and fed back through the control law to adjust the swashplate input to cancel the vibration. Different gain constants were investigated. Since it is possible for the controller signal to exceed the voltage limit of individual actuators, an anti-windup mechanism was added to each feedback structure. Results showed that in some flight conditions, 4P vibration was almost eliminated (40 db reduction). The other components of the fixed-system loads at 4/rev were also reduced, except for the negligible change in yawing moment. The studies clearly show the potential of active fiber composites in helicopter vibration reduction. The wind tunnel results with heavy gas medium allow the matching of full scale Mach number leading to more realistic results. Studies also showed that considerable reduction in vibration could be obtained even if tip twists were less than the target 1.5–2°.

1.8.5 Smart Composite Rotor Modeling

The studies discussed until now were more focused toward experiments and limited in terms of modeling. Addressing this issue, some researchers discussed detailed modeling of smart composite rotors. Chattopadhyay et al. [156] presented numerical results for a four-bladed bearingless model rotor with self-sensing piezoelectric actuators mounted at 45° on the top and bottom walls of the composite box-beam blade. The composite box beam consisted of graphite/epoxy in stacking sequence of

$[-45/45]_{2s}$ in horizontal walls and $[45]_8$ in vertical walls. The box-beam dimensions were: length = 7.92 m, width = 0.18 m, and height = 0.05 m. The thickness of PZT was 0.196 mm and the blade had a rectangular planform with a linear twist of −10°. Planform area, is defined as the projected area of a wing when viewed from above, perpendicular to flow direction.

First, a comprehensive analysis technique was developed for the analysis of rotor dynamic loads using a composite rotor blade built around an active box beam. Second, the model was used to investigate reduction in hub dynamic loads using closed-loop control. A higher order theory-based approach was used to model the smart composite beam [157]. Higher-order displacement field was developed to model the individual walls of arbitrary thickness, in the presence of eccentricity. The theory approximates the elasticity solution so that the beam cross-sectional properties are not reduced to one-dimensional beam parameters. It considers inplane and out-of-plane warping. Because the relationships between the induced strain due to actuation and the applied electric field are nonlinear at high voltages, the formulation includes these nonlinear induced strain effects for the piezoelectric strain coefficients. For the rotor analysis, an unsteady aerodynamic model was coupled with a rotor blade dynamic model to develop an integrated rotor vibratory load analysis procedure. The rotor dynamic analysis accounted for deformations in the flap, lag, and torsion directions. A finite-state induced flow model was used for predicting the aerodynamic loads. The (4/rev) dynamic forces and moment at the rotor hub were calculated in forward flight. Parametric studies were also done to assess the influence of the number and location of actuators on the vibratory load reduction at the hub.

Strain rate feedback was used for structural vibration control. The feedback actuator voltage was defined with the help of the current developed from the electric charge. A finite element formulation was used to implement the coupled theory. After the induced-velocity coefficients were obtained, the aerodynamic, inertial, and centrifugal forces along the blade span were integrated to calculate the rotor vibratory loads.

For maximum efficiency, the distribution of the actuator locations must be closely related to the region of high strain rate of the vibration modes. Elements 1 (root area where maximum strain rate occurs, blade length is divided into 10 elements) and 5 represent optimal actuator locations for the control of the second flapping mode and elements 1, 3, 4, 9, and 10 represent the most effective locations for the third flapping mode. Because seven modes are involved in the rotor loads analysis, it was difficult to predict the optimal locations of the actuators for the control of all hub loads. Thus, two different cases were studied here. The first configuration involved 10 pairs of self-sensing PZT actuators surface bonded to the horizontal walls of the box beam (elements 1–10) and the second configuration involved 4 pairs of self-sensing PZT actuators placed on elements 1, 3, 5, and 7. Blade natural frequencies were calculated for both the configurations. Differences were observed in the modal frequencies and mode shapes of both the configurations, because of the variation in stiffness arising due to the different number of actuators in both cases.

Significant reductions were observed in all the hub forces and moments using active control. A reduction in vertical shear force of about 50 % was achieved. In general, according to the authors, among all six dynamic hub loads, the pitching moment, and the rolling moment contribute most significantly to vehicle vibration. A reduction of about 33 % was obtained in the pitching moment and 30 % in the rolling moment. On comparing both the configurations, a reduction of about 50 % in vertical shear force was achieved in the first configuration (10 pairs) and about 30 % in the second configuration (4 pairs). The power consumed was found to be 60 % lower in the second configuration. This indicated the existence of a nonlinear tradeoff between actuator power and vibration reduction. Cesnik and Shin [158] performed detailed modeling of the active twist rotors with active fiber composites. The analysis was based on an asymptotic formulation for the two-cell thin-walled anisotropic active beam [159, 160]. This formulation stems from shell theory, and the displacement field (including out-of-plane warping) is not assumed a priori, but results from an asymptotical approach. It was observed in case of two-celled box beams that increasing the stiffness in the active members always decreased the actuation. Two cases were analyzed; one had a symmetric distribution of active layers (top and bottom) and other had an asymmetric distribution. Wall 1 was the front wall, 2 the middle wall and 3 the back wall of the two-cell box beam. For the asymmetric distribution; adding passive stiffness at wall 3 decreased the twist actuation, which supports the common view that twist actuation will decrease with an increase in torsional stiffness. A change in the stiffness of wall 2 did not affect the twist actuation. However, by adding passive stiffness at wall 1, the twist actuation was found to increase. But, for symmetric distribution, twist actuation was found to be insensitive to the change in stiffness in walls 1, 2, and 3. For both these cases, significant coupling of active and passive components between both cells was seen. This coupling was noted to diminish for a single-cell section. Hence, the behavior in case of two-cell was complex and more realistic than of single cell because of coupling effects. To illustrate this, two possible single-cell derivatives of the real airfoil-shaped beam were considered. One is the configuration in which trailing-edge (fairing) was of very low or zero stiffness, which is known as D-spar single-cell beam. The other possible model is one without web, and it is simply called the outer-shell beam. It was observed that in none of these cases, comparable results with a full (two-section) model were obtained. Either over- or under-prediction was observed in these cases. By adding plies at the nose and the web, the twist actuation increases, while it decreases when the passive plies are added at the active region and fairing. The torsional stiffness was proportional to the variation in the wall thickness and the bulk active twist moment increased on increasing the stiffness in walls, except in the active regions.

In case of a more representative two-cell active beam with a NACA 0012 airfoil-shape cross-section, active regions were the front spar skins with embedded AFC layers, and all other parts of the blade were passive. An electric field of 1.795 MV/mm was applied in all cases. By adding or removing E-glass plies at each region of the cross-section, the stiffness was varied and a parametric study was performed. They found that there was a possibility of obtaining an increase in twist actuation in the two-cell model, on adding passive plies at regions other than the active one (web and nose). An important conclusion of this study was that the net change in twist actuation in the two-cell active beam depends on the local stiffness variation in the active or passive region. The conventional design belief that torsional stiffness must be reduced to increase twist actuation was not found to be true. It was shown that torsional stiffness can be increased by up to 20 % with an increase in twist actuation of about 5 %. The authors mention that the single-cell cross-section model is incapable of capturing the interaction between active and non-active cross-sectional walls and hence it is insufficient to predict the performance of multi-cell beams.

Buter and Breitbach [161] investigated the active twist concept using an actively controlled tension-torsion-coupling of the structure at the DLR (German Aerospace Laboratory). Tension-torsion coupling is an anisotropic behavior which appears in structural components. It can be realized by properly orienting the blade stiffnesses. The anisotropic material behavior has to be separated from the anisotropic structural behavior resulting from structural elements like ribs or stringers. The actuator was integrated inside a composite helicopter rotor blade. The rotor blade was represented by a tension-torsion coupled thin-walled rectangular beam similar to the BO105 model rotor with a scaling factor of 2.54. A piezoelectric stack actuator was found to be suitable for twisting the blade and a deformation of 1.5° was achieved. However, a disadvantage of the actuator was the high spanwise stiffness of the rotor blade spar. The adaptive twist was achieved using only the outer part of the blade resulting in a comparatively small control effect. An advantage of the actuation at only the outer part of the rotor blade was that the aerodynamic forces were the largest at that location and it was possible to compensate disturbances induced by the flow field.

Kube and Kloppel [162] discuss the role of predictive computer programs for the development of smart helicopter rotor. Accurate aeroelastic simulations were suggested to speed up the development of an adaptive rotor system, since the number of design iterations during the development process needs to be minimized. The key contributions of this research project were (1) reduce rotor induced vibrations by 90 % (2) reduce BVI noise by 6 dB and (3) expand the helicopter flight envelope by 5 %. A comprehensive analysis was developed with accurate modeling of the rotor dynamics, inflow field, and aerodynamics. The code was validated with wind tunnel test data and unsteady Navier-Stokes simulations.

Vibrations can be reduced by means of higher harmonic blade twist variation with NΩ, (N−1)Ω, (N+1)Ω frequencies where N is the number of blades and Ω is the rotor rotational frequency. Also, the rotor efficiency and blade track can be optimized by adjusting the steady part of the blade twist. The tension-torsion coupling method discussed in detail in [161] was used and realized using a helical winding in the blade skin. Laboratory experiments showed that a twist of at least 1° was realizable at the blade tip for a Mach scaled rotor with piezoelectric actuators. The actuator mass was 12 % of the blade mass and was driven with an electric field of approximately 200 V. The BO105 helicopter was equipped with an IBC system and noise was measured at the ground level and on the landing gear. However, these studies with IBC did not use smart material actuators. Ghiringelli et al. [163] developed an analytical procedure for modeling of an active twist helicopter rotor using active fiber composites. The rotor blade was twisted using induced strain actuators distributed into the structure of the blade. Active fiber components made of piezoelectric fibers actuated by inter-digitated electrodes (IDE) were used. Fiberglass was used for the outer skin and placed between the piezoelectric plies. Unidirectional graphite epoxy (T900) was used for the inner part of the spar. The elastic, inertial and piezoelectric properties of the blade section were determined by a dedicated semi-analytical formulation. A 4-bladed, articulated rotor was studied in hover and forward flight. The model was an analytical benchmark full-scale rotor based on the work done by Wilkie et al. [164] and is representative of a large class of medium weight helicopter rotors. Frequencies of the simulations were matched with the experimental frequencies. The aeroelastic analysis was conducted using a multibody formulation. The multibody dynamics approach accounts for the possibility of large rotations, nonlinear aeroelastic behavior or nonuniform rotation speeds. The rotor blades were modeled as finite volume beam elements subject to large displacements and rotations. This is shown by Ghiringelli et al. [165]. Pitt et al. [166] used strip theory with dynamic stall, radial flow, and Mach number correction factor along with dynamic inflow modeling. This approach allows a detailed analysis of the kinematics and dynamics of rotorcraft avoiding any undue approximation in the kinematics of the system, and with the same order of refinement of a finite element model in the description of flexibility with reasonable and acceptable computational cost.

The above studies show improved modeling capabilities for the smart composite structure to predict the behavior of active twist rotor blades. The analytical methods provide a way to conduct parametric studies on actuator placement and on the design of the rotor cross-section for obtaining maximum twist actuation. These analyses also allow the development of control algorithms. The studies also show the use of monoliths placed at 45° on the top and bottom of the box-beam, AFC's and stack actuators. It should be noted that all of the above studies use direct piezoelectric coefficients.

1.8.6 Induced Shear Mechanisms

Piezoelectric strains could be defined in terms of two direct strain coefficients d_{31} or d_{33} and one shear strain d_{15} coefficient. All the studies discussed until now used the direct strain coefficients d_{31} or d_{33} piezoelectric coefficients. Some researchers have looked at the use of the shear strain coefficient d_{15} of the piezoelectric actuator for active twist rotors.

Smith [167] analytically evaluated an induced shear piezoelectric tube as an active blade twist actuator. This induced shear actuator had already been investigated for trailing-edge flap models. An effort for developing an active twist rotor was pursued in this study. A finite element model of the induced shear actuator and rotor blade was used to guide the design of the tube actuator for both small-scale and full-scale rotor blade applications. The performance of the induced shear actuator was then compared to the NASA/Army/MIT Active Twist Rotor. One advantage of using shear strain in a piezoelectric actuator is that the piezoelectric coefficient for shear is higher in magnitude than that for bending or axial. In addition, in the design aspects, unlike the trailing-edge flap and active tip blade applications, amplification mechanism is not required for the active twist rotor. One end of the actuator tube is clamped to a rib in the rotor blade while the other end of the tube actuator is attached to the rotor blade tip. A precompressive load must be applied to the tube actuator to alleviate any tension loading due to centrifugal effects. Possible chord-wise locations of the active blade twist actuator include the leading edge of the blade and the 25 % chord location in the D-spar cavity. The leading-edge placement was desirable because this will allow the replacement of the inactive leading-edge weights with active leading-edge weights. Placement at 25 % chord was also desirable because it allows for the most range in the tube radius design and will not shift the blade center of mass location. The twist performance of both these actuators was analyzed and compared.

A torsion finite element model of the rotor blade and induced shear actuator was developed using Lagrange's equation to predict the blade twist angle distribution. Boundary conditions were such that the tube actuator twist is zero at the actuator clamping near the blade root. It was assumed that there was no blade twist at the root and blade twist and tube twist were equal at the tip. Both full-scale and small-scale applications were investigated. Actuator location at 25 % chord was found to be more suitable than the leading edge location as the actuator torque is a function of the tube radius to the fourth power, while the tube twist is related to the tube radius to the first power. Thus, the drop of torque is more significant than the increase in twist that results from shifting the actuator to the leading edge. A full scale active twist tube actuator was designed with a weight of 1.36 kg (5 % rotor blade mass penalty) and a maximum tube radius of 0.89 cm when placed at 25 % chord location. In case of a small-scale active twist rotor the weight of the tube actuator was 0.23 kg and a maximum tube radius of 0.635 cm. The tube was subjected to an electric field of 4 kV/cm.

The induced shear tube actuator runs across 33 % of the blade span while the active fiber packs are embedded from 30 to 98 % radius location. The shear tube actuator has the benefit of generating the blade twist at a lower applied voltage level when compared to the AFC. The shear tube actuator has an advantage of being discrete, while the active fiber design requires to be embedded in the rotor blade airfoil skin. Moreover, a discrete actuator is more accessible for maintenance, inspection, and replacement than an actuator embedded in the airfoil skin. One drawback of the tube actuator is the larger (approximately 1.13 kg) added weight compared to AFC (approximately 0.127 kg). A key result of this study was that the finite element analysis indicated that a 121.92 cm long tube actuator generates a tip twist of $\pm 1.10°$ in a full scale MD900 blade and a 45.72 cm long tube actuator generates a tip twist of $\pm 1.40°$ for a small-scale blade.

1.9 Suppression of Dynamic Stall-Induced Vibration

The aerodynamic loads on the rotor are mainly due to the time varying relative airflow during the forward flight, wake and the dynamic deflection of the rotor blades. At high forward speed and thrust coefficient, very high pitching moments are generated due to retreating blade-stall. Stall occurs due to flow separation over the airfoil leading to a sudden loss in lift and increase in drag and pitching moment. Static stall occurs due to a slow increase of the angle of attack (α) experienced by the airfoil. In contrast, dynamic stall occurs when the airfoil also undergoes heave and pitch rate motions which leads to separation and reattachment of flow on the airfoil. Dynamic stall results in high values of lift and the delay of the fall of lift to much higher angles of attack compared to static stall. However, dynamic stall also generates very large increases in pitching moment and drag coefficients. This in turn stress the pitch links, decrease fatigue life of components and induce excessive vibration. Another problem is because of the blade tips penetrating into supercritical and transonic flow regimes inducing strong shock waves. Rapid increase in blade torsion moments and vibration because of dynamic stall on the retreating side and compressibility effects on the advancing side is the primary limiting factor for helicopters. Modeling dynamic stall requires improved aerodynamic analysis contributing to the final goal of achieving accurate aeroelastic analysis of the rotor [168, 169]. It is also important to mention that in general, for a 4-bladed rotor 2/rev control is good for reducing stall area and 3, 4, and 5/rev control for reducing vibration.

The highest levels of vibration can occur at very high-speed flight ($\mu > 0.3$) due to flow separation and dynamic stall. Reducing vibration induced by dynamic stall can allow the helicopter to fly at higher forward speeds and therefore expand the flight envelope of the helicopter. Active control of helicopter blade-stall was addressed by Nguyen and Chopra [170, 171] using HHC. The rotor model used for this study is a four-bladed hingeless rotor of the BO-105. A blade pretwist of $-10°$ was used. Forward speed upto $\mu = 0.35$ were investigated for analysis. Finite element analysis including unsteady aerodynamic effects and a nonuniform inflow model is used. Stall suppression is based on a transfer-matrix approach. Each element of the transfer matrix represents the sensitivity of the controlled parameter to each harmonic of the blade root actuation. Transfer matrix was computed using a finite difference method in which each harmonic of the control input is perturbed individually. Stall index was used as the measure for identifying the severity of the stall. Stall index was defined in terms of the excess lift over the stall area. Open-and closed-loop controllers were investigated. Both performance improvement and reduction of stall index were addressed in this study. It was observed that stall could be suppressed effectively with higher harmonic control at both cruise- and high-speed flight with a control amplitude of around 1°. It was also observed that at low speed flight, stall index was fairly insensitive to higher harmonic and multi harmonic input. Reduction in stall index was less with closed-loop multi-pharmonic input but a gain performance was achieved. However, the studies by Nguyen do not consider smart material actuation.

Wilkie et al. used the Active Fiber Composite (AFC) approach to suppress dynamic stall. AFC laminae were embedded in the blade structure at alternating $\pm 45°$ orientation angles. Classical laminated plate theory was used for deriving structural and piezoelectric actuation properties. Numerical results were obtained using the PETRA active twist rotor aeroelasticity computer code. Primary constraints on the

AFC blade design were to maintain the blade mass and bending stiffness approximately equal to that of a conventional helicopter rotor blade structure. Blade torsional stiffness was reduced as much as possible to maximize active twist. A blade twist of -8° was used. Flight speed upto $\mu = 0.3$ were investigated. In this study, 2/rev control was used to suppress the stall area on the rotor disk. Active fiber composite plies were used to improve the dynamic stall response characteristics of the typical rotor. It was shown that with active control a delay in dynamic stall could be achieved resulting in reduced rotor power and an expansion of the flight envelope of the helicopter. Trade-off studies between torsional stiffness and actuation capabilities were also conducted. However, this study does not directly address vibration suppression which results from a complex interaction of nonuniform wake, unsteady aerodynamics, flow separation, dynamic stall, and blade motion.

Later, Depailler and Friedmann [172] demonstrated the suppression of dynamic stall using actively controlled trailing edge flap. The Onera dynamic stall model was used in conjunction with a rational function approximation for unsteady aerodynamic loads for attached flow. A parametric study was conducted over single and dual-flap configurations and their efficiency in terms of flap deflection was studied. With the help of these configurations, reduction in vibration for cruise and high-speed flight conditions was conducted. Using a single flap all the vibratory loads were reduced by 70–85 % with the exception of the vertical hub shear. The dual-flap configuration reduces all loads by 70–95 % and is 40 % more effective than single flap.

The above-mentioned approaches and models have been devised and proposed for suppressing the dynamic stall induced vibration. The above studies mostly focussed on reducing vibration at various flight conditions with the help of either higher harmonic control or active control approach.

1.10 Organization of the Book

Chapter 2 represents the mathematical modeling for the smart helicopter rotor. For the active twist rotor, the derivations need to be modified to include smart structures terms. For the active flap rotor, the rotor equations are not changed and the trailing-edge flap is included through an aerodynamic model and a piezoelectric stack actuator model. Chapters 3–7 develop the trailing-edge flap-based smart rotor concepts through a series of numerical simulations which uncover the physics of the problem. Aspects of controller development, flap placement, and actuator hysteresis compensation are highlighted in these chapters. Chapters 8–11 discuss the active twist rotor using the piezoceramic-induced stear concepts. The use of the active twist concepts for dynamic stall suppression is brought out. The book illustrates a variety of algorithms, concepts, and methods for the development of smart helicopter rotor systems.

References

1. Friedmann, P.P.: Renaissance of aeroelasticity and its future. J. Aircr. **36**(1), 105–121 (1999)
2. Bielawa, R.: Rotary Wing Structural Dynamics and Aeroelasticity. AIAA, Washington (1992)
3. Reichert, G.: Helicopter vibration control—a survey. Vertica **5**(1), 1–20 (1981)
4. Loewy, R.G.: Helicopter vibrations: a technological perspective. J. Am. Helicopter Soc. **29**(4), 4–30 (1984)
5. Viswanathan, S.P., Myers, A.W.: Reduction in helicopter vibration through control of hub impedance. J. Am. Helicopter Soc. **25**(4), 3–12 (1980)
6. Schuett, E.: Application of passive rotor isolation for alleviation of rotor induced vibration. J. Am. Helicopter Soc. **14**(2), 34–48 (1969)
7. Braun, D.: Development of antiresonant force isolators for helicopter vibration reduction. J. Am. Helicopter Soc. **27**(4), 37–44 (1982)
8. Smollen, L.E., Marshall, P., Gabel, R.: Active vibration isolation of helicopter rotors. J. Am. Helicopter Soc. **7**(2), 42–55 (1962)
9. Kuczynski, W.A., Madden, J.: The RSRA active isolation/rotor balance system. J. Am. Helicopter Soc. **25**(2), 17–25 (1980)
10. Amer, K.B., Neff, J.R.: Verical-plane pendulum absorbers for minimizing helicopter vibratory loads. J. Am. Helicopter Soc. **19**(4), 44–48 (1974)
11. Paul, W.F.: Development and evaluation of the main rotor bifilar absorber. In: Proceedings of the 25th Annual National Forum of the American Helicopter Society, Washington, D.C., USA (1969)
12. Friedmann, P.P., Shanthakumaran, P.: Optimum design of rotor blades for vibration reduction in forward flight. J. Am. Helicopter Soc. **29**(4), 70–80 (1984)
13. Celi, R., Friedmann, P.P.: Structural optimization with aeroelastic constraints of rotor blades with straight and swept tips. AIAA J. **28**(5), 928–936 (1990)
14. Chattopadhyay, A., Walsh, J.L., Riley, M.F.: Integrated aerodynamic load/dynamic optimization of helicopter rotor blades. J. Aircr. **28**(1), 58–65 (1991)
15. Chattopadhyay, A., McCarthy, T.R.: A multidisciplinary optimization approach for vibration reduction in helicopter rotor blades. Comput. Math. Appl. **25**(2), 59–72 (1993)
16. Ganguli, R., Chopra, I.: Aeroelastic optimization of a helicopter rotor with composite coupling. J. Aircr. **32**(6), 1326–1334 (1995)
17. Ganguli, R., Chopra, I.: Aeroelastic optimization of a helicopter rotor to reduce vibration and dynamic stresses. J. Aircr. **33**(4), 808–815 (1996)
18. Yuan, K.A., Friedmann, P.P.: Structural optimization for vibratory loads reduction of composite helicopter rotor blades with advanced geometry tips. J. Am. Helicopter Soc. **43**(3), 246–256 (1998)
19. Miura, H.: Applications of numerical optimization methods in helicopter rotor design problems—a survey. Vertica **9**(2), 141–154 (1985)
20. Glaz, B., Friedmann, P.P., Liu, L., Bain, J., Sankar, L.: A surrogate-based approach to reduced-order dynamic stall modeling. J. Am. Helicopter Soc. **57**(2), 1–9 (2012)
21. Friedmann, P.P.: Helicopter vibration reduction using structural optimizationwith aeroelastic/multidisciplinary constraints—a survey. J. Aircr. **28**(1), 8–21 (1991)
22. Celi, R.: Recent applications of design optimization to rotorcraft—a survey. J. Aircr. **36**(1), 176–189 (1999)
23. Ganguli, R.: Survey of recent developments in rotorcraft design optimization. J. Aircr. **41**(3), 493–510 (2004)
24. Murugan, S., Ganguli, R.: Aeroelastic stability enhancement and vibration suppressionin a composite helicopter rotor. J. Aircr. **42**(4), 1013–1024 (2005)
25. Murugan, S., Ganguli, R., Harursampat, D.: Surrogate based design optimizationof composite airfoil cross-section for helicopter vibration reduction. Aeronaut. J. **116**(1181), 709–726 (2012)
26. Ganguli, R.: Optimal design of composite structures: a historical review. J. Indian Inst. Sci. **93**(4), 557–570 (2013)

27. Kumar, D., Cesnik, C.E.S.: New strategy for designing composite rotor blades with active flaps. J. Intell. Mater. Syst. Struct. **53**(2), 436–448 (2014)
28. Lim, S., Park, S., Kim, K.: AI vibration control of high-speed rotor systems using electrorheological fluid. J. Sound Vib. **284**(3–5), 685–703 (2005)
29. Wang, J., Meng, G.: Study of the vibration control of a rotor system using a magnetorheological fluid damper. J. Vib. Control **11**(2), 263–276 (2005)
30. Zhu, C.: A disk-type magneto-rheological fluid damper for rotor system vibration control. J. Sound Vib. **283**(3–5), 1051–1069 (2005)
31. Yong, C., Zimcik, D.G., Wickramasinghe, V.K., Nitzsche, F.: Development of the smart spring for active vibration control of helicopter blades. J. Intell. Mater. Syst. Struct. **15**(1), 37–47 (2004)
32. Chen, Y., Wickramasinghe, V., Zimcik, D.: Smart spring impedance control algorithm for helicopter blade harmonic vibration suppression. J. Vib. Control. **11**(4), 543–560 (2005)
33. Nitzsche, F., Oxley, G.: Smart spring control of vibration and noise in helicopter blades. In: Proceedings of the 46th AIAA/ASME/ASCE/AHS/ASC Structures, Structural Dynamics, and Materials Conference, No. AIAA-2005-2270, Austin, Texas, USA (2005)
34. Anusonti-Inthra, P., Gandhi, F.: Optimal control of helicopter vibration through cyclic variations in blade root stiffness. Smart Mater. Struct. **10**(1), 86–95 (2001)
35. Staple, A.E.: An evaluation of active control of structural response as a means of reducing helicopter vibration. In: Proceedings of the 15th European Rotorcraft Forum, Amsterdam, The Netherlands, pp. 51.1-51.18 (1989)
36. Chiu, T., Friedmann, P.P.: An analytical model for ACSR approach to vibration reduction in helicopter rotor- flexible fuselage system. Aeronaut. J. **101**(1009), 399–408 (1997)
37. Pearson, J.T., Goodall, R.M., Lyndon, I.: Active control of helicopter vibration. Comput. Control Eng. J. **5**(5), 277–284 (1994)
38. Payne, P.R.: Higher harmonic rotor control. Aircr. Eng. **30**(354), 222–226 (1958)
39. Arcidiacono, P.: Theoretical performance of a helicopter having second and higher harmonic feathering control. J. Am. Helicopter Soc. **6**(2), 8–19 (1961)
40. Molusis, J.A., Hammond, C.E., Cline, J.H.: A unified approach to optimal design of adaptive and gain scheduled controllers to achieve minimum helicopter rotor vibration. J. Am. Helicopter Soc. **28**(2), 9–18 (1983)
41. Davis, M.W.: Refinement and evaluation of helicopter real-time self adaptive active vibration controller algorithm. NASA CR-3821 (1984)
42. Robinson, L.H., Friedmann, P.P.: Aeroelastic simulation of higher harmonic control NASA. CR-4623 (1994)
43. Nguyen, K., Chopra, I.: Application of higher harmonic control to hingeless rotor systems. Vertica **14**, 545–556 (1990)
44. Hammond, C.E.: Wind tunnel results showing rotor vibratory loads reduction using higher harmonic blade pitch. J. Am. Helicopter Soc. **28**(1), 10–15 (1983)
45. Shaw, J., Albion, N., Hanker, E.J., Teal, R.S.: Higher harmonic control: wind tunnel demonstration of fully effective vibratory hub force suppression. J. Am. Helicopter Soc. **34**(1), 14–25 (1989)
46. Wood, E.R., Powers, R., Cline, J.H., Hammond, C.E.: On developing and flight testing a higher harmonic control system. J. Am. Helicopter Soc. **30**(1), 3–20 (1985)
47. Polychroniadis, M., Achache, M.: Higher harmonic control: flight tests of an experimental system on SA 349 research Gazelle. In: Proceedings of the 42nd Annual Forum of the American Helicopter Society, Washington, D.C., USA, pp. 811–820 (1986)
48. Miao, W., Kottapalli, S.B.R., Frye, H.M.: Flight demonstration of higher harmonic control (HHC) on S-76. In: Proceedings of the 42nd Annual Forum of the American Helicopter Society, Washington, D.C., USA, pp. 777–791 (1986)
49. Mura, R., Ghalamzan Esfahani, A.M., Lovera, M.: Robust harmonic control for helicopter vibration attenuation. In: American Control Conference (ACC), pp. 3850–3855 (2014)
50. Kretz, M.: Research in multicyclic and active control of rotary-wings. Vertica **1**(1/2), 95–105 (1976)

51. Ham, N.D.: Helicopter gust alleviation, attitude stabilization and vibration alleviation using individual blade control through a conventional swashplate. In: Proceedings of the 11th European Rotorcraft Forum, London, UK, pp. 75.1–75.5 (1985)
52. Ham, N.D.: Helicopter individual blade control research at MIT 1977–1985. Vertica **11**(1/2), 109–122 (1987)
53. Richter, P., Eisbrecher, H.D., Kloppel, V.: Design and first tests of individual blade control actuators. In: Proceedings of the 16th European Rotorcraft Forum, Glasgow, Scotland, UK (1990)
54. Wilbur, M.L., Mirick, P.H., Yeager, W.T., Langston, C.W., Cesnik, C.E.S., Shin, S.: Vibratory loads reduction testing of the NASA/Army/MIT active twist rotor. J. Am. Helicopter Soc. **47**(2), 123–133 (2002)
55. Friedmann, P.P., de Terlizzi, M., Myrtle, T.F.: New developments in vibration reduction with actively controlled trailing edge flaps. Math. Comput. Model. **33**(10–11), 1055–1083 (2001)
56. Jacklin, S.A., Leyland, J.A., Blaas, A.: Full-scale wind tunnel investigation of a helicopter individual blade control system. In: Proceedings of the 34th AIAA/ASME/ASCE/AHS/ASC Structures, Structural Dynamics, and Materials Conference, No. AIAA-1993-1361, La Jolla, CA, USA (1993)
57. Jacklin, S.A., Blaas, A., Teves, D., Kube, R.: Reduction of helicopter BVI noise, vibration and power consumption through individual blade control. In: Proceedings of the 51st Annual Forum of the American Helicopter Society, Fort Worth, Texas, USA (1995)
58. Arnold, U.T.P., Muller, M., Richter, P.: Theoretical and experimental prediction of individual blade control benefits. In: Proceedings of the 23rd European Rotorcraft Forum, Dresden, Germany (1997)
59. Morbitzer, D., Arnold, U.T.P., Muller, M.: Vibration and noise reduction through individual blade control: experimental and theoretical results. In: Proceedings of the 24th European Rotorcraft Forum, Marseille, France (1998)
60. Schimke, D., Arnold, U.T.P., Kube, R.: Individual blade root control demonstration—evaluation of recent flight tests. In: Proceedings of the 54th Annual Forum of the American Helicopter Society, Washington, D.C., USA (1998)
61. Chopra, I.: Review of state of art of smart structures and integrated systems. J. Am. Inst. Aeronaut. Astronaut. **40**(11), 2145–2187 (2002)
62. Chopra, I.: Status of application of smart structures technology to rotorcraft systems. J. Am. Helicopter Soc. **45**(4), 228–252 (2000)
63. Chopra, I., Sirohi, J.: Smart Structures Theory, vol. 35. Cambridge University Press, Cambridge (2013)
64. Irschik, H.: A review on static and dynamic shape control of structures by piezoelectric actuation. Eng. Struct. **24**(1), 5–11 (2002)
65. Kamath, G.M., Hurt, M.K., Wereley, N.M.: Analysis and testing of Bingham plastic behavior in semi-active electrorheological fluid dampers. Smart Mater. Struct. **5**(5), 576 (1996)
66. Wereley, N. (ed.).: Magnetorheology: Advances and Applications, vol. 6. Royal Society of Chemistry, Cambridge (2013)
67. Wereley, N.M., Pang, L., Kamath, G.M.: Idealized hysteresis modeling of electrorheological and magnetorheological dampers. J. Intell. Mater. Syst. Struct. **9**(8), 642–649 (1998)
68. Singh, H.J., Wereley, N.M.: Adaptive magnetorheological shock isolation mounts for drop-induced impacts. Smart Mater. Struct. **22**(12), 122001 (2013)
69. Damjanovic, D.: Ferroelectric, dielectric and piezoelectric properties of ferroelectric thin films and ceramics. Rep. Prog. Phys. **61**(9), 1267–1323 (1998)
70. Kazanci, M.: A review of polymeric smart materials for biomedical applications. Mater. Technol. **18**(2), 87–93 (2003)
71. Valliappan, S., Qi, K.: Review of seismic vibration control using smart materials. Struct. Eng. Mech. **11**(6), 617–636 (2000)
72. Trindade, A., Benjeddou, A.: Hybrid active-passive damping treatments using viscoelastic and piezoelectric materials: review and assessment. J. Vib. Control **8**(6), 699–745 (2002)

73. Park, S., Shrout, T.: Ultrahigh strain and piezoelectric behavior in relaxer ferroelectric single crystals. J. Appl. Phys. **4**(82), 1804–1811 (1997)
74. Bergami, L.: Smart Rotor Modeling: Aero-Servo-Elastic Modeling of a Smart Rotor with Adaptive Trailing Edge Flaps. Springer, New York (2014)
75. Zhang, J., Smith, E., Wang, K.: Active-passive hybrid optimization of rotor blades with trailing edge flaps. J. Am. Helicopter Soc. **49**(1), 54–65 (2004)
76. Fenn, R., Downer, J., Bushko, D., Gondhalekar, V., Ham, N.: Terfenol-D driven flaps for helicopter vibration reduction. Smart Mater. Struct. **5**(1), 49–57 (1996)
77. Hassan, A., Sankar, L., Tadghighi, H.: Effects of leading and trailing edge flaps on the aerodynamics of airfoil/vortex interactions. J. Am. Helicopter Soc. **39**(2), 35–46 (1994)
78. Li, N., Balas, M.J.: Aeroelastic control of wind turbine blade using trailing edge flap. Wind Eng. **38**(5), 549–560 (2014)
79. Malgaca, L., Al-Qahtani, H., Sunar, M.: Vibration control of rotating blades using root-embedded piezoelectric materials. Arab. J. Sci. Eng. **40**(5), 1487–1495 (2015)
80. Spangler, R.L., Hall, S.R.: Piezoelectric actuators for helicopter rotor control. In: Proceedings of the 31st AIAA/ASME/ASCE/AHS/ASC Structures, Structural Dynamics, and Materials Conference, No. AIAA-1990-1076, Long Beach, California, USA, 2–4 April 1990
81. Hall, S.R., Prechtl, E.F.: Development of a piezoelectric servo flap for helicopter rotor control. Smart Mater. Struct. **5**(1), 26–34 (1996)
82. Samak, D.K., Chopra, I.: A feasibility study to build a smart rotor: trailing edge flap actuation. SPIŠs North American Symposium on Smart Structures and Materials, Albuquerque, New Mexico, USA (1993)
83. Walz, C., Chopra, I.: Design and testing of a helicopter rotor model with smart trailing edge flaps. In: Proceedings of the 35th AIAA/ASME Structures, Structural Dynamics, and Materials Conference, Adaptive Structures Forum, No. AIAA-1994-1767- CP, Hilton Head, South Carolina, USA, pp. 309–319 (1994)
84. Ben-Zeev, O., Chopra, I.: Advances in the development of an intelligent helicopter rotor employing smart trailing-edge flaps. Smart Mater. Struct. **5**(1), 11–25 (1996)
85. Koratkar, N.A., Chopra, I.: Analysis and testing of a Froude-scaled helicopter rotor with piezoelectric bender actuated trailing edge flaps. J. Intell. Mater. Syst. Struct. **8**(7), 555–570 (1997)
86. Koratkar, N.A., Chopra, I.: Wind tunnel testing of a Mach-scaled rotor model with trailing-edge flaps. Smart Mater. Struct. **10**(1), 1–14 (2001)
87. Fulton, M.V., Ormiston, R.A.: Hover testing of a small-scale rotor with on-blade elevons. In: Proceedings of the 53rd Annual Forum of the American Helicopter Society, Virginia Beach, Virginia, USA, April 29–May 1 1997
88. Fulton, M.V., Ormiston, R.A.: Small-scale rotor experiments with on-blade elevons to reduce vibratory loads in forward flight. In: Proceedings of the 54th Annual Forum of the American Helicopter Society, Washington, D.C., pp. 433–450 (1998)
89. Samak, D.K., Chopra, I.: Desing of high force, high displacement actuators for helicopter rotors. Smart Mater. Struct. **5**(1), 58–67 (1996)
90. Chandra, R., Chopra, I.: Actuation of trailing edge flap in a wing model using piezostack device. J. Intell. Mater. Syst. Struct. **9**(10), 847–853 (1998)
91. Prechtl, E.F., Hall, S.R.: Design of a high-efficiency discrete servo-flap actuator for helicopter rotor control. In: Regelbrugge, M.E. (ed.) Proceedings of SPIE, vol. 3041, pp. 158–182 (1997)
92. Hall, S.R., Prechtl, E.F.: Preliminary testing of a Mach-scaled active rotor blade with a trailing-edge servo flap. In: Wereley, N.M. (ed.) Proceedings of SPIE, vol. 3668, pp. 14–21 (1999)
93. Lee, T., Chopra, I.: Design and spin testing of an active trailing edge flap actuated with piezostacks. In: Proceedings of the 40th AIAA/ASME/ASCE/AHS/ASC Structures, Structural Dynamics, and Materials Conference, No. AIAA-1999-1502, St.Louis, Missouri, USA, 12–15 April 1999
94. Lee, T., Chopra, I.: Development and validation of a refined piezostack-actuated trailing-edge flap actuator for a helicopter rotor. In: Wereley, N.M. (ed.) Proceedings of SPIE, vol. 3668, pp. 22–36 (1999)

95. Straub, F.K., Ngo, H.T., Anand, V., Domzalski, D.B.: Development of a piezoelectric actuator for trailing edge flap control of full scale rotor blades. Smart Mater. Struct. **10**(1), 25–34 (2001)
96. Janker, P., Kleoppel, V., Hermle, T., Lorkowski, T., Storm, S., Christmann, M., Wettemann, M.: Development and evaluation of a hybrid piezoelectric actuator for advanced flap control technology. In: Proceedings of the 25th European Rotorcraft Forum, Rome, Italy, 14–16 September 1999
97. Shaner, M.C., Chopra, I.: Design and testing of a piezostack-actuated leadingedge flap. In: Wereley, N.M. (ed.) Proceedings of SPIE, vol. 3668, pp. 50–59 (1999)
98. Wang, Q., Xu, B., Kugel, V., Cross, L.: Characteristics of shear mode piezoelectric actuators. IEEE Trans. Ultrason. Ferroelectr. 767–770 (1996)
99. Benjeddou, A., Trindade, M., Ohayon, R.: New shear actuated smart structure beam finite element. J. Am. Inst. Aeronaut. Astronaut. **37**, 378–383 (1999)
100. Trindade, M.A., Benjeddou, A.: Finite element homogenization technique for the characterization of d15 shear piezoelectric macro-fibre composites. Smart Mater. Struct. 20(7), 075012(2011)
101. Berik, P., Benjeddou, A.: Piezoelectric d15 shear response-based torsion actuation mechanism: an experimental benchmark and its 3D finite element simulation. Int. J. Smart Nano Mater. **1**(3), 224–235 (2010)
102. Kranz, B., Benjeddou, A., Drossel, W.G.: Numerical and experimental characterizationsof longitudinally polarized piezoelectric d 15 shear macro-fiber composites. Acta Mech. **224**(11), 2471–2487 (2013)
103. Viswamurthy, S.R., Ganguli, R.: An optimization approach to vibration reduction in helicopter rotors with multiple active trailing edge flaps. Aerosp. Sci. Technol. **8**(3), 185–194 (2004)
104. Lim, K.B.: Method for optimal actuator and sensor placement for large flexible structures. J. Guid. Control Dyn. **15**(1), 49–57 (1992)
105. Dhingra, A.K., Lee, B.H.: Optimal placement of actuators in actively controlled structures. Eng. Optim. **23**(2), 99–118 (1994)
106. Varadan, V.V., Kim, J., Varadan, V.K.: Optimal placement of piezoelectric actuators for active noise control. AIAA J. **35**(3), 526–533 (1997)
107. Bhargava, A., Chaudhry, Z., Liang, C., Rogers, C.A.: Experimental verification of optimal actuator location and configuration based on actuator power factor. J. Intell. Mater. Syst. Struct. 6(3), 411–418 ((1995)
108. Wang, Q., Wang, C.M.: Optimal placement and size of piezoelectric patches on beams from the controllability perspective. Smart Mater. Struct. **9**(4), 558–567 (2000)
109. Liang, C., Sun, F.P., Rogers, C.A.: Determination of design of optimal actuator location and configuration based on actuator power factor. J. Intell. Mater. Syst. Struct. **6**(4), 456–464 (1995)
110. Kang, Y.K., Park, H.C., Hwang, W., Han, K.S.: Optimum placement of piezoelectric sensor/actuator for vibration control of laminated beams. AIAA J. **34**(9), 1921–1926 (1996)
111. Liang, C., Sun, F.P., Rogers, C.A.: Determination of design of optimal actuator location and configuration based on actuator power factor. J. Intell. Mater. Syst. Struct. **6**(4), 456–464 (1995)
112. Li, Y., Onoda, J., Minesugi, K.: Simultaneous optimization of piezoelectric actuator placement and feedback for vibration suppression. Acta Astronaut. **50**(6), 335–341 (2002)
113. Damaren, C.J.: Optimal location of collocated piezo-actuator/sensor combinations in spacecraft box structures. Smart Mater. Struct. **12**(3), 494–499 (2003)
114. Wang, S.Y., Quek, S.T., Ang, K.K.: Optimal placement of piezoelectric sensor/actuator pairs for vibration control of composite plates. In: Rao, V.S. (ed.) Proceedings of SPIE, vol. 4693, p. 461471 (2002)
115. Dutta, R., Ganguli, R., Mani, V.: Swarm intelligence algorithms for integrated optimization of piezoelectric actuator and sensor placement and feedback gains. Smart Mater. Struct. **20**(10), 105018 (2011)
116. Lee, I., Han, J.H.: Optimal placement of piezoelectric actuators in intelligent structures using genetic algorithms. In: Gobin, P.F., Tatibouet, J. (eds.) Proceedings of SPIE, vol. 2779, pp. 872–875 (1996)

117. Han, J.H., Lee, I.: Optimal placement of piezoelectric sensors and actuators for vibration control of a composite plate using genetic algorithms. Smart Mater. Struct. **8**(2), 257–267 (1999)
118. Sadri, A.M., Wright, J.R., Wynne, R.J.: Modelling and optimal placement of piezoelectric actuators in isotropic plates using genetic algorithms. Smart Mater. Struct. **8**(4), 490–498 (1999)
119. Sheng, L., Kapania, R.K.: Genetic algorithms for optimization of piezoelectric actuator locations. AIAA J. **39**(9), 1818–1822 (2001)
120. Li, Q.S., Liu, D.K., Tang, J., Zhang, N., Tam, C.M.: Combinatorial optimal design of number and positions of actuators in actively controlled structures using genetic algorithms. J. Sound Vib. **270**(4–5), 611–624 (2004)
121. Kapania, R.K., Sheng, L.: Toward more effective genetic algorithms for the optimization of piezoelectric actuator locations. AIAA J. **40**(6), 1246–1250 (2002)
122. Liu, D.K., Yang, Y.L., Li, Q.S.: Optimum positioning of actuators in tall buildings using genetic algorithm. Comput. Struct. **81**(32), 2823–2827 (2003)
123. Goldberg, D.E.: Genetic Algorithms in Search, Optimization and Machine Learning. Addison-Wesley, Reading (1989)
124. Ganguli, R.: Engineering Optimization: A Modern Approach. Universities Press, Boca Raton (2012)
125. Heverly, D.E., Wang, K.W., Smith, E.C.: An optimal actuator placement methodology for active control of helicopter airframe vibrations. J. Am. Helicopter Soc. **46**(4), 251–261 (2001)
126. Padula, S.L., Kincaid, R.K.: Optimization strategies for sensor and actuator placement. NASA/TM-1999-209126 (1999)
127. Ge, P., Jouaneh, M.: Modeling hysteresis in piezoceramic actuators. Precis. Eng. **17**(3), 211–221 (1995)
128. Hughes, D., Wen, J.T.: Preisach modeling of piezoceramic and shape memory alloy hysteresis. Smart Mater. Struct. **6**(3), 287–300 (1997)
129. Kurdila, A.J., Li, J., Strganac, T., Webb, G.: Nonlinear control methodologies for hysteresis in PZT actuated on-blade elevons. J. Aerosp. Eng. **16**(4), 167–176 (2003)
130. Muir, E.R., Liu, L., Friedmann, P.P., Kumar, D.: Effect of piezoceramic actuator hysteresis on helicopter vibration and noise reduction. J. Guid. Control Dyn. **35**(4), 1299–1311 (2012)
131. Ganguli, R., Mallick, R., Bhat, M.S.: An experimental and numerical study of piezoceramic actuator hysteresis in helicopter active vibration control. In: Proceedings of the Institution of Mechanical Engineers, Part G: Journal of Aerospace Engineering (2013). 0954410013478254
132. Krasnoselskii, M.A., Pokrovskii, A.V.: Systems with Hysteresis. Springer, Berlin (1983)
133. Mayergoyz, I.D.: Hysteresis models from the mathematical and control theory points of view. J. Appl. Phys. **57**(1), 3803–3805 (1985)
134. Doong, T., Mayergoyz, I.D.: On numerical implementation of hysteresis models. IEEE Trans. Magn. **21**(5), 1853–1855 (1985)
135. Sreeram, P.N., Naganathan, N.G.: Hysteresis prediction for a piezoceramic material system. In: Proceedings of 1993 ASME Winter Annual Meeting, New Orleans, Louisiana, USA, pp. 35–42 (1993)
136. Yu, Y., Naganathan, N., Dukkipati, R.: Preisach modeling of hysteresis for piezoceramic actuator system. Mech. Mach. Theory. **37**(1), 49–59 (2002)
137. Hu, H., Mrad, R.B.: A discrete-time compensation algorithm for hysteresis in piezoceramic actuators. Mech. Syst. Signal Process. **18**(1), 169–185 (2004)
138. Natale, C., Velardi, F., Visone, C.: Identification and compensation of Preisach hysteresis models for magnetostrictive actuators. Phys. B. **306**(1–4), 161–165 (2001)
139. Park, Y.W., Lim, M.C., Kim, D.Y.: Control of a magnetostrictive actuator with feedforwarding inverted hysteresis. J. Magn. Magn. Mater. 272–276, pp. e1757–e1759, Supplement 1 (2004)
140. Mayergoyz, I.D.: Dynamic Preisach models of hysteresis. IEEE Trans. Magn. **24**(6), 2925–2927 (1988)
141. Bertotti, G.: Dynamic generalization of the scalar Preisach model of hysteresis. IEEE Trans. Magn. **28**(5), 2599–2601 (1992)

142. Song, D., Li, C.J.: Modeling of piezo actuatorsŠs nonlinear and frequency dependent dynamics. Mechatronics **94**, 391–410 (1999)
143. Dang, X., Tan, Y.: Neural networks dynamic hysteresis model for piezoceramic actuator based on hysteresis operator of first-order differential equation. Phys. B. **365**(1–4), 173–184 (2005)
144. Nitzsche, F., Breitbach, E.: Using adaptive structures to attenuate rotary wing aeroelastic response. J. Aircr. **31**(5), 1178–1188 (1994)
145. Chen, P., Chopra, I.: Induced strain actuation of composite beams and rotor blades with embedded piezoceramic elements. Smart Mater. Struct. **5**, 35–48 (1996)
146. Chen, P., Chopra, I.: Hover testing of smart rotor with induced-strain actuation of blade twist. J. Am. Inst. Aeronaut. Astronaut. **35**(1), 6–16 (1997)
147. Chen, P., Chopra, I.: Wind tunnel test of a smart rotor model with individual blade twist control. J. Intell. Mater. Syst. Struct. **8**(5), 414–423 (1997)
148. Derham, R., Hagood, N.: Rotor design using smart materials to actively twist blades. Presented at 52nd Annual forum, American Helicopter Society, Washington, DC, USA (1996)
149. Bernhardm, A., Chopra, I.: Analysis of a bending-torsion coupled actuator for a smart rotor with active blade tips. Smart Mater. Struct. **10**(1), 35–52 (1998)
150. Chandra, R., Chopra, I.: Structural modeling of composite beams with induced strain actuations. J. Am. Inst. Aeronaut. Astronaut. **31**(9), 1692–1701 (1993)
151. Vlasov, V.: Thin-Walled Elastic Beams. Wiley-Interscience, New York (1961)
152. Rodgers, J., Hagood, N.: Design, manufacture, and testing of an integral twist actuated rotor blade. Presented at 8th International Conference on Adaptive Structures and Technology, Wakayama, Japan (1997)
153. Wilkie, W.K., Wilbur, M.L., Mirick, P.H., Cesnik, C.E.S., Shin, S.J.: Aeroelastic analysis of the NASA/Army/MIT active twist rotor. In: Proceedings of the 55th Annual Forum of the American Helicopter Society, Montreal, Canada, May 25–27, pp. 545–557 (1999)
154. Johnson, W.: Rotorcraft aerodynamics models for a comprehensive analysis. Presented at 54th Annual forum, American Helicopter Society, Washington, DC, USA (1998)
155. Shin, S., Cesnik, C., Hall, S.R.: Nasa/army/mit active twist rotor closed loop control test for vibration reduction. In: Proceedings of 29th European Rotorcraft Forum, Friedrichshafen, Germany (2003)
156. Chattopadhyay, A., Lin, Q., Gu, H.: Vibration reduction in rotor blades using active composite box beam. J. Am. Inst. Aeronaut. Astronaut. **38**(7), 1125–1131 (2000)
157. Chattopadhyay, A., Gu, H., Lin, Q.: Modeling of smart composite box beams with nonlinear induced strain. J. Compos. 30(B), 603–612 (1999)
158. Cesnik, C., Shin, S.: On the twist performance of a multiple-cell active helicopter blade. Smart Mater. Struct. **10**, 53–61 (1999)
159. Cesnik, C., Shin, S.: On the modeling of integrally actuated helicopter blades. Int. J. Solids Struct. **38**(10–13), 1765–1789 (2001)
160. Cesnik, C., Shin, S.: Dynamic response of active twist rotor blades. Smart Mater. Struct. **10**, 62–76 (2001)
161. Buter, A., Breitbach, E.: Adaptive blade twist–calculations and experimental results. Aerosp. Sci. Technol. **4**(5), 309–319 (2000)
162. Kube, R., Kloppel, K.: On the role of prediction tools for adaptive rotor system developments. Smart Mater. Struct. **10**(1), 137–144 (2001)
163. Ghiringhelli, G., Masarati, P., Mantegazza, P.: Analysis of an actively twisted rotor by multibody global modeling. Compos. Struct. **52**(1), 113–122 (2001)
164. Wilkie, W., Park, K., Belvin, W.: Helicopter dynamic stall suppression using piezoelectric active fiber composite rotor blades. AIAA/ASME/AHS Structures, Structural Dynamics and Materials Conference, Long Beach, USA (1998)
165. Ghiringhelli, G., Masarati, P., Mantegazza, P.: A multi-body implementation of finite volume beams. J. Am. Inst. Aeronaut. **38**(1), 131–138 (2000)
166. Pitt, D., Peters, D.: Theoretical prediction of dynamic inflow derivatives. Vertica **5**(1), 21–34 (1981)

167. Smith, E.: Induced-shear piezoelectric actuators for active twist rotor blades. In: 43rd AIAA/ASME/ASCE/AHS/ASC Structures, Structural Dynamics, and Materials Conference, AIAA Paper No. 2002–1446-CP (Denver, Colorado) (2002)
168. Gessow, A., Myers, G.: Aerodynamics of the Helicopter. College Park Press, College Park (1985)
169. Leishmann, J.: Principles of Helicopter Aerodynamics. Cambridge University Press, Cambridge (2000)
170. Nguyen, K., Chopra, I.: Application of higher harmonic control to rotors operating at high speed and thrust. J. Am. Helicopter Soc. **35**(3), 78–89 (1990)
171. Nguyen, K.: Active control of helicopter blade stall. J. Aircr. **35**(1), 91–98 (1998)
172. Depailler, G., and Friedmann, P.: Reduction of vibrations due to dynamic stall in helicopters using an actively controlled flap. AIAA Paper No., AIAA-2002-1431 (2002)

Chapter 2
Mathematical Modeling

The governing equations of the helicopter rotor blade enabling piezoceramic-axial, bending, and shear actuation are derived in this chapter and an outline of the aeroelastic analysis used in this book is given. A background on piezoelectric materials used in this book is also provided. Section 2.1 begins with an introduction to piezoelectric materials and Sect. 2.2 explains the piezoceramic actuation concept. Section 2.3 provides an introduction to terminology used in the helicopter field. In Sect. 2.4, structural modeling is explained. Section 2.5 explains the aerodynamic model used for the aeroelastic analysis. Section 2.6 presents the blade and hub loads. Section 2.7 explains the aeroelastic analysis of a rotor. Section 2.8 gives the summary of this chapter.

2.1 Piezoelectric Materials

Piezoelectricity stems from the Greek word *piezein*, meaning to press or squeeze. It was found out that quartz changed its dimensions when subjected to an electric field and conversely generated an electric charge when it was pressed. A piezoelectric material develops a potential across its boundaries when subjected to a mechanical stress (or pressure), and vice versa, when an electric field is applied to the material, a mechanical deformation ensues. Piezoelectric materials therefore fall in the class of smart materials which are typically controllable using some physical variable such as electric or magnetic fields or temperature. Ferroelectricity is a subgroup of piezoelectricity, where a spontaneous polarization exists that can be reoriented by application of an AC electric field. Two types of piezoceramics exist: Soft (Ex: PZT-5H, PZT-5A) and hard (Ex: PZT-4D, PZT-8).

Soft piezoceramics are donor-doped PZT (Lead zirconate titanate) and hard piezoceramics are considered to be acceptor-doped PZT's. Material research says that soft

R. Ganguli et al., *Smart Helicopter Rotors*, Advances in Industrial Control,
DOI 10.1007/978-3-319-24768-7_2

Table 2.1 Differences between soft and hard piezoceramics

Property	Soft PZT	Hard PZT
Piezoelectric constants	Larger	Smaller
Permittivity	Higher	Lower
Electromechanical coupling factors	Higher	Lower
Resistance	High	Lower
Linearity	Poor	Better
Polarization/Depolarization	Easier	Difficult
Hysteresis	Greater	Lower

PZTs have an exceptionally high-domain wall mobility, while the hard piezoceramics suppress the domain wall response. Regions in the piezoceramic with uniformly oriented spontaneous polarization are called domains and the region between two such domains is called a domain wall. This characteristic leads to the following differences between soft and hard piezoceramics (Table 2.1).

When manufactured, a piezoelectric material has electric dipoles arranged in random directions. In order to avoid a random response from each of these dipoles, alignment of dipoles is required so that a uniform response is achieved when an external stimuli is applied as shown in Fig. 2.1. For poling (aligning), the material is heated above its Curie temperature and a strong electric field is applied. The direction in which the field is applied is the polarization direction, resulting in the alignment of the dipoles in that direction. For fixing the poling direction permanently, the material is cooled below its Curie temperature. Every piezoelectric material has a specific Curie temperature. Polarization never results into a full/perfect alignment of all dipoles to give the ideal piezoelectric effect; however, the polycrystalline ceramic exhibits a large piezoelectric effect. While the polarization is being carried out, the dipoles are properly aligned but as soon as the external stimuli are removed, there is an introduction of an extremely small amount of randomness in dipoles once again. This process is depicted in Fig. 2.2. During this process, there is a very small expansion of the material along the poling axis and a contraction in both

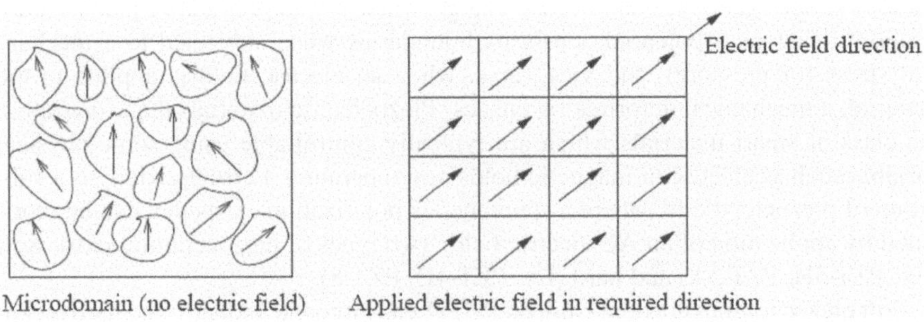

Fig. 2.1 Polarization in a piezoelectric domain

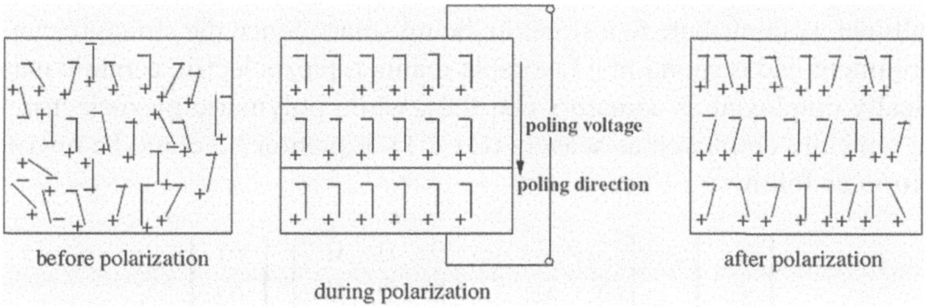

Fig. 2.2 Detailed polarization process

directions perpendicular to it. These dipoles after poling respond collectively to the external stimuli [1]. Application of an electric field (voltage) to a poled PZT results in its deformation (straining). Depolarization of the piezoelectric ceramic can result if it is exposed to excessive heat, electrical drive, or mechanical stress or any combination thereof. The temperature at which piezoelectric ceramic will be totally depoled is known as the "Curie point." The constitutive equations are based on the assumption that the total strain is the sum of the mechanical strain induced by stress, the thermal strain induced by temperature, and the actuation strain dependent on applied electric field. The relation between strain and voltage is linear in the first order at relatively low electric fields and low mechanical stress levels. In general, piezoelectric materials are relatively linear at low electric fields and bipolar in nature. At high levels of electric fields, they exhibit nonlinearity to quite an extent. Although under an applied electric field they generate very low strains, they cover a wide range of actuation frequency [2].

Coupled electromechanical constitutive relations [3] for a piezoceramic are given as follows:

$$D_i = e^{\sigma}_{ij} E_j + d^{d}_{im} \sigma_m + a_i \Delta T \qquad i = 1..3,\ j = 1..3, \tag{2.1}$$

$$\epsilon_k = d^{c}_{jk} E_j + S^{E}_{km} \sigma_m + \alpha_{tck} \Delta T \qquad k = 1..6,\ m = 1..6. \tag{2.2}$$

Piezoelectric materials have the capability of undergoing strain on the application of an electric field. This concept is known as the converse piezoelectric effect (Eq. 2.2) and is exploited in actuators. Examples for the use of converse piezoelectric effect are sonar, ultrasound generation, positioners, motors, vibration cancelation, ultrasonic surgery, and so on. It allows the alteration of system characteristics as well as the system response. The same material (PZT) generates an electric charge when subjected to a mechanical force or deformation. This is known as the direct piezoelectric effect (Eq. 2.1) and is used in sensors. Examples for the use of direct piezoelectric effect are accelerometers, hydrophones, ultrasonic transducers, etc. The piezoelectric converse and direct effect is explained in Fig. 2.3. The microprocessors analyze the response from the sensors and use distributed parameter control theory to command the actuators to apply localized stresses to minimize the response of the system. Hence, piezoelectric materials are utilized either as actuators or sensors which ultimately contribute to a structure being smart. Since the structure can sense its environment and respond in a favorable manner, piezoelectric ceramic materials are typically employed as actuators (Eq. 2.2), while polymeric piezoelectric materials are typically employed as sensors (Eq. 2.1). Equation 2.2 could be rewritten in matrix form as follows:

$$\begin{Bmatrix} \epsilon_1 \\ \epsilon_2 \\ \epsilon_3 \\ \epsilon_{23} \\ \epsilon_{31} \\ \epsilon_{12} \end{Bmatrix} = \begin{bmatrix} S_{11} & S_{12} & S_{13} & 0 & 0 & 0 \\ S_{12} & S_{11} & S_{13} & 0 & 0 & 0 \\ S_{13} & S_{13} & S_{33} & 0 & 0 & 0 \\ 0 & 0 & 0 & S_{44} & 0 & 0 \\ 0 & 0 & 0 & 0 & S_{44} & 0 \\ 0 & 0 & 0 & 0 & 0 & S_{66} \end{bmatrix} \begin{Bmatrix} \sigma_1 \\ \sigma_2 \\ \sigma_3 \\ \tau_{23} \\ \tau_{31} \\ \tau_{12} \end{Bmatrix} + \begin{bmatrix} 0 & 0 & d_{31} \\ 0 & 0 & d_{32} \\ 0 & 0 & d_{33} \\ 0 & d_{15} & 0 \\ d_{15} & 0 & 0 \\ 0 & 0 & 0 \end{bmatrix} \begin{Bmatrix} E_1 \\ E_2 \\ E_3 \end{Bmatrix} + \begin{Bmatrix} \alpha_{tc1} \\ \alpha_{tc2} \\ \alpha_{tc3} \\ \alpha_{tc4} \\ \alpha_{tc5} \\ \alpha_{tc6} \end{Bmatrix} \Delta T. \tag{2.3}$$

Extension/Contraction in a piezoceramic element depends on the polarity of the electric filed and the poling direction of the piezoceramic. The piezoceramic will shorten if a voltage of opposite polarity to the poling direction is applied to the electrode. If the applied voltage has same polarity as the poling voltage, the piezoceramic will lengthen. But if an AC voltage is applied to the electrodes, the piezoceramic will grow and shrink at the same frequency as that of the applied voltage. These phenomena are depicted in Fig. 2.4. Piezoelectric solid-state transducers are characterized by high forces in the range of kilo newtons, with reaction time of the order of a few milliseconds, along with a positioning accuracy of the order of a few nanometers. They have been successfully used over many years in a wide range of applications such as ultrasonic transducers, accelerometers, gramophones, resonators, filters, inkjet printers, and as various kinds of sensors in structural problems. Reviews on aerospace applications are given by [4]. Regarding the availability, there are various manufacturers who supply the characteristics of their products. Piezoelectrics are available in the form of sheets and films.

The advantages of piezoelectric materials being used as sensors and actuators include ease of integration into existing structures, easy control by voltage, low weight, low power requirements, low-field linearity, and high bandwidth (allowing large range of applications). In general, piezoelectric materials can be broadly classified into two groups: piezoceramics and piezopolymers. The most common piezoceramic is lead zirconate titanate (PZT) and the most common piezopolymer is Polyvinylidene Fluoride (PVDF). PZT is a ceramic and has high stiffness, while the PVDF polymer is more flexible and has low stiffness and high damping. The high stiffness of the PZT makes it a suitable actuator because of its high actuation authority and fast actuation response. In contrast, the flexibility and low stiffness of PVDF

Fig. 2.3 Direct and converse piezoelectric effect

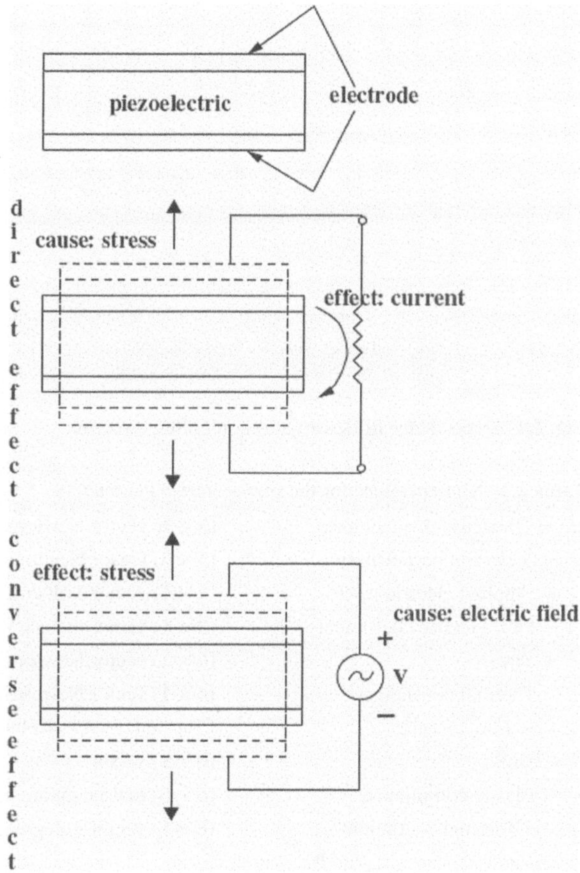

makes it a better sensor. From these two types of piezoelectric material, there exist a variety of configurations in which they can be manufactured to be used as sensors and actuators. In terms of handling and practicality, the brittleness of piezoceramics places a restriction on its minimum thickness. Also, the attachment of piezoelectric materials to the host structure has to be performed with proper electrical insulation (Table 2.2).

2.1.1 Single Crystals

In general, relaxor-based ferroelectric single crystals have a different crystal composition and orientation as compared to conventional soft or hard piezoceramics. Physicists and metallurgists consider this structure to be very complex. Nonstoichiometric doping proved to be effective in creating instabilities in the material to

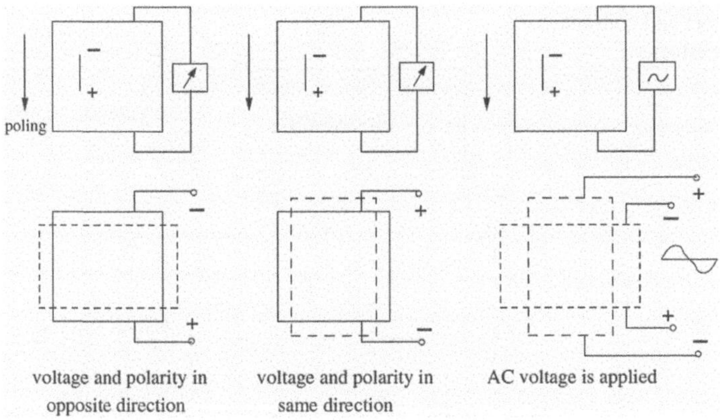

Fig. 2.4 Dependence of deformation on poling direction

Table 2.2 Nomenclature for the piezoceramic parameters

D = Dielectric displacement	(3 × 1) vector Newtons/milliVolts
e = Dielectric permittivity	(3 × 3) tensor Newtons/square Volt
E = Applied electric field	(3 × 1) vector Volts/meter
d = Piezoelectric coefficients	(3 × 1) vector meters/Volt
σ = Stress	(6 × 1) vector Newtons/square meter
a = Thermal constants	(6 × 1) vector Newtons/Volt-meter-degrees Kelvin
ΔT = Temperature	Constant degrees Kelvin
ϵ = Strain	(6 × 1) vector
S = Elastic compliance	(6 × 6) matrix square meters/Newton
α_{tc} = Thermal coefficient	(6 × 1) vector 1/degrees Kelvin

generate large responses, making the material "smarter." The lead zirconate–lead titanate solid solution (PZT) system is one such successful example. The use of Nb doping can dramatically increase the value of the piezoelectric property of PZT. Of this category are $(1-x)Pb(Zn_{1/3}Nb_{2/3})O_3 - xPbTiO_3$ (PZN–PT), and lead zinc niobate–lead titanate crystals. Here x is the percentage of $PbTiO_3$ in the crystal.

Domain engineered single-crystal systems (like PZN–PT) exhibit superior electromechanical properties compared to the conventional PZT ceramics. It is natural to ask the question, "Why the electromechanical properties have such values?" The answer to this lies in the crystallographic orientation and phase of the structure. It has been found that near the morphotropic phase boundary (MPB) between the rhombohedral and tetragonal phases, the levels of strain and the piezoelectric coefficients rise anomalously. In summary, without going into greater detail regarding the material configuration, this is the key to the high piezoelectric coefficients in single crystals like PZN–8 %PT.

Table 2.3 Comparison of the electromechanical properties

	Max strain %	Elastic energy density J/cm^3	Coupling efficiency (k^2) %	Relative speed full cycle
PZT	0.2	0.10	52	Fast
PZN–PT	1–1.7	1.0	81	Fast

Single-crystal materials are grown in crucibles at high, very tightly controlled temperatures using a starting seed to determine the crystallographic orientation. Orientations can be tailored to optimize response to longitudinal, transverse, and shear excitations commonly used in transducer designs. A significant reorientation of domains is achieved during polarization of single-crystal materials, resulting in higher electromechanical coupling factors (applied electrical energy is converted into mechanical or acoustic energy) of 90 % or more, as compared to 70 % for polarized conventional piezoceramics. Single-crystal piezoelectric materials also offer higher bandwidths, up to 135 %, as compared to 40–45 % for PZT. In actuator applications, single-crystal materials can achieve field-induced strains three times greater than those obtained in PZT. The dielectric losses in single crystal are much less than 1 %, as compared to 2 % for PZT. A quantitative comparison of the properties for general PZT and PZN–PT ceramics is made in Table 2.3 [5].

According to Park and Shrout [5], in designing an actuator, the maximum strain energy density should be as high as possible. Generally, the induced strain due to the applied field is the most important parameter for an actuators performance. This is also reflected in the strain energy density function which can be defined as below:

$$e_{max} = \frac{1}{\rho}\left(\frac{1}{4}\right)\left(\frac{1}{2}\right)E(s_{max}^2), \tag{2.4}$$

where e_{max} is the strain energy density, E is the actuators elastic modulus, s_{max} is the maximum field-induced strain, and ρ is the actuators density. The factor $1/4$ in Eq. 2.4 is appropriate for an actuator impedance related to its surroundings.

Piezoelectric properties of PZN–PT-type single crystals with <001>-orientation have been found to posses large direct piezoelectric coefficients (d_{33}, d_{31}) which are almost of the order of 10–15 times that of a soft piezoceramic such as PZT-5H. However, the shear coefficient d_{15} of these materials is relatively poor as compared to soft piezoceramics such as PZT-5H. Liu et al. [6] reported that the piezoelectric shear coefficient d_{15} of PZN–4.5 %PT and PZN–8 %PT crystals oriented in <111> is very high as compared to the conventional soft piezoceramic materials. Here, <001> and <111> orientations are directions in a crystal. These directions are shown in Fig. 2.5. Such single crystals are now commercially available and should therefore be evaluated for applications.

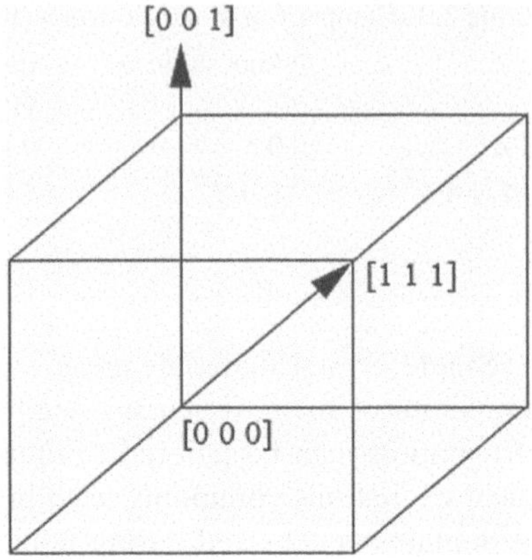

Fig. 2.5 The <001> and <111> orientations in a crystal

2.2 Piezoceramic Actuation

The relationship between induced strain and the applied electric field for piezoelectric materials can be understood with help of the piezoelectric constitutive equations. Modeling of induced strain actuation was addressed initially by Crawley [7] and later was discussed and validated in detail by Hong and Chopra [8]. The constitutive equations for piezoelectric material are 'poling-direction' dependent; therefore, appropriate care should be taken when using them. For a piezoelectric segment, there are three directions in which poling can be established.

- If axial and bending actuation is desired, poling direction should be selected as '3.' For a piezo segment shown in Fig. 2.6, poling is in the 3 direction and the constitutive equations can be written as

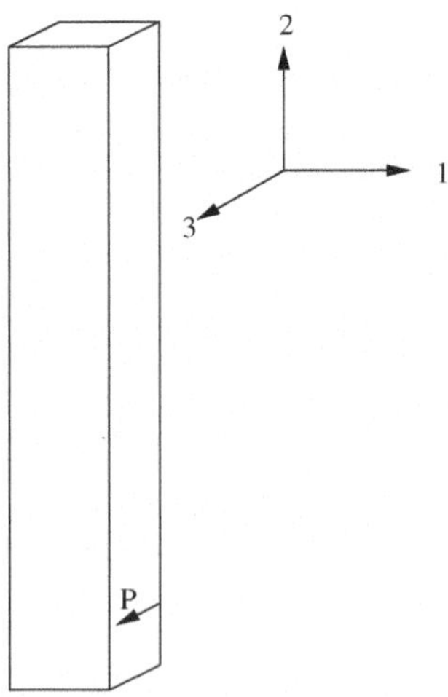

Fig. 2.6 Configuration of a piezo segment with poling direction 3

$$\begin{Bmatrix} \lambda_{11} \\ \lambda_{22} \\ \lambda_{33} \\ \lambda_{23} \\ \lambda_{31} \\ \lambda_{12} \end{Bmatrix} = \begin{bmatrix} 0 & 0 & d_{31} \\ 0 & 0 & d_{32} \\ 0 & 0 & d_{33} \\ 0 & d_{24} & 0 \\ d_{15} & 0 & 0 \\ 0 & 0 & 0 \end{bmatrix} \begin{Bmatrix} E_1 \\ E_2 \\ E_3 \end{Bmatrix}. \tag{2.5}$$

This relationship is found in most literature where poling in 3 direction is assumed. Here, λ is the actuation strain and in general, the first subscript of the d-constant d_{jk} gives the "electrical" direction (field or dielectric displacement) and the second gives the component of mechanical deformation or stress. The convention is to define the poling direction as the 3-axis, and the shear planes are indicated by the subscripts 4, 5, and 6 and are perpendicular to directions 1, 2, and 3, respectively. The planar isotropy of poled ceramics is expressed in their piezoelectric constants by the equalities $d_{32} = d_{31}$ (an electric field parallel to the poling axis 3 interacts in the same way with axial stress along either the 2-axis or the 1-axis) and $d_{24} = d_{15}$. Similar relations hold for the elastic constants because of the isotropy in the plane perpendicular to the polar axis. In general, the piezoelectric strain coefficients d_{15} and d_{24} contribute to shear strains and others to normal strains.

- If shear/torsional actuation is desired, then the poling direction should be selected in the length direction, which is '2' as shown in Fig. 2.7. For such a piezo segment, the constitutive equations can be written as

$$\begin{Bmatrix} \lambda_{11} \\ \lambda_{22} \\ \lambda_{33} \\ \lambda_{23} \\ \lambda_{31} \\ \lambda_{12} \end{Bmatrix} = \begin{bmatrix} 0 & d_{31} & 0 \\ 0 & d_{33} & 0 \\ 0 & d_{32} & 0 \\ 0 & 0 & d_{24} \\ 0 & 0 & 0 \\ d_{15} & 0 & 0 \end{bmatrix} \begin{Bmatrix} E_1 \\ E_2 \\ E_3 \end{Bmatrix}. \tag{2.6}$$

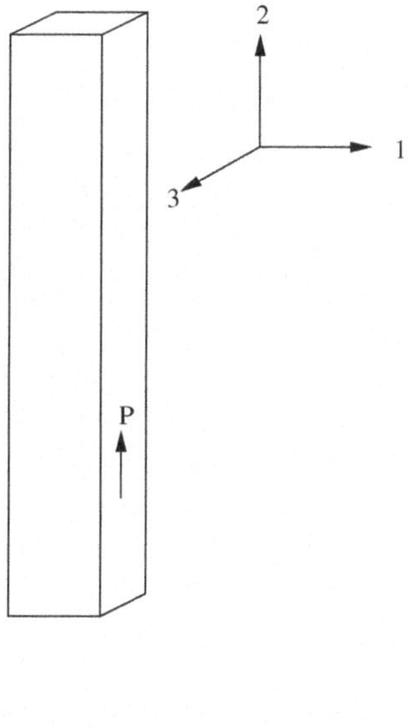

Fig. 2.7 Configuration of a piezo segment with poling direction 2

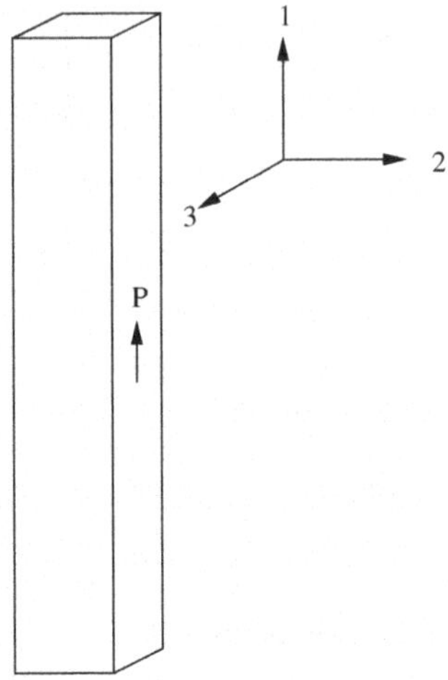

Fig. 2.8 Configuration of a piezo segment with poling direction 1

One can select the case which corresponds to the poling in the span/length direction. But the coordinate system that we refer to is different from that mentioned in the above cases. Thus, we transform case 2 to the following coordinate system

$$\begin{Bmatrix} \lambda_{11} \\ \lambda_{22} \\ \lambda_{33} \\ \lambda_{23} \\ \lambda_{31} \\ \lambda_{12} \end{Bmatrix} = \begin{bmatrix} d_{33} & 0 & 0 \\ d_{31} & 0 & 0 \\ d_{32} & 0 & 0 \\ 0 & 0 & 0 \\ 0 & 0 & d_{24} \\ 0 & d_{15} & 0 \end{bmatrix} \begin{Bmatrix} E_1 \\ E_2 \\ E_3 \end{Bmatrix}. \tag{2.7}$$

For the piezoelectric segment shown in Fig. 2.8, the span and poling direction is the '1' direction. It can be seen from Eq. 2.7 that both λ_{31} and λ_{12} utilize d_{15}. In Chaps. 8–11 of this book, the shear strain λ_{12} is employed to produce twist in the structure.

2.3 Key Terminologies and Nondimensional Parameters

Since this book deals with helicopter main rotor, some key terminologies used in this book are explained first. Figure 2.9 shows the rotor disk. Any point on the rotor disk can be located based on the polar coordinates (r, ψ). Here, r is the radial location and varies from $r = 0$ at the hub to $r = R$ at the tip, where R is the blade radius. The rotor rotates counterclockwise with a constant rotation speed of Ω and moves forward with a speed of V. When the blade is at $\psi = 90°$, it is at the advancing side of the rotor. When the blade is at $\psi = 270°$, it is at the retreating side of the rotor.

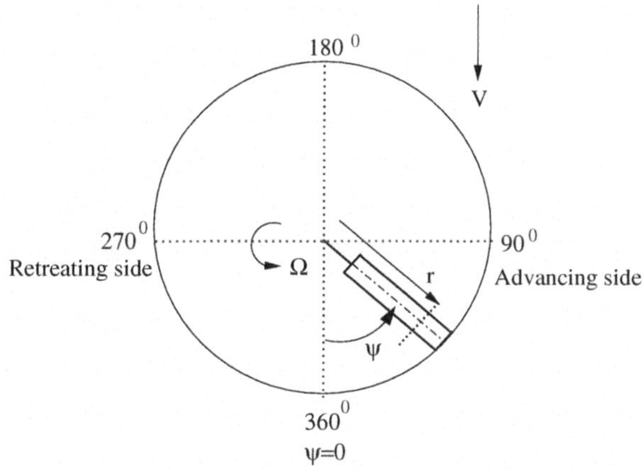

Fig. 2.9 Rotor disk of the helicopter

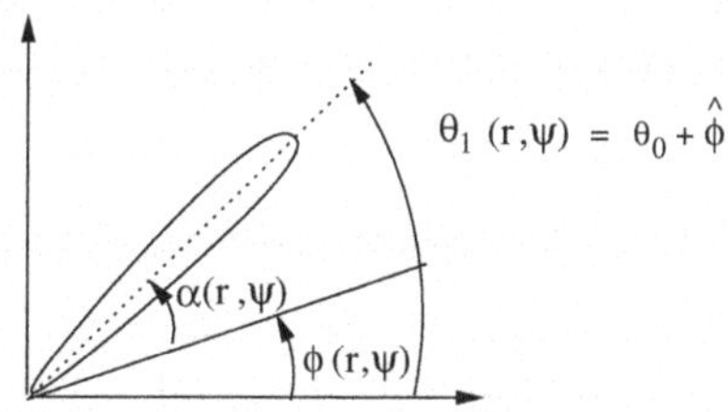

Fig. 2.10 Blade cross section

At any section (r,ψ), the blade section is an airfoil as shown in Fig. 2.10. The angle of attack at any blade section can be written as

$$\alpha(r, \psi) = \theta_1(r, \psi) - \phi(r, \psi), \tag{2.8}$$

where $\theta_1(r, \psi)$ is the local blade pitch and $\phi(r, \psi)$ is the local inflow angle. The inflow angle is usually obtained from a wake model. Wake models for helicopters range from momentum theory which assumes a uniform inflow to free wake models which account for the nonuniformity in inflow which is characteristic of real rotors. Blade section loads are calculated using the angle of attack, $\alpha(r, \psi)$. The local blade pitch $\theta_1(r, \psi)$ includes the effect of both pilot control input θ_0 and elastic twist $\hat{\phi}$.

The nondimensional forward speed (μ), thrust coefficient (C_T), power coefficient (C_P), lock number, and solidity of a rotor are defined below. Here, μ is defined as the ratio of the forward speed of the helicopter to the rotor tip rotation speed (ΩR). Lock number γ is defined as the ratio of aerodynamic forces to the inertial forces and solidity (σ) is defined as the ratio of rotor blade area to rotor disk area.

$$\mu = \frac{V}{\Omega R},$$
$$C_T = \frac{T}{\frac{1}{2}\rho A(\Omega R)^2},$$
$$C_P = \frac{P}{\frac{1}{2}\rho A(\Omega R)^3},$$
$$\gamma = \frac{\rho C_{l\alpha} c R^4}{I_b},$$
$$\sigma = \frac{N_b\, c}{\pi R}.$$

Here, V is the forward speed of the helicopter, ρ is the density of air, Ω is the rotor rotational speed, T is the thrust, P is the power, $C_{l\alpha}$ is the lift curve slope, c is the blade chord, and I_b is the mass moment of inertia about the flapping hinge. Also, A is the rotor disk area and N_b is the number of blades.

2.3.1 Nondimensionalization

The entire formulation and all computations are carried out in nondimensional form. In addition to increase the generality of the analysis, working with nondimensional quantities can help avoid scaling problems while computing. It should be noted that the following physical quantities are nondimensionalized by the given reference parameters. Only nondimensional quantities are used in the analysis (Table 2.4).

Here, m_0 is defined as the reference mass per unit length of an equivalent uniform blade which has the same flap inertia as the actual (nonuniform) blade. Using this definition, m_0 can be defined as follows:

$$m_0 = \frac{3\int_0^R mr^2 dr}{R^3}. \tag{2.9}$$

Table 2.4 Nondimensionalization

Physical quantity	Reference parameter
Length	R
Time	$1/\Omega$
Mass/length	m_0
Velocity	ΩR
Acceleration	$\Omega^2 R$
Force	$m_0\Omega^2 R^2$
Moment	$m_0\Omega^2 R^3$
Energy or work	$m_0\Omega^2 R^3$

2.4 Structural Modeling

2.4.1 Hamilton's Principle

The generic governing differential equations are derived for a smart beam/rotor blade undergoing axial, chord-wise bending, span-wise bending and elastic twist and supporting axial, bending, and torsion actuation. A point P on the undeformed elastic axis undergoes deflection u, v, w in the x, y, z and moves to a point P' as shown in Fig. 2.11. X and Y are the hub coordinate axes and blade is at a precone angle β_p from the hub axes. The cross section undergoes a rotation θ_1 as shown in Fig. 2.12. Here θ_0 is given as

$$\theta_0 = \theta_{75} + \theta_{tw}\left(\frac{x}{R} - 0.75\right) + \theta_{1c}\cos\psi + \theta_{1s}\sin\psi, \tag{2.10}$$

where θ_{75} is the blade pitch at 75 % span of blade, θ_{tw} is the blade linear pretwist, and θ_{1s} and θ_{1c} are the cyclic pitch controls.

The constitutive equations of an isotropic beam plate [9] can be written as follows:

$$\begin{Bmatrix} \epsilon_{xx} \\ \epsilon_{\eta\eta} \\ \epsilon_{\zeta\zeta} \\ \epsilon_{x\eta} \\ \epsilon_{\eta\zeta} \\ \epsilon_{x\zeta} \end{Bmatrix} = \begin{bmatrix} 1/E & -\nu/E & -\nu/E & 0 & 0 & 0 \\ -\nu/E & 1/E & -\nu/E & 0 & 0 & 0 \\ -\nu/E & -\nu/E & 1/E & 0 & 0 & 0 \\ 0 & 0 & 0 & 1/G & 0 & 0 \\ 0 & 0 & 0 & 0 & 1/G & 0 \\ 0 & 0 & 0 & 0 & 0 & 1/G \end{bmatrix} \begin{Bmatrix} \sigma_{xx} \\ \sigma_{\eta\eta} \\ \sigma_{\zeta\zeta} \\ \tau_{x\eta} \\ \tau_{\eta\zeta} \\ \tau_{x\zeta} \end{Bmatrix}. \tag{2.11}$$

The following fundamental assumptions are made for the analysis [10]:

(i) Mid-line of a plate segment does not deform in its own plane;
(ii) the normal stress in the contour direction, $\sigma_{\eta\eta}$, is neglected relative to the normal axial stress σ_{xx}; and

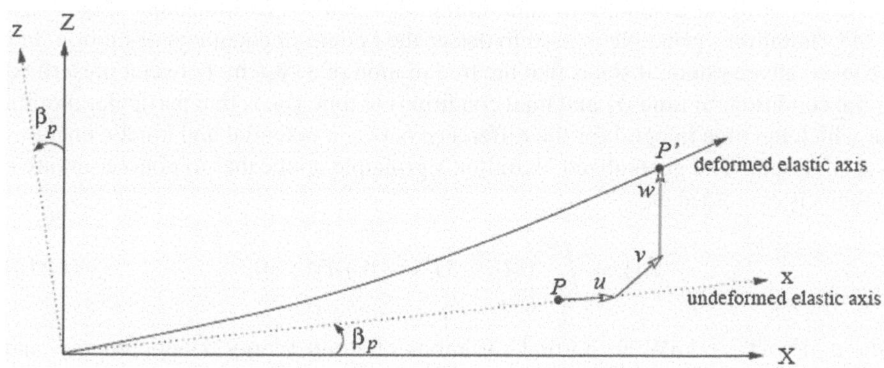

Fig. 2.11 Deformation of rotor blade

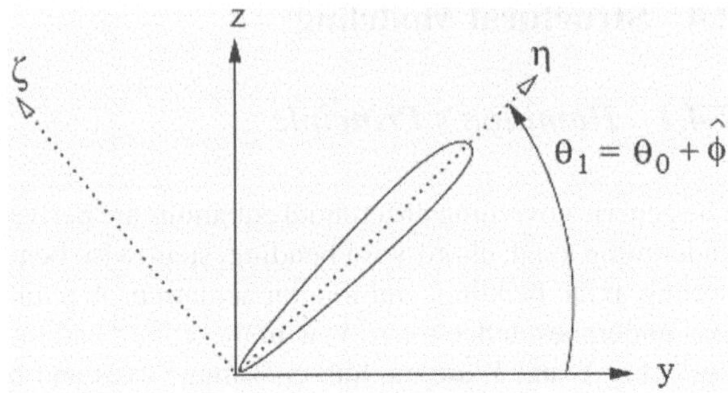

Fig. 2.12 Deformation of cross section

(iii) rotor blade is a long slender beam and hence the uniaxial stress assumptions can be made; $\sigma_{\eta\eta} = 0$, $\sigma_{\zeta\zeta} = 0$ and $\tau_{\eta\zeta} = 0$.

The strain displacement field (accurate up to second order and accounting for moderate deflections) is defined as

$$\begin{aligned}\epsilon_{xx} &= u' + \frac{v'}{2} + \frac{w'^2}{2} - \lambda_T \phi'' + (\eta^2 + \zeta^2)(\theta_0 \phi' + (\phi')^2/2) \\ &- v''[\eta \cos(\theta_0 + \hat{\phi}) - \zeta \sin(\theta_0 + \hat{\phi})] \\ &- w''[\eta \sin(\theta_0 + \hat{\phi}) + \zeta \cos(\theta_0 + \hat{\phi})],\end{aligned} \tag{2.12}$$

$$\epsilon_{x\eta} = -(\zeta + \lambda_{T,\eta})\phi' = -\hat{\zeta}\phi', \tag{2.13}$$

$$\epsilon_{x\zeta} = (\eta - \lambda_{T,\zeta})\phi' = \hat{\eta}\phi'. \tag{2.14}$$

This results into equations being geometrically nonlinear. Here, v, w and $\hat{\phi}$ are axial, chord-wise bending, span-wise bending, and elastic twist, respectively; θ_0 is the blade pitch; η and ζ are the cross-sectional coordinates; and λ_T is the cross-sectional warping function. Also, $\hat{\phi}' = \phi' - w'v''$ and $u = u_e - \frac{1}{2}\int_0^x (v'^2 + w'^2)dx$.

Based on these assumptions and strain-displacement relationships (Eqs. 2.12–2.14), Hamilton's principle is used to derive the system of equations of motion. For a conservative system, it states that the true motion of a system, between prescribed initial conditions at time ψ_1 and final conditions at time ψ_2, is that particular motion for which the time integral for the difference between potential and kinetic energies is a minimum. The generalized Hamilton's principle applicable to nonconservative systems is expressed as

$$\delta\Pi = \int_{\psi_1}^{\psi_2} (\delta U - \delta T - \delta W)\, d\psi = 0, \tag{2.15}$$

where δU, δT, and δW are virtual variations of strain energy, kinetic energy, and virtual work done by external force, respectively, and $\delta\Pi$ represents the total potential of the system.

The strain energy of the system can be written as follows:

$$U = \frac{1}{2}\int_0^R \int_A \int (\sigma_{xx}\epsilon_{xx} + \sigma_{x\eta}\epsilon_{x\eta} + \sigma_{x\zeta}\epsilon_{x\zeta})\, d\eta\, d\zeta\, dx. \tag{2.16}$$

On addition of the smart layer, both the strain and kinetic energy of the existing structure are modified but the virtual work done due to applied forces remains unchanged. The kinetic and strain energy terms of the isotropic beam are defined in Appendix A. The nonlinear equations listed in Appendix A have been derived by Hodges and Dowell for elastic bending and torsion of twisted nonuniform rotor blades. Appendix B lists the section properties used in Appendix A. This theory was used by Epps and Chandra [11] to validate the rotating frequencies obtained by experiments. Also, Ganguli et al. [12] have correlated the analysis with vibration data and found that the analysis predicts vibration trends properly.

2.4.2 *Smart Structure Expressions*

The use of piezoceramic actuators changes the strain energy of the rotor blade. In the framework of above-mentioned assumptions, the equations of motion for a rotor blade/beam with surface-bonded piezoceramic actuators are derived. This derivation is generic for axial, bending, and torsion actuation and could be used as per the requirement of the application. The equations are used in Chaps. 8–11 for the active twist rotor. On using the variation of strain energy in the Hamilton's principle, we get

$$\delta U = \frac{1}{2}\int_0^R \int_A \int (\sigma_{xx}\delta\epsilon_{xx} + \sigma_{x\eta}\delta\epsilon_{x\eta} + \sigma_{x\zeta}\delta\epsilon_{x\zeta})\, d\eta\, d\zeta\, dx. \tag{2.17}$$

The induced strains due to the bonding of piezoelectric layer were discussed in Eq. 2.7 and now superposition of these two constitutive equations will yield the derivation of smart terms.

$$U = \frac{1}{2}\int_0^R \int_A \int \{\sigma_{xx}(\epsilon_{xx} - \lambda_{11}) + \sigma_{x\eta}(\epsilon_{x\eta} - \lambda_{12}) + \sigma_{x\zeta}(\epsilon_{x\zeta} - \lambda_{31})\}\, d\eta\, d\zeta\, dx. \tag{2.18}$$

Here, we utilize only λ_{11} and λ_{12} in the above equation. The required bending and axial actuation can be obtained from λ_{11} and shear/torsion actuation is achievable from λ_{12}.

where
$$\begin{aligned}\delta\epsilon_{xx} &= \delta u' + v'\delta v' + w'\delta w' + (\eta^2 + \zeta^2)(\theta_0{}' + \phi')\delta\phi' - \lambda_T\delta\phi'' \\ &\quad - [\eta\cos(\theta_0 + \hat{\phi}) - \zeta\sin(\theta_0 + \hat{\phi})](\delta v'' + w''\delta\hat{\phi}) \\ &\quad - [\eta\sin(\theta_0 + \hat{\phi}) + \zeta\cos(\theta_0 + \hat{\phi})](\delta w'' - v''\delta\hat{\phi}),\end{aligned} \tag{2.19}$$

$$\delta\epsilon_{x\eta} = -\hat{\zeta}\ \delta\phi^{'}, \tag{2.20}$$

$$\delta\epsilon_{x\zeta} = \hat{\eta}\ \delta\phi^{'}, \tag{2.21}$$

where $\hat{\zeta} = \zeta + \frac{\partial\lambda_T}{\partial\eta}$ and $\hat{\eta} = \eta - \frac{\partial\lambda_T}{\partial\zeta}$.

Performing the variation in strains and then nondimensionalizing with respect to 'R' (Radius of rotor), the following smart strain energy contributions are obtained:

$$U_{u'e} = \frac{1}{2}\int_0^1 A_0 dx, \tag{2.22}$$

$$U_{\hat{\phi}'} = \frac{1}{2}\int_0^1 \{A_1\ (\theta_0{}' + \hat{\phi}' + w'\ v'') - A_4\}\ dx, \tag{2.23}$$

$$U_{v''} = \frac{1}{2}\int_0^1 \{-A_2(\cos\theta_0 - \hat{\phi}\sin\theta_0)\ + A_3(\sin\theta_0 + \hat{\phi}\cos\theta_0) + A_4 w'\}dx,$$

$$U_{\hat{\phi}''} = \frac{1}{2}\int_0^1 -A_5\ dx, \tag{2.24}$$

$$U_{\hat{\phi}} = \frac{1}{2}\int_0^1 \{[A_2(\sin\theta_0 + \hat{\phi}\cos\theta_0)\ +\ A_3(\cos\theta_0 - \hat{\phi}\sin\theta_0)]\ v''$$

$$+ [-A_2(\cos\theta_0 - \hat{\phi}\sin\theta_0)\ +\ A_3(\sin\theta_0 + \hat{\phi}\cos\theta_0)]\ w''\}dx, \tag{2.25}$$

$$U_{w''} = \frac{1}{2}\int_0^1 [-A_2(\sin\theta_0 + \hat{\phi}\cos\theta_0)]\ -\ A_3(\cos\theta_0 - \hat{\phi}\sin\theta_0)]\ dx. \tag{2.26}$$

Hence, evaluating and deriving the new terms which arise due to surface bonding of the smart layer with appropriate substitution by section integrals yields

$$\begin{aligned}\frac{\delta U_{smart}}{m_0\Omega^2 R^3} = \int_0^1 & [A_0\ (\delta u_e')\ + A_1\ (\theta_0{}' + \phi')\delta\phi' \\ & + (-A_2\ \cos(\theta_0 + \hat{\phi}) + A_3\ \sin(\theta_0 + \hat{\phi}))\ (\delta v'' + w''\delta\hat{\phi}) \\ & + (-A_2\ \sin(\theta_0 + \hat{\phi}) - A_3\ \cos(\theta_0 + \hat{\phi}))\ (\delta w'' - v''\ \delta\hat{\phi}) \\ & - (A_4\ \delta\phi') - (A_5\ \delta\hat{\phi}'')]\ dx,\end{aligned} \tag{2.27}$$

where A_0, A_1, A_2, A_3, A_4 and A_5 are the section properties due to piezoceramic actuation and are defined in Appendix B. The above derivation is generic for a beam undergoing displacements (axial, transverse bending, inplane bending, and torsion); also, it is generalized for axial, bending, and torsional actuation. A_0 and A_1 contribute to axial actuation, A_3 to bending actuation, and A_4 to shear actuation.

2.4.3 Identification of Smart Terms

The governing equations derived in Eq. 2.27 and shown in Appendix A contain a very large number of terms and are complicated to understand as they are derived for a generic cross-sectional beam with pretwist. To explain the physical significance, a symmetric section is taken as an example with θ_0 and β_p assumed to be zero. This simulates a rotating beam. In this case, the equations become simpler and hence the following element stiffness, damping, and mass matrices are obtained. The linear and nonlinear element force vectors are also shown for the same case.

$$\mathbf{K} = \frac{1}{2}\left[\begin{array}{c|c|c|c} \int_0^1 EAH_u{'}^T H_u{'} ds & 0 & 0 & 0 \\ \hline 0 & \begin{array}{c} \int_0^1 EI_z H'^T H' ds \\ +\int_0^1 F_A H'^T H' ds \\ -\int_0^1 mH^T H ds \end{array} & \int_0^1 A_4 H''^T H' ds & +\int_0^1 A_3 H''^T H_{\hat{\phi}} ds \\ \hline 0 & \int_0^1 A_4 H''^T H' ds & \begin{array}{c} \int_0^1 EI_y H'^T H' ds \\ +\int_0^1 F_A H'^T H' ds \end{array} & 0 \\ \hline 0 & -\int_0^1 A_3 H''^T H_{\hat{\phi}} ds & 0 & \begin{array}{c} \int_0^1 m(k_{m2}{}^2 - k_{m1}{}^2) H_{\hat{\phi}}^T H_{\hat{\phi}} ds \\ +\int_0^1 GJ H_{\hat{\phi}}'^T H_{\hat{\phi}}' ds \\ +\int_0^1 A_1 H_{\hat{\phi}}'^T H_{\hat{\phi}}' ds \end{array} \end{array}\right], \tag{2.28}$$

$$\mathbf{C} = \frac{1}{2}\left[\begin{array}{c|c|c|c} 0 & -\int_0^1 2mH_u{}^T H_u ds & 0 & 0 \\ \hline \int_0^1 2mH_u{}^T H_u ds & 0 & 0 & 0 \\ \hline 0 & 0 & 0 & 0 \\ \hline 0 & 0 & 0 & 0 \end{array}\right], \tag{2.29}$$

$$\mathbf{M} = \frac{1}{2}\left[\begin{array}{c|c|c|c} \int_0^1 mH_u{}^T H_u ds & 0 & 0 & 0 \\ \hline 0 & \int_0^1 mH^T H ds & 0 & 0 \\ \hline 0 & 0 & \int_0^1 mH^T H ds & 0 \\ \hline 0 & 0 & 0 & \int_0^1 mk_m{}^2 H_{\hat{\phi}}^T H_{\hat{\phi}} ds \end{array}\right], \tag{2.30}$$

$$\mathbf{F_L} = \frac{1}{2}\left[\begin{array}{c} \int_0^1 mxH_u{}^T ds + \int_0^1 A_0 H_u{}^T ds \\ \hline 0 \\ \hline -\int_0^1 A_3 H''^T ds \\ \hline +\int_0^1 A_4 H_{\hat{\phi}}'^T \end{array}\right], \tag{2.31}$$

$$
\mathbf{F_{NL}} = \frac{1}{2}\left[\begin{array}{c}
-\int_0^1 EAk_A{}^2\frac{\hat{\phi}^2}{2}H_u{}'^T ds \\
\hline
-\int_0^1 GJ\hat{\phi}'w'H''^T ds - \int_0^1 (EI_z - EI_y)w''\hat{\phi}H''^T ds \\
+\int_0^1 A_1(\hat{\phi}w' + w'^2v'')H''^T ds + \int_0^1 2m[\int_0^x (v'\dot{v}' + w'\dot{w}')d\xi]H^T ds \\
-\int_0^1 (2v'\int_x^1 m\dot{v}d\xi)H'^T ds \\
\hline
-\int_0^1 GJ\hat{\phi}'v''H'^T ds - \int_0^1 (2w'\int_x^1 m\dot{v}d\xi)H'^T ds \\
\hline
-\int_0^1 (EI_z - EI_y)v''w''H_{\hat{\phi}}^T ds \\
-\int_0^1 (EAk_A{}^2\hat{\phi}'u_e{}' + GJw'v'')H_{\hat{\phi}}'^T ds \\
+\int_0^1 A_1w'v''H_{\hat{\phi}}'^T ds + \int_0^1 A_3w''\hat{\phi}H_{\hat{\phi}}^T ds
\end{array}\right]. \quad (2.32)
$$

The first, second, third, and fourth row and column in the matrix **K** and rows in **F** represent the axial, lag, flap, and torsion displacements. The off-diagonal terms in **K** introduce couplings between the displacements. Depending on the type of actuation given, the smart terms get activated. For a symmetric section, A_0 and A_1 get activated in extension actuation, A_3 in bending actuation, and A_4 in torsion actuation. It should be noted that A_2 and A_5 are zero for a symmetric section. For bending actuation (A_3), lag bending–torsion and flap bending–torsion couplings are introduced through the stiffness matrix. Linear forcing is introduced into flap and lag bending equations and nonlinear forcing in the torsion equation. For torsion actuation (A_4), lag–flap bending coupling is introduced through the stiffness matrix and a linear forcing in the torsion equation.

Here, $F_A(x) = \int_1^x mx\, d\xi$ is the centrifugal axial force due to rotation. We observe that smart actuation terms cause inplane bending–transverse bending coupling and inplane bending–torsion coupling through the stiffness terms. The mass matrix does not contain any specific smart structure terms. The damping matrix shows the presence of the antisymmetric Coriolis effect, but no influence of smart terms. The linear force vector shows the influence of smart actuation on the axial, transverse bending, and torsion directions. Nonlinear smart terms are present in the force vector for inplane bending and torsion forces. The above equations of motion can be used to identify terms that can be tailored by designing a beam cross section to maximize the smart actuation effect and minimize the centrifugal stiffening effect.

2.4.4 Finite Element—Spatial Discretization

Each of the N beam elements has fifteen degrees of freedom as shown in Fig. 2.13. These degrees of freedom are distributed over five element nodes (2 boundary nodes and 3 interior nodes). There are six degrees of freedom at each element boundary node. These six degrees of freedom correspond to u, v, v', w, w', and $\widehat{\phi}$. There are two internal nodes for axial deflection u and one internal node for elastic twist $\widehat{\phi}$. Between elements, there is continuity of displacements and slope for flap and lag bending deflections, and continuity of displacement for elastic twist and axial deflections. These elements ensure physically consistent linear variations of bending

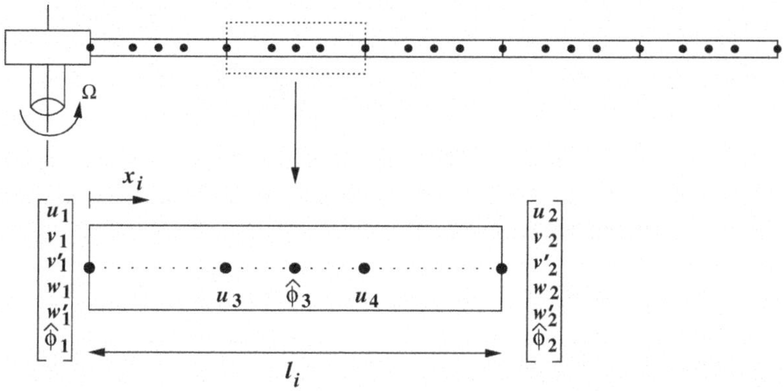

Fig. 2.13 The 15-DOF finite element used for spatial discretization of rotor blade

moments and torsional moment, and quadratic variation of axial force within each element. Using the interpolating polynomials, the distribution of deflections over a beam element is expressed in terms of the elemental nodal displacements $\mathbf{q}_i$. The shape functions used are Hermite polynomials for lag and flap bending, and Lagrange polynomials for axial and torsion deflection. For the ith beam element, the elemental nodal displacement vector is defined as

$$\mathbf{q}_i^T = [u_1, u_2, u_3, u_4, v_1, v_1', v_2, v_2', w_1, w_1', w_2, w_2', \hat{\phi}_1, \hat{\phi}_2, \hat{\phi}_3]. \tag{2.33}$$

Assembling the blade finite element equations and applying boundary conditions results in Eq. (2.16) becoming,

$$\mathbf{M}\ddot{\mathbf{q}}(\psi) + \mathbf{C}\dot{\mathbf{q}}(\psi) + \mathbf{K}\mathbf{q}(\psi) = \mathbf{F}(\mathbf{q}, \dot{\mathbf{q}}, \psi). \tag{2.34}$$

The nodal displacement vector $\mathbf{q}$ is a function of time and all nonlinear terms have been moved into the force vector on the right-hand side. The spatial functionality has been removed using finite element discretization and partial differential equations have been converted into ordinary differential equations.

2.4.5 Normal Mode Transformation

The finite element equations representing each rotor blade are transformed to normal mode space for efficient solution of blade response using modal expansion. Typically, 6–10 modes are used. The displacements are expressed in terms of normal modes as

$$\mathbf{q} = \Phi\mathbf{p}. \tag{2.35}$$

Substituting Eq. 2.35 into Eq. 2.34 leads to the normal mode equations of the following form:

$$\bar{\mathbf{M}}\ddot{\mathbf{p}}(\psi) + \bar{\mathbf{C}}\dot{\mathbf{p}}(\psi) + \bar{\mathbf{K}}\mathbf{p}(\psi) = \bar{\mathbf{F}}(\mathbf{p}, \dot{\mathbf{p}}, \psi). \tag{2.36}$$

These equations are nonlinear ODEs but their dimensions are much reduced compared to the full finite element Eq. 2.34. The normal mode mass, stiffness and damping matrix, and force vectors are defined as

$$\bar{\mathbf{M}} = \Phi^T \mathbf{M}\Phi, \quad \bar{\mathbf{C}} = \Phi^T \mathbf{C}\Phi, \quad \bar{\mathbf{K}} = \Phi^T \mathbf{K}\Phi, \quad \bar{\mathbf{F}} = \Phi^T \mathbf{F}. \tag{2.37}$$

The mode shapes or eigen vectors in Eqs. 2.35 and 2.37 are obtained by solving the general eigenvalue problem:

$$\mathbf{K}_s \Phi = \omega^2 \mathbf{M}_s \Phi. \tag{2.38}$$

Here, ω are the natural frequencies of the rotating blade.

2.4.6 Finite Element—Temporal Discretization

The blade normal mode Eq. 2.36 can be written in the following variational form:

$$\int_0^{2\pi} \delta\mathbf{p}^{\mathrm{T}} \left(\bar{\mathbf{M}}\ddot{\mathbf{p}}(\psi) + \bar{\mathbf{C}}\dot{\mathbf{p}}(\psi) + \bar{\mathbf{K}}\mathbf{p}(\psi) - \bar{\mathbf{F}}(\mathbf{p}, \dot{\mathbf{p}}, \psi) \right) \mathrm{d}\psi \;=\; 0. \tag{2.39}$$

Integrating Eq. 2.39 by parts, we obtain

$$\int_0^{2\pi} \begin{Bmatrix} \delta\mathbf{p} \\ \delta\dot{\mathbf{p}} \end{Bmatrix}^T \begin{Bmatrix} \bar{\mathbf{F}} - \bar{\mathbf{C}}\dot{\mathbf{p}} - \bar{\mathbf{K}}\mathbf{p} \\ \bar{\mathbf{M}}\dot{\mathbf{p}} \end{Bmatrix} d\psi = \begin{Bmatrix} \delta\mathbf{p} \\ \delta\dot{\mathbf{p}} \end{Bmatrix}^T \begin{Bmatrix} \bar{\mathbf{M}}\dot{\mathbf{p}} \\ 0 \end{Bmatrix} \Bigg|_0^{2\pi}. \tag{2.40}$$

Since the helicopter rotor is a periodic system with a time period of one revolution, we have $\dot{\mathbf{p}}(0) = \dot{\mathbf{p}}(2\pi)$. Imposing periodic boundary conditions on Eq. 2.40 results in the right-hand side becoming zero and yields the following system of first-order ordinary differential equations:

$$\int_0^{2\pi} \delta\mathbf{y}^T \mathbf{Q} d\psi = 0, \tag{2.41}$$

where

$$\mathbf{y} = \begin{Bmatrix} \mathbf{p} \\ \dot{\mathbf{p}} \end{Bmatrix}, \quad \mathbf{Q} = \begin{Bmatrix} \bar{\mathbf{F}} - \bar{\mathbf{C}}\dot{\mathbf{p}} - \bar{\mathbf{K}}\mathbf{p} \\ \bar{\mathbf{M}}\dot{\mathbf{p}} \end{Bmatrix}.$$

The nonlinear, periodic, ordinary differential equations are then solved for blade steady response using the finite element in time in conjunction with Newton–Raphson method [13]. Discretizing Eq. 2.41 over N_t time elements around the rotor disk (where $\psi_1 = 0$, $\psi_{N_t+1} = 2\pi$) and taking first-order Taylor series expansion about the steady-state value $\mathbf{y}_0 = [\mathbf{p}_0^{\mathrm{T}} \ \dot{\mathbf{p}}_0^{\mathrm{T}}]^{\mathrm{T}}$ yields the following algebraic equations.

$$\sum_{i=1}^{N_t} \int_{\psi_i}^{\psi_{i+1}} \delta\mathbf{y}_i^T \mathbf{Q}_i(\mathbf{y}_0 + \Delta\mathbf{y}) d\psi = \sum_{i=1}^{N_t} \int_{\psi_i}^{\psi_{i+1}} \delta\mathbf{y}_i^T [\mathbf{Q}_i(\mathbf{y}_0) + \mathbf{K}_{ti}(\mathbf{y}_0)\Delta\mathbf{y}] d\psi = 0, \tag{2.42}$$

where

$$\mathbf{K}_{ti} = \begin{bmatrix} \frac{\partial \bar{\mathbf{F}}}{\partial \mathbf{p}} - \bar{\mathbf{K}} & \frac{\partial \bar{\mathbf{F}}}{\partial \dot{\mathbf{p}}} - \bar{\mathbf{C}} \\ 0 & \bar{\mathbf{M}} \end{bmatrix}_i .$$

Here, $\mathbf{K}_{ti}$ is the tangential stiffness matrix for time element i and $\mathbf{Q}_i$ is the load vector. Behavior of the modal displacement vector can be approximated in terms of shape functions and a vector of temporal nodal coordinates as follows:

$$\mathbf{p}_i(\psi) = \mathbf{H}(s)\mathbf{r}_i, \tag{2.43}$$

where $\mathbf{H(s)}$ are time shape functions (in terms of the element coordinate s) used for approximating the normal mode coordinate $\mathbf{p}$. Here $\mathbf{r}$ is the temporal nodal coordinate. In this book, mixed Lagrange–Hermite polynomials are used for interpolation within the time element [14]. Substituting Eq. 2.43 and its derivative into Eq. 2.42 yields the time discretized blade response.

$$\mathbf{Q}^G + \mathbf{K}_t^G \Delta\mathbf{r}^G = 0, \tag{2.44}$$

where

$$\mathbf{Q}^G = \sum_{i=1}^{N_t} \int_{\psi_i}^{\psi_{i+1}} \mathbf{H}^T \mathbf{Q}_i d\psi,$$

$$\mathbf{K}_t^G = \sum_{i=1}^{N_t} \int_{\psi_i}^{\psi_{i+1}} \mathbf{H}^T \begin{bmatrix} \frac{\partial \bar{F}}{\partial p} - \bar{\mathbf{K}} & \frac{\partial \bar{F}}{\partial \dot{p}} - \bar{\mathbf{C}} \\ 0 & \bar{\mathbf{M}} \end{bmatrix}_i d\psi,$$

$$\Delta\mathbf{r}^G = \sum_{i=1}^{N_t} \Delta\mathbf{r}_i .$$

Solving the above equations iteratively yields the blade steady response.

2.5 Aerodynamic Models

The aerodynamic environment of a helicopter rotor in forward flight is extremely complex involving transonic flow on advancing blades and reversed flow on retreating blades. The blade is also exposed to unsteady variations in angle of attack and free-stream velocity. Because of the complexity of the flow, accurate modeling of the unsteady flow field on the blade requires a sophisticated analysis. This section deals with aerodynamic loads due to blade/airfoil motion only. The aerodynamic loads enter the blade governing equations through their contribution to the external virtual work in Hamilton's principle.

The aerodynamic modeling of a rotor can be divided into two parts: a local blade element model and a global wake model. In this book, an unsteady aerodynamic model proposed by Leishman and Beddoes [15] is used to estimate the aerodynamic loads due to blade motion. The flow is assumed to be under a linear attached potential flow regime. Separated flow and dynamic stall effects are not modeled.

2.5.1 Attached Flow Formulation

The attached flow formulation is based on the work of Leishman and Beddoes [16]. This model implicitly includes the effects of compressibility. Also, this model requires few empirical constants which can be derived from static airfoil data. In this formulation, the unsteady lift, drag, and pitching moment are assumed to consist of circulatory and impulsive (noncirculatory) components. These airloads are calculated using an indicial response representation implemented in the form of a finite-difference discretization of integral equations. This method determines the aerodynamic loadings due to a step change in the downwash at the three-quarter chord position. The circulatory loads include the effects of the near shed wake. The impulsive loads are due to the presence of propagating pressure waves.

2.5.2 Reverse Flow

In forward flight, an inboard region of the retreating side of the rotor disk experiences reverse flow. In this reverse flow region, the forward speed component of the total velocity relative to the blade becomes larger than the velocity component due to the rotational speed. Therefore, the velocity relative to the blade is directed from the trailing edge to the leading edge. If the blade dynamics are neglected, the reverse flow region is given by Eq. 2.45, which represents a circular region (Fig. 2.14).

$$r \leq -\mu \sin\psi . \tag{2.45}$$

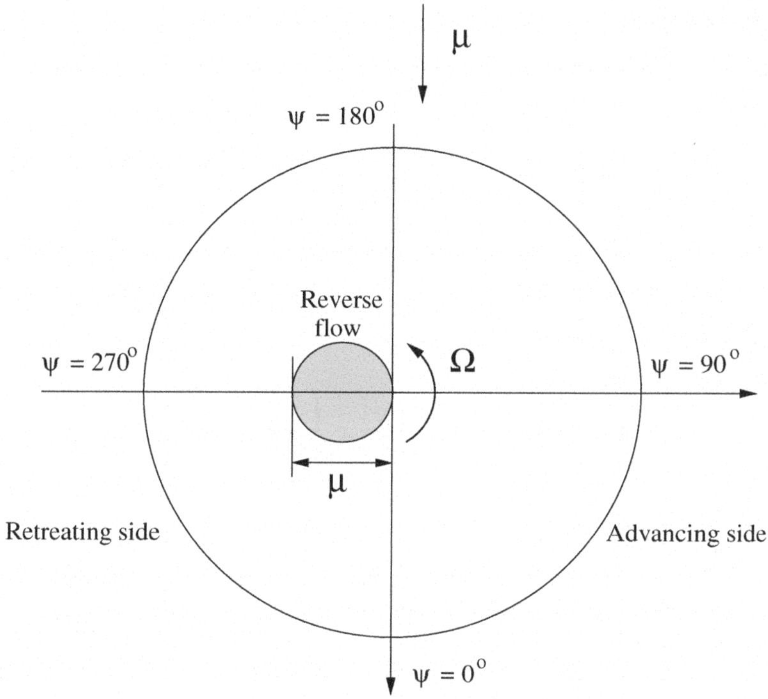

Fig. 2.14 Reverse flow region

The reverse flow boundary is thus a circle of diameter μ, centered at $r = \mu/2$ for the azimuth station of 270° on the retreating side. Reverse flow causes the aerodynamic center of the airfoil to shift from x_{ac} to $(c - x_{ac})$, where x_{ac} is the aerodynamic center of the airfoil.

2.5.3 *Inflow and Rotor Wake Modeling*

The wake behind the rotor disk determines the induced inflow distribution over the disk and plays a very important role in the prediction of blade response, vibration, and rotor performance. An accurate modeling of the induced inflow is essential, especially at low-speed flight condition. At low speed, the wake stays close to the disk and has a dominating influence on blade airloads. There are many wake models available with varying levels of complexity and accuracy. Two inflow models are used. One is the simple linear inflow model and the other is the elaborate free wake model.

Linear Inflow Model

In a linear inflow model, the induced velocities are assumed to vary linearly across the rotor disk:

$$\lambda = \mu\ \tan\alpha_s + \lambda_i = \mu\ \tan\alpha_s + \frac{C_T/2}{\sqrt{\lambda^2 + \mu^2}}(1 + \kappa_x x\ \cos\psi + \kappa_y x\ \sin\psi), \quad (2.46)$$

where λ is the total inflow ratio, λ_i is the induced inflow ratio, and α_s is the forward tilt of the rotor disk plane. There are many forms of this model available in literature. One form is called Drees [17] model, where κ_x and κ_y are defined as

$$\kappa_x = \frac{4}{3}\left[(1 - 1.8\mu^2)\sqrt{1 + \left(\frac{\lambda}{\mu}\right)^2} - \frac{\lambda}{\mu}\right],$$
$$\kappa_y = -2\mu. \quad (2.47)$$

This simple inflow model captures the global effects of rotor wake and is usually satisfactory for high-speed flight condition, more so for rotor performance and stability predictions. However, this model becomes much less accurate at low-speed flight or hover condition when the inflow distribution becomes highly nonuniform over the rotor disk. The linear inflow model can be easily implemented in rotor aeroelastic analysis due to its simplicity.

Free Wake Model

The free wake model used was developed by Bagai and Leishman [17]. The wake is divided into three distinct regions: near wake, rolling-up wake, and far wake. The near wake is of prime importance to its generating blade due to its proximity. The most difficult part of the wake analysis is to calculate the wake geometry. In the present wake model, the wake is modeled as a finite number of vortex filaments, which are trailed and shed into the wake. These filaments are convected with the local flow velocity, which consists of the free-stream velocity and the wake self-induced velocity. Since the self-induced velocity is, in turn, a function of the wake geometry, the analysis is highly nonlinear in nature. A rigid wake geometry, which consists of a simple helix convected by the free stream and the mean inflow, is used to initialize the calculation, and then the wake geometry evolves until it is stabilized. The vorticity strength is determined by the bound circulation which in turn is calculated from the lift distribution along the rotor blade. Once the vorticity strength and wake geometry are known, the induced velocity can be evaluated using the Biot–Savart law. Free wake models can capture nonuniform inflow distributions expected at low forward speeds but at the cost of much greater computational effort compared to linear inflow models.

2.5.4 Dynamic Stall Formulation

This is characterized by separation at the leading edge and the shedding of concentrated vorticity from the leading edge region. The vorticity is swept downstream chord-wise and significantly affects the load distribution on the airfoil. The vortex convection rate is calculated using a nondimensional time parameter based on the distance traveled by the airfoil in terms of semi-chord length. By monitoring the excess lift and the center of pressure movement, the vortex-induced pitching moment is computed. As the vortex reaches the trailing edge, the model assumes rapid decay of the increment lift.

Inflow distribution over the rotor disk can be computed using either a simple uniform linear inflow model or a free wake model. Typically, the actual wake of a rotor is quite nonuniform and requires a free wake model for accurate vibratory load predictions.

2.6 Blade and Hub Loads

Once the blade steady-state response is obtained, the loads acting at any section of the blade are calculated by summation of the aerodynamic and inertial loads. The blade root loads are given by

$$\begin{Bmatrix} F_{xR} \\ F_{yR} \\ F_{zR} \\ M_{xR} \\ M_{yR} \\ M_{zR} \end{Bmatrix} = \int_0^R \begin{Bmatrix} L_u \\ L_v \\ L_w \\ -L_v w + L_w v + M_u \\ L_u w - L_w(x+u) + M_v \\ -L_u v + L_v(x+u) + M_w \end{Bmatrix} dx. \tag{2.48}$$

Here, L_u, L_v, and L_w are the forces acting on element dx along the x, y, and z directions and M_u, M_v, and M_w are the moments along the x, y, and z directions, respectively. The section loads are in the rotating frame and are directed toward the undeformed blade axes. In the fixed frame, the rotor hub loads can be expressed as

$$\begin{aligned} F_X^H(\psi) &= \sum_{m=1}^{N_b} (F_x^m \cos\psi_m - F_y^m \sin\psi_m - F_z^m \cos\psi_m \beta_p), \\ F_Y^H(\psi) &= \sum_{m=1}^{N_b} (F_x^m \sin\psi_m + F_y^m \cos\psi_m - F_z^m \sin\psi_m \beta_p), \\ F_Z^H(\psi) &= \sum_{m=1}^{N_b} (F_x^m \beta_p + F_z^m), \end{aligned} \tag{2.49}$$

$$M_X^H(\psi) = \sum_{m=1}^{N_b}(M_x^m \cos\psi_m - M_y^m \sin\psi_m - M_z^m \cos\psi_m \beta_p),$$

$$M_Y^H(\psi) = \sum_{m=1}^{N_b}(M_x^m \sin\psi_m + M_y^m \cos\psi_m - M_z^m \sin\psi_m \beta_p),$$

$$M_Z^H(\psi) = \sum_{m=1}^{N_b}(M_x^m \beta_p + M_z^m).$$

The hub loads are expanded in a Fourier series and the steady and harmonic components are calculated. The steady hub loads are required to calculate the steady rotor forces which are in turn utilized for trim equilibrium equations to obtain the trim controls. After trim controls have been calculated, the blade response for the current trim solution is calculated using an iterative process. For a N_b-bladed rotor, the fixed frame hub loads contain only steady and integer multiples of N_b/rev harmonics. In general, the N_b/rev harmonics are the dominant vibratory components. The analysis of hub loads is needed for both trim and vibration predictions.

2.6.1 Coupled Trim

Once the hub loads are obtained, the helicopter needs to be trimmed. Trim is defined as the condition where the steady forces and moments acting on the helicopter sum to zero and simulate the condition for steady level flight. The trim solution for the helicopter involves finding the pilot control angles θ at which the six steady forces and moments acting on the helicopter are zeros:

$$\mathbf{F}(\theta) = \mathbf{0}, \tag{2.50}$$

where

$$F^T = [F_1, F_2, F_3, F_4, F_5, F_6]. \tag{2.51}$$

The following equations are derived from the force equilibrium of a helicopter in steady level flight.

$$\begin{aligned}
F_1 &= D_F \cos\theta_{FP} + H\cos\alpha_s - T\sin\alpha_s, \\
F_2 &= Y_F + Y\cos\phi_s + T\sin\phi_s + T_{tr}, \\
F_3 &= T\cos\alpha_s\cos\phi_s - D_F\sin\theta_{FP} + H\sin\alpha_s - Y\sin\phi_s - W - L_{ht}, \\
F_4 &= M_{xR} + M_{xF} + Y_F(\bar{h}\cos\phi_s + y_{cg}\sin\phi_s) + W(\bar{h}\sin\phi_s - y_{cg}\cos\phi_s) + T_{tr}(\bar{h} - z_{tr}), \\
F_5 &= M_{yR} + M_{yF} + W(\bar{h}\sin\alpha_s - x_{cg}\cos\alpha_s) - D_F(\bar{h}\cos(\alpha_s + \theta_{FP}) \\
&\quad + x_{cg}\sin(\alpha_s + \theta_{FP})) + L_{ht}(x_{ht} - x_{cg}), \\
F_6 &= M_{zR} + M_{zF} + T_{tr}(x_{tr} - x_{cg}) - D_F Y_{cg}\cos\alpha_s - Y x_{cg}\cos\phi_s,
\end{aligned} \tag{2.52}$$

where F_1, F_2, and F_3 are, respectively, the vehicle force equilibrium residuals in the *X*, *Y*, and *Z* directions in the fuselage axes; and F_4, F_5, and F_6 are the vehicle rolling, pitching, and yawing moment equilibrium residuals about the vehicle cg, respectively. Also, *H*, *Y*, and *T* are, respectively, rotor drag, side force, and thrust; and D_F, Y_F, and *W* are, respectively, fuselage drag, side force, and gross weight. The terms T_{tr}, x_{tr}, and z_{tr} denote the tail rotor thrust, the distance of the tail rotor hub behind the vehicle cg, and the distance of the tail rotor hub above the vehicle cg. The horizontal tail is located at a distance x_{ht} behind the vehicle cg. The terms in the three moment, M_{xR} and M_{xF}, denote the rotor and fuselage moments, respectively. The forces act on the rotor hub and the moments act about the rotor hub. In addition, x_{cg} and y_{cg} and *h* are, respectively, the relative location of the rotor hub center with respect to the vehicle center of gravity in the X_F, Y_F, and Z_F directions; α_s (positive for forward flight) and ϕ_s (positive advancing down) are the longitudinal and lateral shaft tilts, respectively, and θ_{FP} is the flight path angle relative to an axis perpendicular to the gravity vector.

The forces acting on the fuselage are calculated using

$$\begin{aligned} D_F &= \frac{\gamma N_b}{6a\sigma}\frac{f}{A}\mu^2, \\ Y_F &= \frac{\gamma N_b}{6a\sigma}C_{yF}\mu^2, \\ M_{xF} &= \frac{\gamma N_b}{6a\sigma}C_{mxF}\mu^2, \\ M_{yF} &= \frac{\gamma N_b}{6a\sigma}C_{myF}\mu^2, \\ M_{zF} &= \frac{\gamma N_b}{6a\sigma}C_{mzF}\mu^2, \\ W &= \frac{\gamma N_b}{3a\sigma}(C_W - C_{lF}\mu^2/2)\mu^2, \end{aligned} \tag{2.53}$$

where γ is the lock number, N_b is the number of blades, *a* is the reference lift curve slope, σ is the solidity, *f/A* is the fuselage flat plate area, C_{yF} is the fuselage side force coefficient, C_{mxF} is the fuselage yawing moment coefficient, C_W is the helicopter weight coefficient, and C_{lF} is the lift coefficient.

For propulsive trim, the unknown quantities to be determined from the equilibrium equations are

$$\theta^T = [\alpha_s, \phi_s, \theta_0, \theta_{1c}, \theta_{1s}, \theta_{tail}], \tag{2.54}$$

where the rotor trim parameters θ_0, θ_{1c}, and θ_{1s} govern the main rotor blade.

The trim equations are solved iteratively using a Newton–Raphson procedure. A coupled trim procedure is carried out to solve the blade response, pilot input trim controls, and vehicle orientation, simultaneously. This procedure is called coupled trim since the blade response Eq. 2.44 and trim Eq. 2.50 are simultaneously solved thereby accounting for the influence of elastic blade deflections on the rotor steady forces:

$$\Delta\theta_i = -\left[\frac{\partial \mathbf{F}}{\partial \theta}\right]^{-1}_{\theta=\theta_0} \mathbf{F}(\theta_i). \tag{2.55}$$

For a converged solution, the $\Delta\theta$ and F are zero. The controls are updated as follows:

$$\theta_{i+1} = \theta_i + \Delta\theta_i. \tag{2.56}$$

The trim Jacobian, $[\partial F/\partial\theta]$, is calculated using a forward finite-difference approximation at $\theta = \theta_0$, and it is generally held constant throughout the analysis to save computational time.

The coupled trim is solved iteratively until convergence. The coupled trim procedure is essential for elastically coupled blades since elastic deflections play an important role in the net steady forces and moments generated by the rotor.

2.7 Organization of Aeroelastic Analysis

Figure 2.15 shows the flow of the aeroelastic formulation and analysis which was discussed in previous sections. Starting with the input to the analysis, structural modeling is carried out as the first step. The governing differential equations thus obtained are space and time dependent. These equations are nonlinear and periodic in nature. These equations are discretized using FEM (Finite element method) in space and rotating blade natural frequencies and mode shapes are found. Normal mode transformation is done to reduce the number of degrees of freedom and thus reduce the computational time. Aerodynamic modeling is the next major step. Here, the type of aerodynamic model is decided and the governing equations are discretized in time. Thus, both space and time dependencies are removed, resulting into nonlinear, algebraic equations which are solved using the Newton–Raphson method. Blade loads and the helicopter steady loads are calculated which are in turn required for performing trim calculations. Blade steady-state response, vehicle orientation, and trim control angles are solved for iteratively using a coupled trim procedure.

2.8 Summary

The baseline aeroelastic analysis for a helicopter rotor is discussed in this chapter along with the basic physics of piezoelectric materials. The approaches for determining the blade response, blade and hub loads, and vehicle trim are highlighted. The influence of piezoelectric material on the smart rotor equations is highlighted. These equations are used for the active twist rotor. For the active flap rotor, the 'non-smart' equations will be used and the piezoelectric actuator is addressed separately.

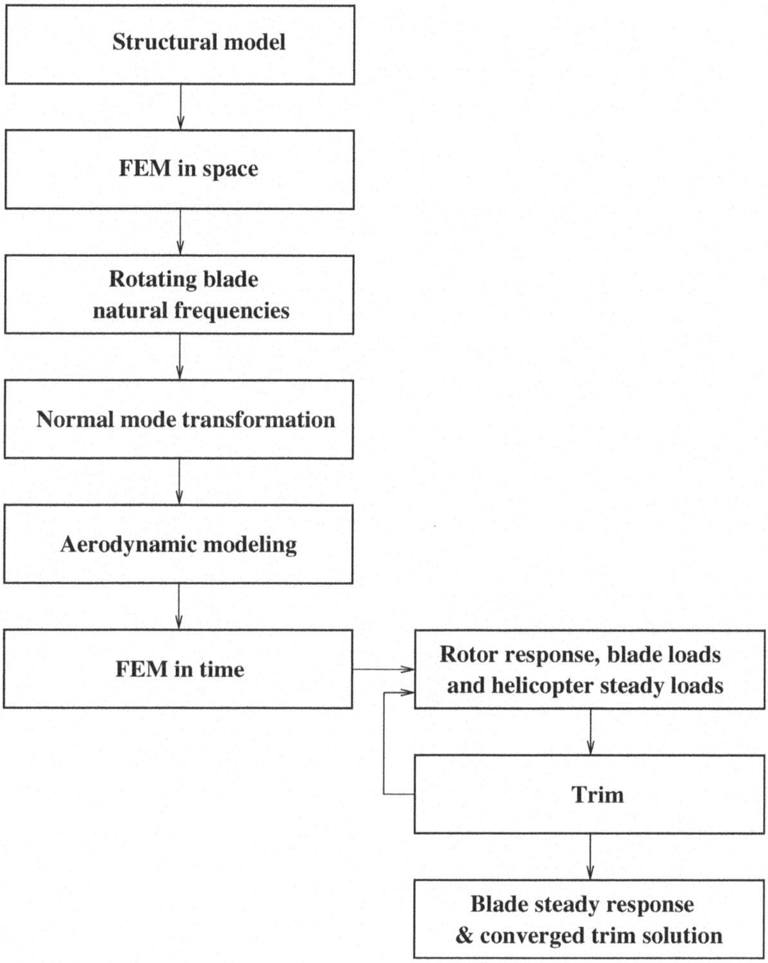

Fig. 2.15 Organization of aeroelastic analysis

References

1. Durr, J., Schmidt, U., Zaglauer, W.: On the integration of piezoceramic actuators in composite structures for aerospace applications. J. Intell. Mater. Syst. Struct. **10**, 880–889 (1999)
2. Srinivasan, A., McFarland, M.: Smart Structures: Analysis and Design. Cambridge University Press, Cambridge (2000)
3. Crawley, E.: Intelligent structures for aerospace: a technology overview and assessment. J. Am. Inst. Aeronaut. Astronaut. **32**(8), 1689–1699 (1994)
4. Loewy, R.: Recent developments in smart structures with aeronautical applications. Smart Mater. Struct. **6**(5), 11–42 (1997)
5. Park, S., Shrout, T.: Relaxor based ferroelectric single crystals for electromechanical actuators. Mater. Res. Innov. **1**, 20–25 (1997)

6. Liu, S., Ren, W., Mukherjee, B.: The piezoelectric shear strain coefficient of <111> piezocrystals. Appl. Phys. Lett. **83**(14), 2886–2888 (2003)
7. Crawley, E., Anderson, E.: Detailed models of piezoceramic actuation of beams. In: Proceedings of the 30th AIAA/ASMI/ASCH/AHS/ASC Structures, Structural Dynamics Conference, Washington, DC (1989)
8. Hong, C., Chopra, I.: Modeling and validation of induced strain actuation of composite coupled plates. J. Am. Inst. Aeronaut. Astronaut. **37**(3), 372–377 (1999)
9. Chen, W., Saleeb, A.: Constitutive Equations for Engineering Materials: Elasticity and Modeling, vol. 1. Wiley-Interscience, New York (1952)
10. Hodges, D., Dowell, E.: Non-linear equations of motion for the elastic bending and torsion of twisted nonuniform rotor blades. Technical report NASA TN D-7818 (1974)
11. Epps, J., Chandra, R.: The natural frequencies of rotating composite beams with tip sweep. J. Am. Helicopter Soc. **41**, 29–36 (1996)
12. Ganguli, R., Chopra, I., Weller, W.: Comparison of calculated vibratory rotor hub loads with experimental data. J. Am. Helicopter Soc. **43**, 312–318 (1998)
13. Chopra, I., Sivaneri, T.: Aeroelastic stability of rotor blades using finite element analysis. Technical report NASA CR 166389 (1982)
14. Leishman, J.G., Beddoes, T.S.: A generalised model for airfoil unsteady aerodynamic behaviour and dynamic stall using the indicial method. In: Proceedings of the 42nd Annual Forum of the American Helicopter Society, Washington, DC, June, pp. 243–265 (1986)
15. Leishman, J.G., Beddoes, T.S.: A semi-empirical model for dynamic stall. J. Am. Helicopter Soc. **34**(3), 3–17 (1989); Leishman, J.G.: Principles of Helicopter Aerodynamics. Cambridge University Press, New York (2000)
16. Drees, J.M.: A theory of air flow through rotors and its application to some helicopter problems. J. Helicopter Assoc. G. B. **3**(2), 79–104 (1949)
17. Bagai, A., Leishman, J.G.: Rotor free-wake modeling using a pseudo-implicit technique—including comparisons with experimental data. J. Am. Helicopter Soc. **40**(3), 29–41 (1995)

Chapter 3
Preliminary Studies with Active Flaps

There are primarily two types of flaps that are suitable to be mounted on helicopter blades: plain flaps and servo-flaps. The servo-flap design consists of auxiliary airfoil sections that are located aft of the trailing edge of the main blades, thereby resulting in high drag due to the exposure of hinges and supporting structure. In comparison, plain flaps are aerodynamically efficient since the flap is integrated into the blade structure, but are located closer to the blade elastic axis and their capability to generate pitching moments is correspondingly reduced. The plain flap configuration is used in this book. In this chapter, a simple aerodynamic model of the trailing-edge flap is used to obtain an idea about the advantages of 1, 2, and 4 trailing-edge flaps. Section 3.1 gives a brief description of the trailing-edge flap aerodynamic model and the flap control law. Section 3.2 explains formulation of optimization problem. Section 3.3 gives the numerical results. Section 3.4 gives the summary of this chapter.

3.1 Trailing-Edge Flap Model

In this chapter, the effects of trailing-edge flaps are included by changing the blade element airfoil properties [1], using experimental data from Abbott and von Doenhoff [2]. This formulation accounts only for the quasisteady effects of the trailing-edge flap. The effects of the rotor wake are captured using the Drees linear inflow model.

In this chapter, we study the effect of multiple trailing-edge flaps, which are used to introduce control input directly in the rotating reference frame. It is observed by researchers that 3/rev, 4/rev, and 5/rev harmonics are most influential in reducing vibration, while 2/rev excitation is useful for BVI noise reduction. These harmonics can be applied individually or simultaneously using a harmonic control law. All flaps in the current study are deflected at 3/rev, 4/rev, and 5/rev harmonics in the rotating reference frame. The following control law is used for the ith flap:

R. Ganguli et al., *Smart Helicopter Rotors*, Advances in Industrial Control,
DOI 10.1007/978-3-319-24768-7_3

$$\delta_i(\psi) = \delta_{i-3c}\cos(3\psi) + \delta_{i-3s}\sin(3\psi) + \delta_{i-4c}\cos(4\psi) + \delta_{i-4s}\sin(4\psi) + \delta_{i-5c}\cos(5\psi) + \delta_{i-5s}\sin(5\psi). \quad (3.1)$$

Three cases are considered:

(a) Single flap of 20 % span and one-fifth chord, centered at about 90 % span position.
(b) Two flaps of 10 % span each and one-fifth chord; the first flap centered about 95 % span position and second about the 85 % span position.
(c) Four flaps of 5 % span each and one-fifth chord; the first flap centered about 97.5 % span position, second about the 92.5 % span position, third about the 87.5 % span position, and fourth about the 82.5 % span position.

3.2 Formulation of Optimization Problem

For a helicopter rotor with N_b blades, an important source of vibration is the N_b/rev hub forces and moments transmitted by the rotor to the fuselage. For a 4-bladed rotor, the vibration objective function J_v is chosen as a combination of the magnitudes of the nondimensional 4/rev harmonics of the three forces (longitudinal, lateral, and vertical) and three moments (pitch, roll, and yaw) at the hub. This objective function has been used in several helicopter rotor optimization studies.

$$J_v = \frac{\sqrt{(F_{xH}^{4p})^2 + (F_{yH}^{4p})^2 + (F_{zH}^{4p})^2}}{F_{zH}^0} + \frac{\sqrt{(M_{xH}^{4p})^2 + (M_{yH}^{4p})^2 + (M_{zH}^{4p})^2}}{M_{zH}^0}. \quad (3.2)$$

Our aim is to reduce J_v by as large an amount as possible. However, the power required to obtain this reduction depends on the motion of the flaps. Adaptive materials which are used in the practical implementation of the actively controlled flap (ACF) are limited in their force and stroke producing capability, leading to fairly small angular displacements. From a control perspective, this leads to *saturation*, which introduces serious problems for vibration control. Therefore, it is essential to limit flap deflections to practically achievable limits. Cribbs and Friedmann [3] have shown that intuitive limits such as scaling or clipping of the optimal control deflection to a given maximum value cause severe degradation in the vibration reduction effectiveness of the ACF system. It was also shown that, by modifying the weighting of the control effort, the flap deflection could be limited to a desired value without any significant degradation in performance. Hence, we define a flap deflection objective function J_f as follows:

$$J_f = \frac{1}{n}\sum_{i=1}^{n}\left[\sqrt{\delta_{i-3c}^2 + \delta_{i-3s}^2 + \delta_{i-4c}^2 + \delta_{i-4s}^2 + \delta_{i-5c}^2 + \delta_{i-5s}^2}\right], \quad (3.3)$$

where n is the number of flaps and the flap deflection harmonics are in radians. Simultaneous minimization of J_v and J_f is highly desirable. It is unlikely that the two objectives would be minimized by the same design variable choices. Hence, some trade-off between the criteria is needed to ensure a satisfactory solution, which brings us to the subject of multicriteria optimization.

A vector is said to be *Pareto optimal* if it is impossible to minimize any objective without allowing an increase in the other objectives. Typically, there is an entire curve or surface of Pareto points, whose shape indicates the nature of the trade-off between different objectives, which in our case are J_v and J_f. Combining the multiple objectives into one scalar objective whose solution is a Pareto optimal point for the original problem can often solve the multi-objective problem. In the present chapter, the standard technique of minimizing a positively weighted convex sum of the objectives is followed, that is,

$$\begin{aligned} \text{Minimize} \quad & J_t(\mathbf{x}) = w_v J_v(\mathbf{x}) + w_f J_f(\mathbf{x}) \quad \text{and} \quad w_v + w_f = 1.0 \\ & \mathbf{x} = \left[\delta_{i-3c}, \delta_{i-3s}, \delta_{i-4c}, \delta_{i-4s}, \delta_{i-5c}, \delta_{i-5s}\right], \quad i = 1 \text{ to } n, \end{aligned} \tag{3.4}$$

where J_t is the total objective function and w_v and w_f are scalar weighting factors. The design variables are the 3/rev, 4/rev, and 5/rev harmonics of the flaps. Therefore, for one flap there are six design variables, for two flaps there are twelve design variables, and for four flaps there are twenty four design variables. A gradient-based optimization technique is used to minimize J_t. The gradients are calculated using the forward finite-difference method.

By minimizing convex sums of the objectives for various settings of the convex weights, we can generate various points in the Pareto set. Although computationally more expensive, this approach gives an idea of the shape of the Pareto surface and provides the user with more information about the trade-off among the various objectives.

3.3 Numerical Results

Numerical results are obtained for a 4-bladed, soft-inplane, uniform hingeless rotor similar to the BO105 rotor. The baseline helicopter rotor properties are shown in Table 3.1. A flap chord ratio of 0.20 is chosen in this book. The rotor blade is divided into five, six, and eight beam finite elements for cases (a), (b), and (c), respectively. Figure 3.1 shows the location of the beam finite elements for a rotor with four trailing-edge flaps. The blade tip (outer 20 %) is divided into four finite elements in this case. In cases (a) and (b), the blade tip (outer 20 %) is represented by one and two finite elements, respectively. Ten modes (four flap, four lag, and two torsion) are used for the trim analysis. The blade steady response is calculated by dividing the azimuth into six time elements with a fifth-order polynomial distribution within each time

Table 3.1 Baseline rotor properties

Number of blades	4
c/R	0.055
Solidity, σ	0.07
Lock number, γ	5.20
C_T/σ	0.07
Blade pretwist	0.0
Precone, β_p	0.0
$EI_y/m_0\Omega^2R^4$	0.0108
$EI_z/m_0\Omega^2R^4$	0.0268
$GJ/m_0\Omega^2R^4$	0.00615
m_0 (kg/m)	6.46
Ω (rpm)	383
R (m)	4.94

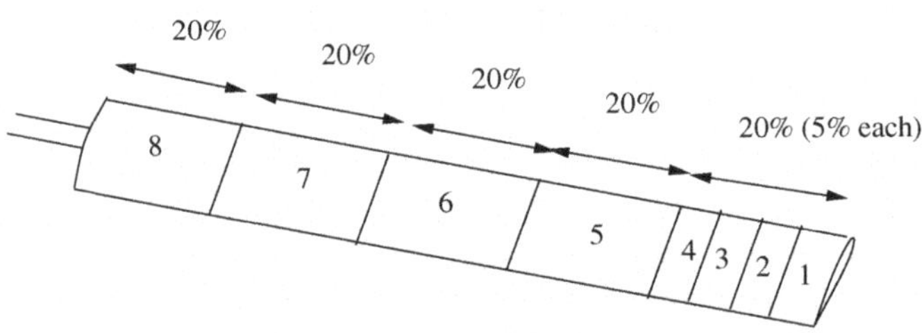

Fig. 3.1 Schematic diagram of a discretized rotor blade with four trailing-edge flaps

element. Results are obtained at a forward speed corresponding to $\mu = 0.30$ and a C_T/σ of 0.07.

Figure 3.2 shows the Pareto optimal curves for the three cases considered. In all three cases, the maximum vibration reduction possible (by setting $w_v = 1$ and $w_f = 0$) was around 72 % from the baseline value. This corresponds to a J_v value of 0.0182 as against a baseline value of 0.0655. The corresponding values of J_f for one, two, and four flaps are 0.0328, 0.0260, and 0.0238, respectively. It is observed that multiple flaps require much lower control effort as compared to a single flap for obtaining the similar reductions in vibration (Fig. 3.3). We also realize that with the same available control deflection power, multiple trailing-edge flaps yield more reduction in vibration as compared to a single flap. Figure 3.2 also indicates that considerable reduction in control effort is obtained when the number of flaps is increased from one to two, whereas increasing the number of flaps from two to four yields only marginal reduction in control effort. This is an important result with practical ramifications.

Now consider only case (c), where the rotor has four trailing-edge flaps. The values of J_v and J_f for the ($w_v = 1$ and $w_f = 0$) case are 0.0182 and 0.0238, respectively.

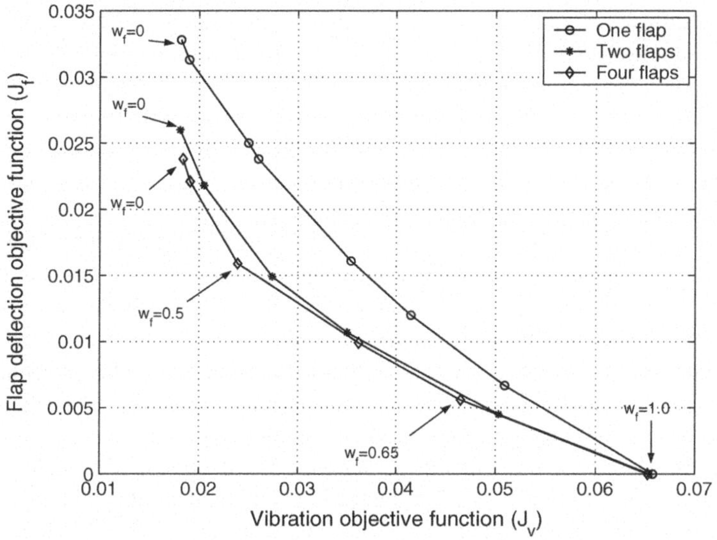

Fig. 3.2 Pareto optimal curve for one, two, and four flaps

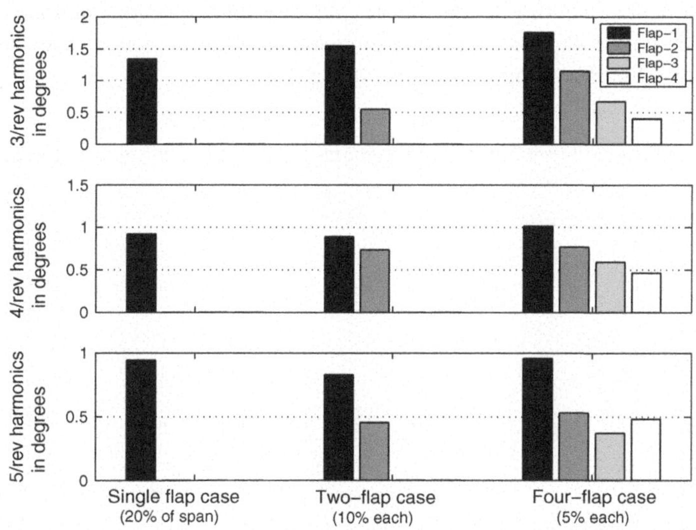

Fig. 3.3 Comparison of flap harmonics for rotor with one, two, and four flaps (vibration minimization only)

Now by setting $w_v = 0.5$ and $w_f = 0.5$, that is, by giving equal weights to both vibration reduction and flap control effort, we obtain another point in the Pareto optimal curve. This corresponds to a J_v value of 0.0239 and J_f value of 0.0159. The vibration reduction is thus 63 % from the baseline value of 0.0655. Now, by setting $w_v = 0.35$ and $w_f = 0.65$, that is, by giving more weight to flap control effort than vibration reduction, we obtain yet another point in the Pareto optimal curve. This corresponds to a J_v value of 0.0465 and J_f value of 0.0056. In this case, we obtain a vibration reduction of 29 % from the baseline value of 0.0655. The above result indicates that it is possible to achieve significant reduction in the control deflections of the flaps compared to the ($w_v = 1$ and $w_f = 0$) case by settling for somewhat lower vibration reduction than obtained before.

The reduction in hub vibratory loads obtained here is due to the creation of new higher harmonic loads by the application of the control law in Eq. 3.1 to the trailing-edge flaps. Due to the change in the blade loads, there is some change in the blade response. The elastic twist or torsion response at the blade tip for a blade with four trailing-edge flaps is shown in Fig. 3.4. The changes in tip axial, flap, and lag response are insignificant and hence not shown. There is a difference in the blade tip torsion response between the three cases shown. Figure 3.5 shows that, as the number of flaps is increased gradually from one to four, the blade response at the tip steadily becomes invariant. This indicates that for a given total span of flaps, rotor blades with more than four flaps are not likely to yield better performance. This can also be observed from the Pareto optimal curve in Fig. 3.2.

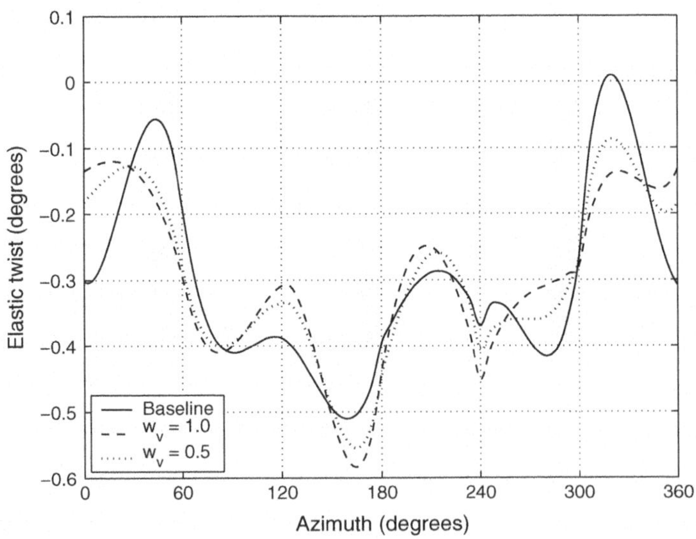

Fig. 3.4 Blade tip torsion response versus azimuth for rotor blade with four trailing-edge flaps

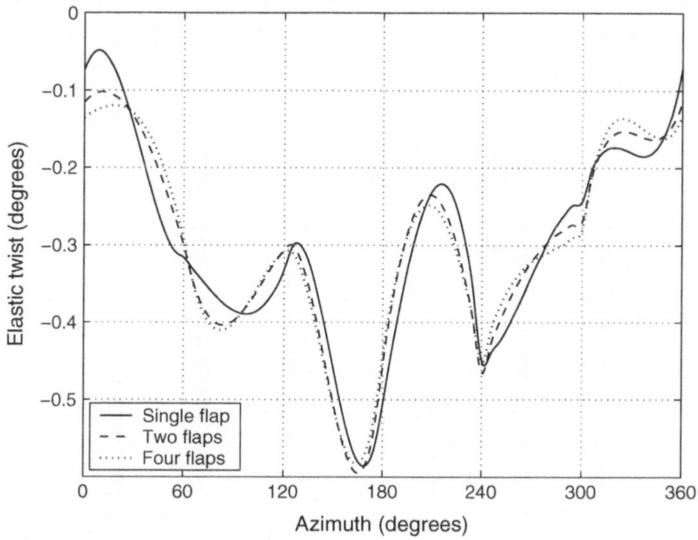

Fig. 3.5 Blade tip torsion response versus azimuth for rotor blade with vibration minimization only ($w_v = 1.0$)

Figure 3.6 shows the 4/rev loads at the hub for a rotor with four trailing-edge flaps. A comparison between selected points on the Pareto curve is also shown. These points correspond to the baseline design (zero flap motion), $w_f = 0.65$, $w_f = 0.50$, and $w_f = 0.0$. For the design point corresponding to $w_f = 0.65$, a lower vibration reduction is accepted to avoid spending too much control power. It is evident that as more and more weight is given for vibration reduction (points $w_f = 0.5$ and $w_f = 0.0$), the vibratory hub moments reduce considerably. However, the vibratory hub forces increase marginally. This is because the magnitudes of vibratory hub moments are higher than that of vibratory hub forces for this configuration, and hence they contribute more to J_v. We observe from Fig. 3.6 that the vibratory hub moments are greatly reduced for the optimum design.

The flap deflection harmonics (in degrees) of a rotor with four trailing-edge flaps for three points on the Pareto optimal curve are shown in Fig. 3.7. It is evident that the ($w_v = 1$ and $w_f = 0$) case requires the highest control deflection angles among the cases shown. From Figs. 3.3 and 3.7, it can be observed that the outermost flap requires a higher deflection than the rest, indicating that it is the most effective flap. The importance of the blade tip region in helicopter rotors is due to the high dynamic pressure experienced by that region. From Fig. 3.7, it can also be observed that the maximum control angles required by the trailing-edge flaps ($\pm 4°$) may be obtained practically using the current state-of-the-art smart materials along with novel amplification mechanisms. The phase difference between the four flaps is shown in Fig. 3.8. Studies show that the flaps are more or less in phase.

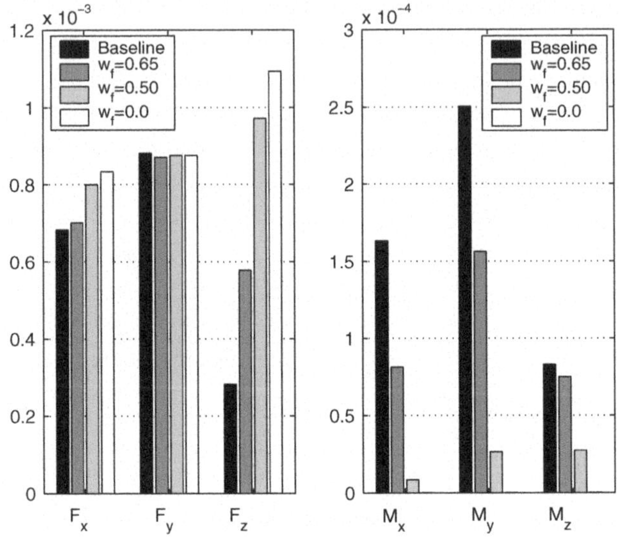

Fig. 3.6 Nondimensional 4/rev hub loads for a rotor with four trailing-edge flaps

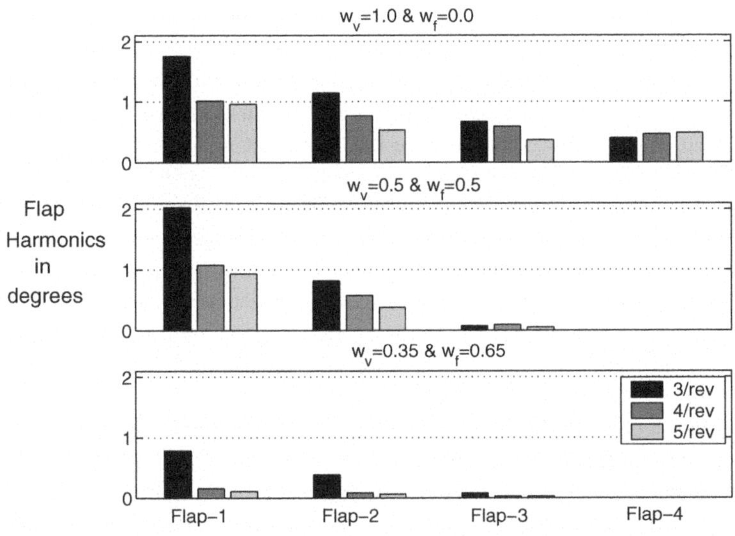

Fig. 3.7 Flap deflection harmonics for a rotor with four flaps

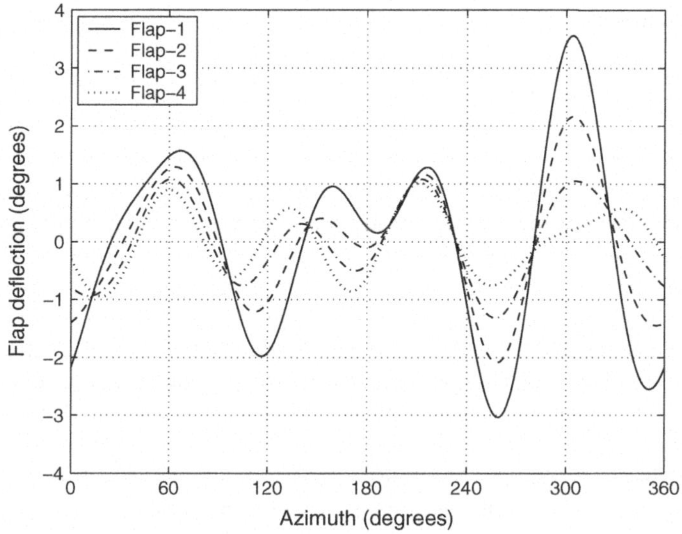

Fig. 3.8 Phase difference between the four flaps for $w_v = 1.0$ and $\mu = 0.30$

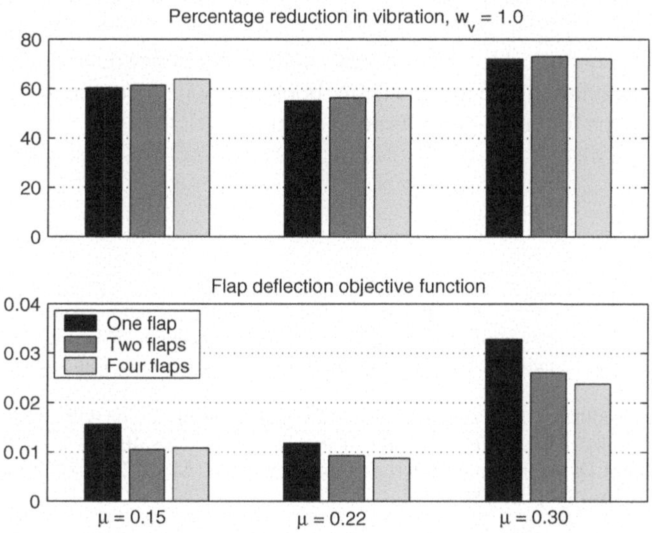

Fig. 3.9 Performance of flaps at different forward speeds, $w_v = 1.0$

The optimization study was performed at two more forward speeds, corresponding to advance ratios of 0.15 and 0.22. Figure 3.9 shows the percentage reduction in vibration and the corresponding value of the flap deflection objective function at the optima for three different forward speeds. Figure 3.9 indicates that multiple flaps perform better than a single flap even at lower advance ratios.

3.4 Summary

In this chapter, we investigate the effect of multiple trailing-edge flaps at the blade tip for a 4-bladed, soft-inplane, hingeless rotor. Flap deflections are modeled by changing the aerodynamic properties at the blade section. All flaps are deflected at 3/rev, 4/rev, and 5/rev harmonics in the rotating reference frame. Compared to the baseline (zero flap motion) case, a reduction of over 72 % is obtained in the vibration cost function comprising vibratory hub loads. By converting the constraints on the flap deflection amplitudes to a flap deflection objective function, a multicriteria optimization study has been performed. The nature of the trade-off between the two objectives (vibration and flap deflection) was obtained by plotting the Pareto optimal curves for rotor blades with single, two, and four flaps. Numerical results show that multiple flaps require much lesser flap deflection angles as compared to a single flap for obtaining the same reduction in vibration and hence require lesser control effort. They also demonstrate that it is possible to achieve significant reduction in the control deflections of the flaps by settling for somewhat lower vibration reduction. It is also shown that rotor blade tips with more than four trailing-edge flaps are unlikely to yield better performance than four trailing-edge flaps. *In fact, dual flaps provide a good level of vibration reduction without the additional complication of four flaps.* Multiple trailing-edge flaps appear attractive for helicopter vibration reduction, since it is easier and economical to move several small trailing-edge devices than one large device. In the remaining chapters of this book, we will focus on single and dual trailing-edge flaps for helicopter vibration control.

References

1. Viswamurthy, S.R., Ganguli, R.: An optimization approach to vibration reduction in helicopter rotors with multiple active trailing edge flaps. Aerosp. Sci. Technol. **8**(3), 185–194 (2004)
2. Abbott, I.H., von Doenhoff, A.E.: Theory of Wing Sections. Dover Publications, New York (1959)
3. Cribbs, R., Friedmann, P.P.: Actuator saturation and its influence on vibration reduction by actively controlled flaps. In: Proceedings of the 42nd AIAA/ASME/ASCE/AHS/ASC Structures, Structural Dynamics, and Materials Conference, No. AIAA-2001-1467, Seattle, Washington, USA, 16–19 April (2001)

Chapter 4
Flap Configuration and Control Law

Preliminary results obtained in Chap. 3 indicate that multiple trailing-edge flaps have great potential in helicopter vibration control. However, the aerodynamic model used in that chapter did not include the unsteady effects due to the motion of trailing-edge flaps. Also, the effects of the rotor wake were captured using a simple linear inflow model. In this chapter, a two-dimensional unsteady aerodynamic model for a flapped airfoil in compressible, subsonic flow is used to calculate the incremental aerodynamic loads due to flap motion. The free wake model described is utilized to obtain the inflow distribution over the rotor disk. Also, in this chapter, a control algorithm based on optimal control theory is used to determine the flap control input. Sections 4.1 and 4.2 give brief description of the flap inertial and aerodynamic load calculations. The flap control algorithm is explained in Sect. 4.3. Section 4.4 provides a comparison of differential versus uniform weighting method. The effect of trailing-edge flap motion on the blade motion is discussed in Sect. 4.5 and its effect on the optimal location of flaps is examined in Sect. 4.6. In Sect. 4.7, trailing-edge flap actuation power is explained. Section 4.8 explains trailing-edge flap hinge loads. Section 4.9 considers the performance of the multiple-flap system in the event of failure of one or more flaps. Section 4.10 gives the summary of this chapter.

4.1 Flap Inertial Loads

The leading edge of the flap is attached to the trailing-edge of the blade through a hinge mechanism. The flap properties are assumed to be symmetrical with respect to its major principal axes and uniform along the span-wise direction. The trailing-edge flap has a chord length of c_f and a span of l_f. The flap sectional mass per unit length is m_f and the first and second mass moments per unit span about the hinge are S_f and I_f, respectively. Also, it is assumed that the trailing-edge flap structural properties

R. Ganguli et al., *Smart Helicopter Rotors*, Advances in Industrial Control,
DOI 10.1007/978-3-319-24768-7_4

are lumped into the properties of the blade itself. The motion of the trailing-edge flaps is prescribed. Therefore, there are no additional degrees of freedom for the trailing-edge flap. The effects of the flaps are represented in the blade equations of motion as incremental contributions to the nonlinear load vector.

Several coordinate systems are defined for describing the motions of blades and active trailing-edge flaps. All coordinate systems are rectangular and right-handed. The following subscripts are used to indicate the reference frame for any vector.

'R' - Hub rotating coordinate system
'U' - Undeformed blade coordinate system
'L' - Local undeformed coordinate system
'D' - Blade deformed coordinate system
'h' - Hinge coordinate system
'f' - Flap coordinate system

The 'R' coordinate system is a rotating Cartesian system with its origin at the hub center and rotates with an angular velocity Ω. The 'U' system also rotates with the blades and is defined as the 'R' system rotated by about $\hat{j}_R$ by the blade precone angle, β_p. This is the coordinate system in which the blade deformations and sectional loads are defined. The 'L' system is a blade-attached coordinate system with its origin at an arbitrary point P with the blade span location of x = r/R. The 'L' system has unit vectors identical to the 'U' system but with the origin at P. The deformed blade is characterized by the 'D' system. The 'h' system has its origin on the trailing-edge flap hinge line with an offset d to the blade elastic axis (Fig. 4.1). The unit vectors in the 'h' system are parallel to the 'D' system. The 'f' system is used to describe the motion of trailing-edge flap. The 'f' system is fixed to the trailing-edge flap and is essentially the 'h' system rotated about the hinge line by the flap deflection angle δ (Fig. 4.1).

The position vector of an elemental mass, dm_f, on the trailing-edge flap can be expressed in the 'U' coordinate system as $\vec{\rho}_U$ via several coordinate transformations. This position vector, $\vec{\rho}_U$, is used in the formulation of the flap inertial loads. Based

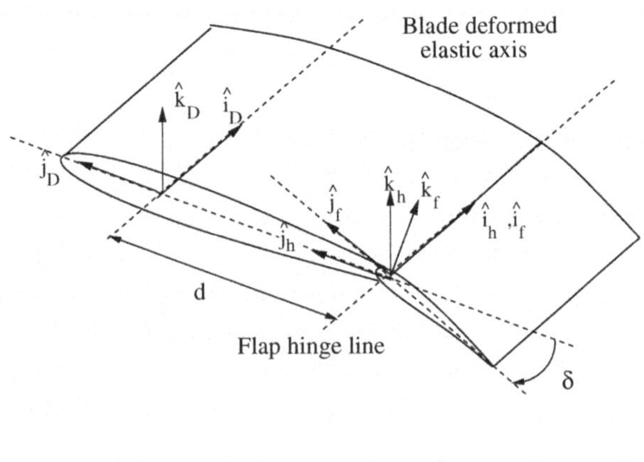

Fig. 4.1 Hinge coordinate system and flap coordinate system

on the Coriolis theorem, the acceleration of the elemental mass on the flap is derived in the 'U' coordinate system.

$$\vec{a} = \dot{\vec{\Omega}} \times \vec{\rho}_U + 2\vec{\Omega} \times \dot{\vec{\rho}}_U + \vec{\Omega} \times \vec{\Omega} \times \vec{\rho}_U + \ddot{\vec{\rho}}_U. \tag{4.1}$$

The acceleration of the elemental mass on the flap can be written in the 'U' coordinate system as

$$\vec{a} = a_{Ux}\hat{i}_U + a_{Uy}\hat{j}_U + a_{Uz}\hat{k}_U. \tag{4.2}$$

If the hub is assumed to have no acceleration and the blade rotational speed, $\vec{\Omega}$, is assumed to be constant, then $\dot{\vec{\Omega}} \times \vec{\rho}_U$ and $\ddot{\vec{\rho}}_U$ can be dropped in Eq. 4.1. From Newton's second law, the sectional inertial forces associated with a trailing-edge flap can be obtained by integrating the elemental contributions over the flap cross section.

$$\begin{Bmatrix} L_u^{I_f} & L_v^{I_f} & L_w^{I_f} \end{Bmatrix} \begin{Bmatrix} \hat{i}_U \\ \hat{j}_U \\ \hat{k}_U \end{Bmatrix} = -\int\int \vec{a}\, dm_f. \tag{4.3}$$

The trailing-edge flap's inertial contributions to the sectional moments about the deformed elastic axis are calculated using

$$\begin{Bmatrix} M_u^{I_f} & M_v^{I_f} & M_w^{I_f} \end{Bmatrix} \begin{Bmatrix} \hat{i}_U \\ \hat{j}_U \\ \hat{k}_U \end{Bmatrix} = -\int\int \vec{\rho}_D \times \vec{a}\, dm_f. \tag{4.4}$$

In this book, only plain flaps are considered and hence the terms independent of trailing-edge flap motion in Eqs. 4.3 and 4.4 are already included in the blade inertial loads. Therefore, only terms associated with trailing-edge flap deflections (δ, $\dot{\delta}$ and $\ddot{\delta}$) need to be considered as the flap inertial contributions. The inertial contribution to the trailing-edge flap hinge moment is calculated as

$$M_h = -\int\int (\vec{\rho}_h \times \vec{a}) .\hat{i}_h\, dm_f. \tag{4.5}$$

4.2 Flap Aerodynamic Loads

The unsteady aerodynamic loads generated due to trailing-edge flap motion are calculated using the formulation developed by Hariharan and Leishman [1]. This formulation is based on indicial concepts and is applicable to thin airfoils with trailing-edge flaps in subsonic compressible flow. This formulation has two main advantages over the classical Theodorsen's theory [2, 3]. Theodorsen's approach gives a solution to the unsteady airloads on a 2-d harmonically oscillating airfoil (with a control sur-

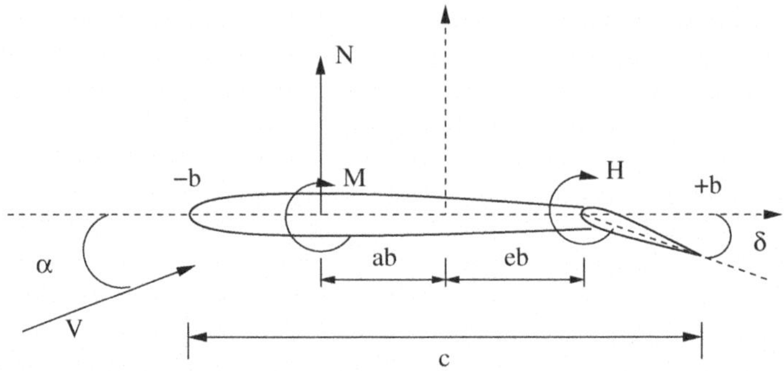

Fig. 4.2 A thin airfoil with trailing-edge flap

face) in inviscid, incompressible flow, and subject to small disturbance assumptions (Fig. 4.2). The present approach is formulated in time domain and hence capable of calculating unsteady airloads due to arbitrary motion of the flap. If the indicial lift/moment response is known, then the unsteady loads due to arbitrary changes in flap deflection can be obtained through the superposition of indicial aerodynamic responses using the Duhamel integral. Also, Theodorsen's theory is applicable only for incompressible flow, while the current approach works well in subsonic compressible flow regime.

Leishman classified the unsteady aerodynamic loads acting on an oscillating airfoil as circulatory and non-circulatory. The non-circulatory loads result from flow acceleration effects and are sometimes referred as impulsive loads. The circulatory loads arise from the creation of circulation about the airfoil and capture the effect of the shed wake. The total airloads acting on the airfoil is found by linear superposition of these two effects. In incompressible flow, non-circulatory loads acting on the airfoil are proportional to the instantaneous flap position since they are related to the propagation and reflection of wave disturbances. The circulatory loads, which include the effect of shed wake, have time-history effects and lag the excitation by a finite value. However, in subsonic compressible flow, both circulatory and non-circulatory loads have time-history effects.

In compressible flow, the initial loading on the airfoil after an indicial (step) input comprises a pressure wave system with a compression wave on one surface of the airfoil and an expansion wave on the other. This initial pressure loading on the surface can be computed directly using piston theory, which is a local wave equation solution for the unsteady airloads. The piston theory gives a result valid for any Mach number, M, but only at the instant after the perturbation has been applied ($s = 0$). Here, s is the distance traveled by the airfoil in semi-chords (b). For indicial flap motion about a hinge located at a distance eb aft of midchord location (Fig. 4.2), the normal force at $s = 0$ from piston theory are

$$\begin{aligned}\Delta C_{N_\delta}(s = 0, M) &= [2(1 - e)/M]\,\Delta\delta,\\ \Delta C_{N_{\dot\delta}}(s = 0, M) &= \left[(1 - e)^2/2M\right]\Delta(\dot\delta c/V).\end{aligned} \tag{4.6}$$

The final values ($s = \infty$) of the indicial response are given by the linearized subsonic theory.

$$\Delta C_{N_\delta}(s = \infty, M) = \frac{2F_{10}}{\beta}\,\Delta\delta, \qquad (4.7)$$
$$\Delta C_{N_{\dot\delta}}(s = \infty, M) = \frac{F_{11}}{2\beta}\,\Delta(\dot\delta c/V),$$

where F_{10} and F_{11} are coefficients defined by Theodorsen. Immediately after the perturbation, the non-circulatory effects dominate the loading, while the final steady-state conditions are dominated by circulatory loading. The intermediate indicial behavior between $s = 0$ and $s = \infty$ is written as

$$\Delta C_{N_\delta}(s, M, e) = \left[\frac{2(1-e)}{M}\phi^i_{N_\delta}(s, M) + \frac{2F_{10}}{\beta}\phi^c_{N_\delta}(s, M)\right]\Delta\delta, \qquad (4.8)$$
$$\Delta C_{N_{\dot\delta}}(s, M, e) = \left[\frac{(1-e)^2}{2M}\phi^i_{N_{\dot\delta}}(s, M) + \frac{F_{11}}{2\beta}\phi^c_{N_{\dot\delta}}(s, M)\right]\Delta(\dot\delta c/V),$$

where the indicial response functions $\phi^i_{N_\delta}$, $\phi^c_{N_\delta}$, $\phi^i_{N_{\dot\delta}}$, and $\phi^c_{N_{\dot\delta}}$ represent the intermediate behavior of the indicial normal force between $s = 0$ and $s = \infty$. The indicial functions can be approximated as

$$\phi^c_{N_\delta} = \phi^c_{N_{\dot\delta}} = 1 - \sum_{i=1}^{N} A_i \exp(-b_i\beta^2 s), \quad \sum A_i = 1, \quad b_i > 0, \qquad (4.9)$$
$$\phi^i_{N_\delta} = \exp\left[\frac{-s}{T'_{N_\delta}(M, e)}\right].$$

The coefficients A_i and b_i are determined using experimental measurements in frequency domain. Also, β is the Glauert compressibility factor and the β^2 term used in the indicial function reflects the effects of compressibility on the circulatory lift. The non-circulatory time constant, T'_{N_δ}, is evaluated using exact solutions for the indicial response in conjunction with aerodynamic reverse flow theorems. Equations 4.8 and 4.9 are substituted in Duhamel's integral to determine the normal force acting on the airfoil resulting from arbitrary flap motion. The airfoil pitching moment and hinge moment resulting from indicial flap deflection and flap rate are derived using a similar approach.

4.3 Flap Control Algorithm

In this chapter, all trailing-edge flaps are deflected at 2/rev, 3/rev, 4/rev, and 5/rev harmonics in the rotating reference frame.

$$\delta_i(\psi) = \delta_i^{2c}\cos(2\psi) + \delta_i^{2s}\sin(2\psi) + \delta_i^{3c}\cos(3\psi) + \delta_i^{3s}\sin(3\psi) \\ + \delta_i^{4c}\cos(4\psi) + \delta_i^{4s}\sin(4\psi) + \delta_i^{5c}\cos(5\psi) + \delta_i^{5s}\sin(5\psi). \quad (4.10)$$

The unknown flap harmonics are determined based on optimal control theory. Since the objective function used in Chap. 3 is not a quadratic form, it is not amenable to the direct application of optimal control theory. Therefore, a control law developed by Johnson is used in this chapter [4]. The periodic nature of the helicopter blade response in steady forward flight allows the transformation of the control problem from time domain to frequency domain. For a helicopter rotor with 4 blades, an important source of vibration is the 4 Ω hub forces and moments transmitted by the rotor to the fuselage. The control algorithm, hence, is based on the minimization of the following quadratic cost functional:

$$J = \mathbf{Z}^T\mathbf{W}_Z\mathbf{Z} + \sum_{i=1}^{N_f} \beta_i \mathbf{u_i}^T\mathbf{u_i}. \quad (4.11)$$

Here, $\mathbf{Z}$ is a vector containing the 4 Ω component of the three hub shears and three hub moments, $\mathbf{u_i}$ is the control input vector containing cosine and sine higher harmonics of the ith flap, and N_f is the number of flaps in one rotor blade. The diagonal matrix $\mathbf{W}_Z$ is a weighting matrix for hub vibratory loads. The first term in the above equation is a scalar quantity that is purely related to the magnitude of the hub vibratory loads and is a measure of hub vibration. The subsequent terms weight the control effort of each of the N_f flaps.

$$\mathbf{Z} = \begin{bmatrix} F_{xH}^{4p} & F_{yH}^{4p} & F_{zH}^{4p} & M_{xH}^{4p} & M_{yH}^{4p} & M_{zH}^{4p} \end{bmatrix}^T,$$
$$\mathbf{u_i} = \begin{bmatrix} \delta_i^{2c} & \delta_i^{2s} & \delta_i^{3c} & \delta_i^{3s} & \delta_i^{4c} & \delta_i^{4s} & \delta_i^{5c} & \delta_i^{5s} \end{bmatrix}^T,$$
$$\mathbf{W}_Z = \left(1 - \sum_{i=1}^{N_f} \beta_i\right) \begin{bmatrix} \alpha\, \mathbf{I}_{3\times 3} & \mathbf{0}_{3\times 3} \\ \mathbf{0}_{3\times 3} & (1-\alpha)\, \mathbf{I}_{3\times 3} \end{bmatrix},$$
$$\sum_{i=1}^{N_f} \beta_i \leq 1. \quad (4.12)$$

The nondimensional weighting parameter α can be adjusted to give more importance to reduce either hub moments or hub shears. By setting α equal to zero, the controller can be made to minimize only the 4/rev hub moments without regard to 4/rev hub shears. If α is set to unity, then the controller tries to minimize 4/rev hub shears only. A value of $\alpha = 0.5$ is used in this chapter which means that all hub loads are weighted equally. The nondimensional weighting parameters β_i can also be changed to establish the relative importance of hub loads ($\mathbf{Z}$) versus flap motions ($\mathbf{u_i}$) in the cost function J. When $\beta_i = 0$, the controller tries to minimize hub loads without regard to the motion of the ith trailing-edge flap. As β_i increases toward unity, the controller will try to reduce the motion of the ith trailing-edge flap to zero. A simple and efficient open-loop global controller is used to determine the optimal flap control input harmonics. The entire range of control input is presumed to be linear. Taylor series expansion about zero control input gives

$$\mathbf{Z} = \mathbf{Z_0} + \sum_{i=1}^{N_f} \mathbf{T_i u_i}, \tag{4.13}$$

where $\mathbf{T_i}$ is the transfer matrix that relates the system response to the motion of ith trailing-edge flap. This transfer matrix is assumed to be constant over the entire range of the control input and hence is calculated only once by perturbing the control harmonics individually around zero control inputs. $\mathbf{Z_0}$ is the vibratory load vector at zero control inputs. Equation 4.13 is substituted into Eq. 4.11 and the following condition is applied for optimality:

$$\frac{\partial J}{\partial \mathbf{u_i}} = 0, \quad i = 1 \text{ to } N_f. \tag{4.14}$$

This yields the following set of linear equations which when solved yield the optimal flap control input harmonics

$$\left(1 - \sum_{i=1}^{N_f} \beta_i\right) \left[\mathbf{Z_0}^T \mathbf{T_i} + \sum_{j=1}^{N_f} \mathbf{u_j} \mathbf{T_j}^T \mathbf{T_i}\right] + \sum_{i=1}^{N_f} \beta_i \mathbf{u_i}^T = 0, \text{ for } i = 1 \text{ to } N_f. \tag{4.15}$$

In this chapter, the values of β_i are adjusted such that *each flap is utilized to full authority* ($\delta_{sat} = 2$ degrees) [5]. Since each flap is weighted to a different extent, this method will be referred to as the *differential weighting* method. Figure 4.3 shows the steps involved in determining the weighting parameters β_i of all flaps. As seen from Fig. 4.3, the weighting parameter β_i is dynamically adjusted to keep the maximum deflection of the ith flap within saturation limits. If the peak deflection of the ith flap is less than the saturation limit, the algorithm automatically decreases the weighting of its control effort. On the other hand, if the peak deflection of the ith flap is more than the saturation limit, the algorithm automatically increases the weighting of its control effort in the next iteration. An initial guess for β_i is used to start the

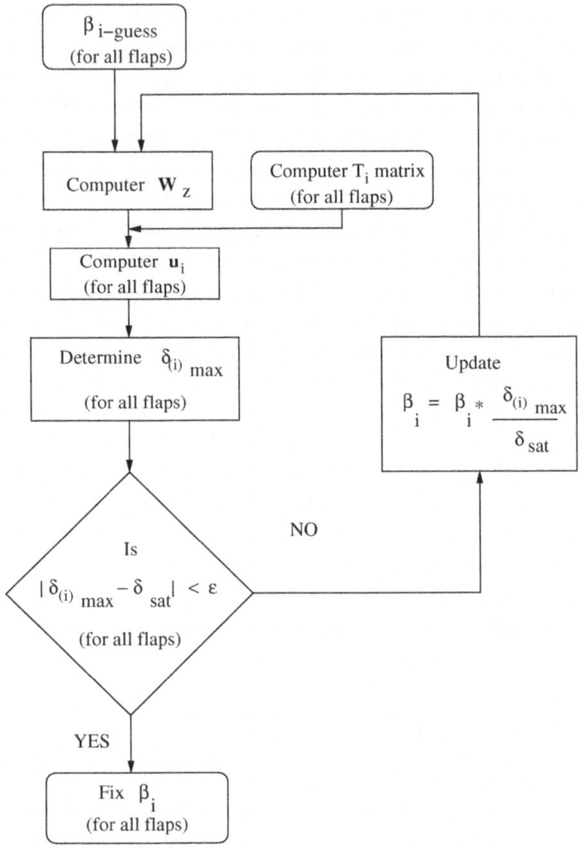

Fig. 4.3 Determining weighting parameters β_i's for differential weighting method

algorithm. It is worth noting that this method to determine the appropriate weighting parameters (β_i's) is numerically efficient and computationally inexpensive compared to the actual helicopter aeroelastic analysis. In earlier studies using multiple flaps, all flaps were weighted equally and the saturation limit was imposed on the flap with maximum deflection. Hence, we shall refer to it as *uniform weighting* method (Fig. 4.4). The drawback of this method is that, typically, only one flap among the multiple flaps is exploited to full authority. The other flap(s) are under-utilized and hence lead to inferior performance as compared to the *differential weighting* method.

The properties of the trailing-edge flap are given in Table 4.1. The flap hinge is assumed to coincide with its leading edge. To avoid effects due to strong tip vortices and blade-tip shape effects, the flaps are placed away from the rotor blade tip. Spacing between the flaps is introduced to reduce inter-flap aerodynamic interaction.

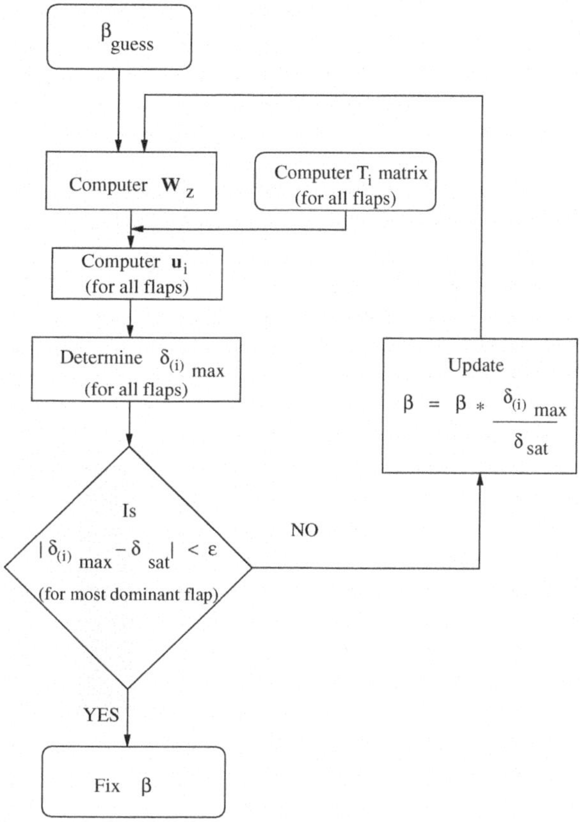

Fig. 4.4 Determining weighting parameter β for uniform weighting method

Table 4.1 Trailing-edge flap properties

c_f/c	0.20
m_f/m_0	0.10
X_g^f/c_f	0.20

4.4 Differential Versus Uniform Weighting Method

Initially, a rotor with three small trailing-edge flaps is considered to demonstrate the advantage of differential weighting method over uniform weighting method (Fig. 4.5). Each flap spans 4 % of the blade span, centered at about 86, 78, and 70 % blade span location. Figure 4.6 compares the vibration reduction achieved at two different flight conditions through uniform weighting and differential weighting of the flap control effort. Numerical studies are conducted at both high ($\mu = 0.30$) and low-speed ($\mu = 0.10$) flight conditions. At $\mu = 0.30$, the differential weighting method gave 68 % reduction in hub vibration as against 57 % reduction obtained

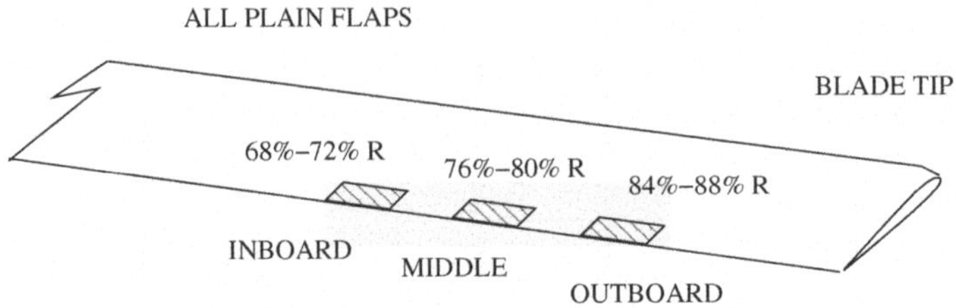

Fig. 4.5 Rotor with three trailing-edge flaps

Fig. 4.6 Comparison of uniform and differential weighting methods

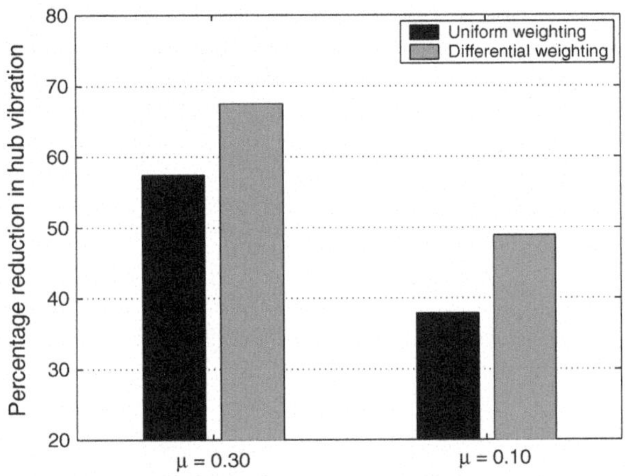

through uniform weighting method. At $\mu = 0.10$, differential weighting and uniform weighting methods gave 49 and 38 % reduction in hub vibrations, respectively. It is seen that there is considerable improvement in performance when every flap is allowed to work to its full authority.

Figures 4.7 and 4.8 show the motion history of all three flaps at $\mu = 0.30$ under uniform and differential weighting methods, respectively. In the uniformly weighted case, only the inboard flap is utilized to full stroke. The peak deflections of the middle and outboard flaps are 1.2 and 0.7 degrees, respectively. However, in the differentially weighted case, all three flaps are completely utilized with a peak deflection of 2 degrees, resulting in better performance in terms of hub vibration reduction. Figures 4.9 and 4.10 show the motion history of all flaps at $\mu = 0.10$. Similar conclusions can be drawn under this low-speed flight condition with only the inboard flap getting completely exploited under uniform weighting and all three flaps being completely used when differentially weighted. It is clear that differential weighting method performs better than uniform weighting method. All subsequent studies are conducted using the differential weighting method.

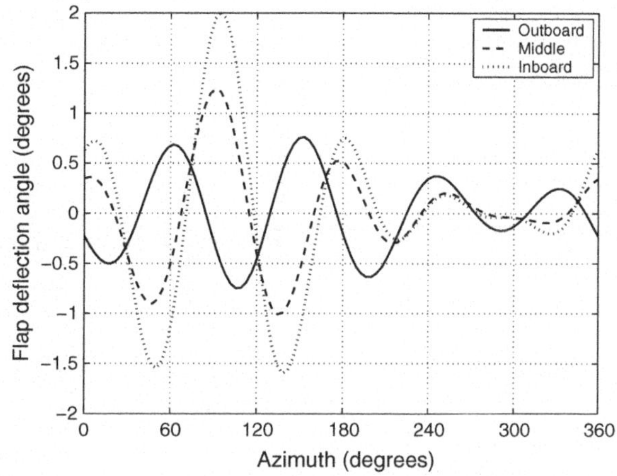

Fig. 4.7 Flap motion history for uniform weighting case, $\mu = 0.30$

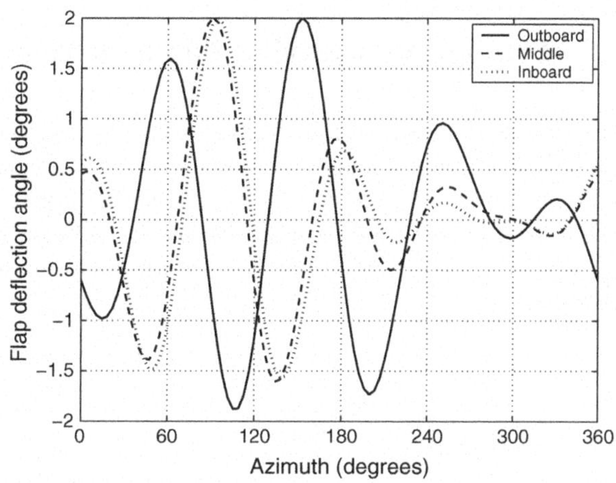

Fig. 4.8 Flap motion history for differential weighting case, $\mu = 0.30$

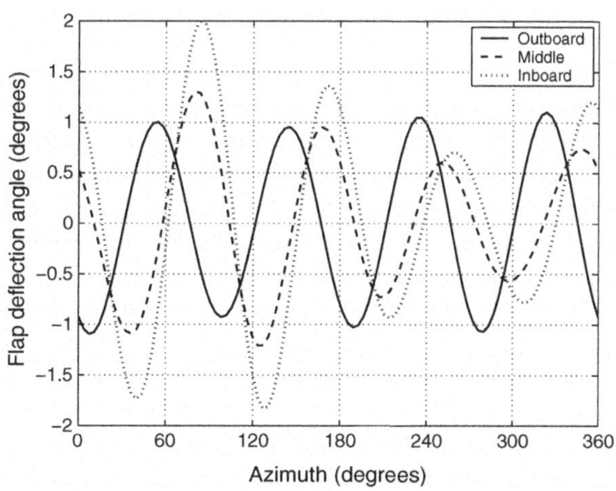

Fig. 4.9 Flap motion history for uniform weighting case, $\mu = 0.10$

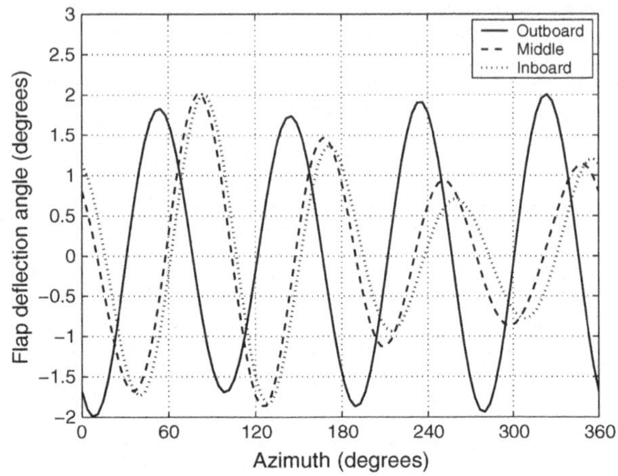

Fig. 4.10 Flap motion history for differential weighting case, $\mu = 0.10$

4.5 Flap Location Studies Based on Modes

The span-wise location of the trailing-edge flaps can have considerable influence on the level of vibration reduction obtained. First, a single trailing-edge flap is used to study the effect of radial flap location. This flap spans 4 % of the blade span and its radial location is varied between 66 and 90 % blade span location. To understand the effect of trailing-edge flap motion on blade motion, we define the following parameters:

$$M_{ib} = \sqrt{\int_0^{2\pi} [m_{ib}(\psi)]^2 \, d\psi},$$

$$D_i = \frac{\sqrt{\int_0^{2\pi} [m_{ic}(\psi) - m_{ib}(\psi)]^2 \, d\psi}}{M_{ib}}. \tag{4.16}$$

Here, $m_{ib}(\psi)$ is the modal participation factor of the ith natural mode of the rotor blade at azimuth ψ under baseline conditions (no flap motion) and $m_{ic}(\psi)$ is the modal participation factor of the ith natural mode of the rotor blade at azimuth ψ when the flap is actuated to obtain maximum reduction in hub vibration. M_{ib} represents the extent to which the ith natural mode contributes to blade motion under baseline conditions. D_i represents the extent to which the actuation of trailing-edge flap affects the contribution of the ith natural mode to blade motion. It is seen from Fig. 4.11 that the natural modes that contribute most to blade motion are first flap bending mode, first lag bending mode, first torsion mode, and second flap bending mode. Of these four dominant modes, only the first torsion mode and second flap bending mode are sufficiently affected by trailing-edge flap motion (see Figs. 4.12 and 4.13). As the location of the flap shifts toward the blade tip, its effect on the first torsion mode increases monotonically. However, the effect of trailing-edge flap

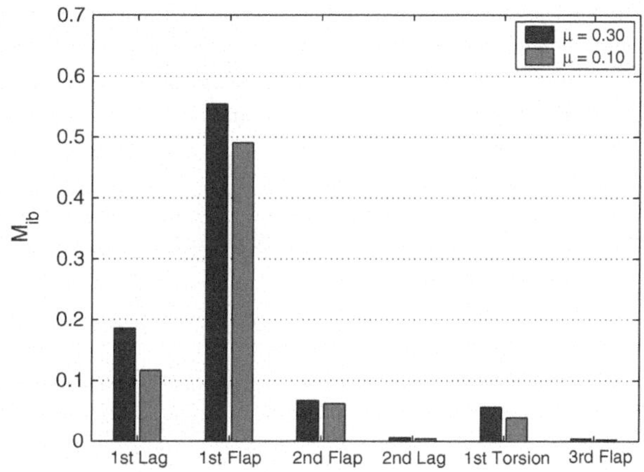

Fig. 4.11 Contribution of natural modes to blade motion under baseline conditions

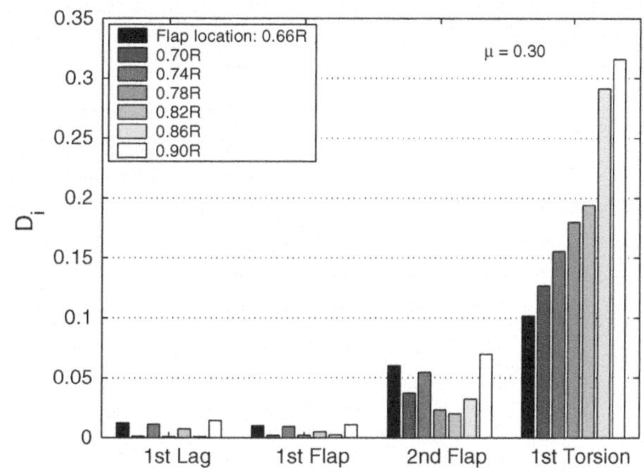

Fig. 4.12 Effect of trailing-edge flap motion on the four dominant modes; $\mu = 0.30$

motion on the second flap bending mode is least when the flap is located near the node of that mode (79 % blade radial location). Figure 4.14 shows the reduction in hub vibration versus the radial location of the small single trailing-edge flap. It is clear that the flap is least effective when it is located near the node of the second flap bending mode. Furthermore, there is a direct correlation between the reduction in hub vibration and the extent to which the contribution of second flap bending mode is affected due to trailing-edge flap motion. The effect of trailing-edge flap motion on the contribution of first blade torsion mode does not seem to affect the extent of hub vibration reduction.

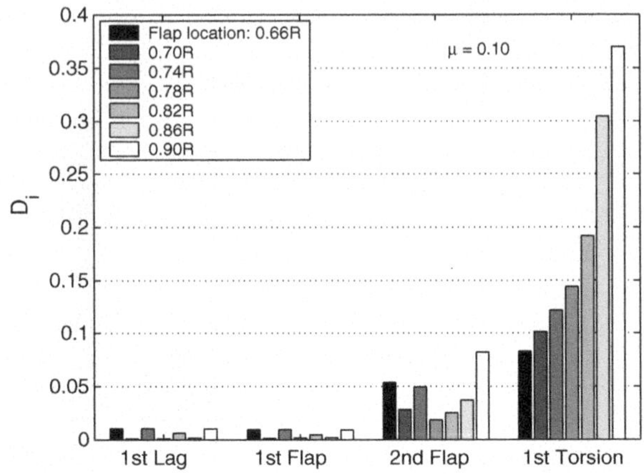

Fig. 4.13 Effect of trailing-edge flap motion on the four dominant modes; $\mu = 0.10$

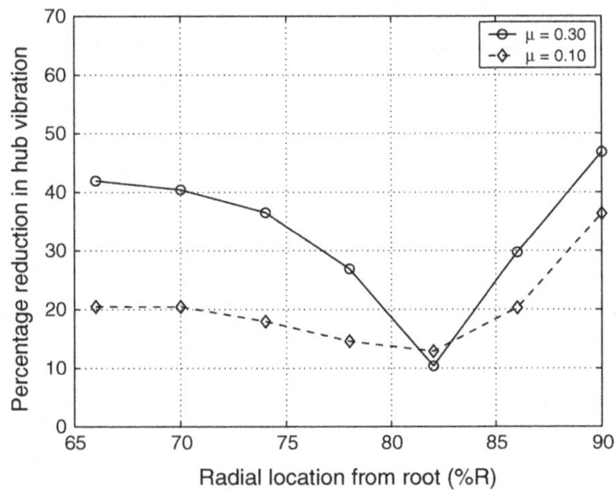

Fig. 4.14 Hub vibration reduction versus trailing-edge flap location

Using the results from the above study, we now try to place single (12 % span), dual (6 % span each), and three (4 % span each) flaps between 64 and 92 % blade radial location to achieve the best performance in terms of hub vibration reduction. It can be seen from Fig. 4.14 that there is a gradual decrease in the vibration reduction achieved as the small flap is moved from inboard toward the node of the second flap bending mode. Moving the flap further outboard toward the tip away from the node results in a rapid rise in the effectiveness of the flap in terms of hub vibration reduction. It is therefore desirable to place the large single flap into the inboard region and away from the node of the second flap bending mode to achieve its best performance. In the dual-flap configuration, the participation of the second flap bending mode can be most affected by placing each flap away from the node on either side and actuating them in an out-of-phase manner. However, one should note that in the dual-flap case, the inboard and outboard flaps (of equal span) produce generalized forces of

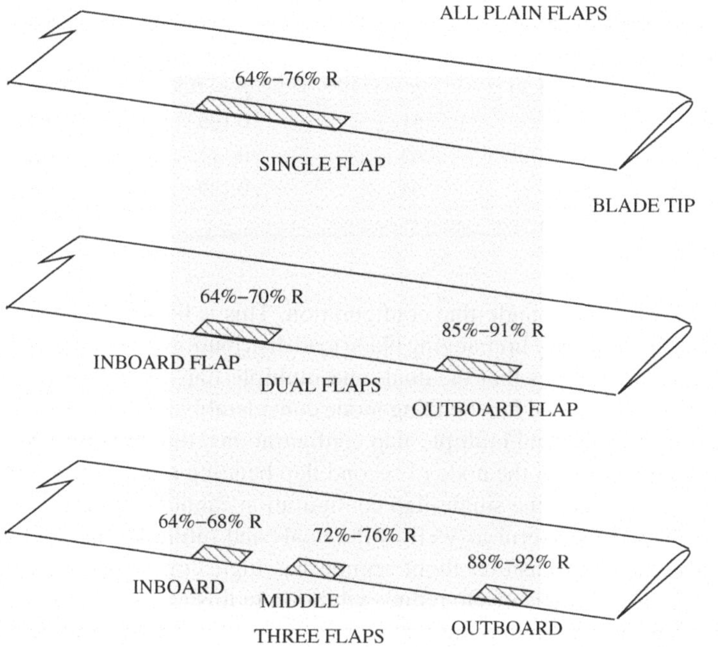

Fig. 4.15 Optimum location of single, dual, and multiple flaps from parametric studies

different magnitudes (for the same deflection) due to their location in lower and higher dynamic pressure region, respectively. This disadvantage can be ameliorated using the three-/multiple-flap configurations. Here too it is advantageous to place flaps on either side of the node of the second flap bending mode but to compensate for the lower dynamic pressure on the inboard region, two small flaps are placed inboard of the node and one flap is placed outboard of it to achieve optimal performance. The optimal locations of the flaps in single-, dual-, and multiple-flap configurations are shown in Fig. 4.15. We should point out that the final flap locations were obtained using a combination of the physics-based reasoning discussed above and extensive numerical experiments simulating an exhaustive search. In Chap. 6, we will look at other more formal approaches based on optimization for finding the best flap locations using optimization methods.

4.6 Physics Behind Modal Location of Flaps

Table 4.2 shows the reduction in hub vibration achieved in the each of the cases shown in Fig. 4.15. It is seen that all three configurations are equally effective at high-speed forward flight. However, at the low-speed flight condition, dual and multiple flaps

Table 4.2 Performance comparison of single-, dual-, and multiple-flap configurations

Flap configuration	Vibration reduction, %		$D_{2nd\ flap\ bending\ mode}$	
	$\mu = 0.30$	$\mu = 0.10$	$\mu = 0.30$	$\mu = 0.10$
Single flap	73	51	0.105	0.081
Dual flaps	75	62	0.111	0.120
Multiple flaps	76	64	0.129	0.129

perform better than the single-flap configuration. This is because dual and multiple flaps have more flexibility in changing blade load distributions as compared to a single flap. Table 4.2 also shows that the dual- and multiple-flap configurations affect the contribution of the second flap bending mode considerably more than the single-flap configuration. In dual- and multiple-flap configurations, this is achieved by placing the flaps on either side of the node of second flap bending mode and actuating them out-of-phase. However, the single-flap configuration cannot exploit this advantage, and hence does not perform as well as the dual- and multiple-flap configurations. Since the low-speed transition flight regime has high vibration levels, the use of multiple flaps is advantageous to reduce vibration in this regime. Figure 4.16 shows the motion of all flaps in the multiple-flap cases at high-speed forward flight. The middle and inboard flaps which are in the lower dynamic pressure region have similar motion trends across the azimuth. The outboard flap, however, is almost out-of-phase with the middle and inboard flap.

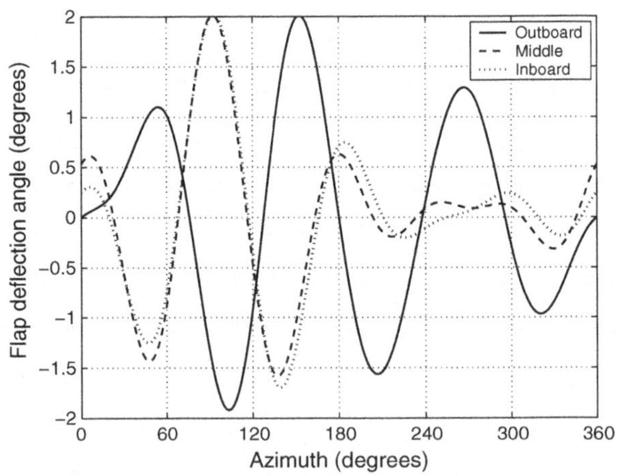

Fig. 4.16 Flap motion history for multiple-flap configurations, $\mu = 0.30$

4.7 Trailing-Edge Flap Actuation Power

The mean power required by the trailing-edge flap (single blade) is obtained by integrating the product of the hinge moment and flap deflection rate over one rotor revolution.

$$P_f = -\frac{1}{2\pi}\int_0^{2\pi} M_h\dot{\delta}\, d\psi. \tag{4.17}$$

The flap power required may change sign over some portions of the azimuth. As the actuator will generally not be able to transfer this power back to the flap actuation power supply, the negative power is neglected. Therefore, the actuation power of the flap system is

$$P_f = \frac{1}{2\pi}\int_0^{2\pi} \max(-M_h\dot{\delta}, 0)\, d\psi. \tag{4.18}$$

The power required by the trailing-edge flap is an important parameter which needs to be considered for practical implementation. Figures 4.17 and 4.18 show the mean and maximum (instantaneous) powers required to drive the flaps at high- and low-speed forward flight conditions, respectively. In high-speed forward flight, the mean power required by each of the dual flaps is about two-thirds of the power required to drive a single large flap. However, in multiple-flap configurations, each small flap requires approximately one-third of the power required to drive a single flap. In low-speed flight regime, the inboard flap in the dual-flap configuration

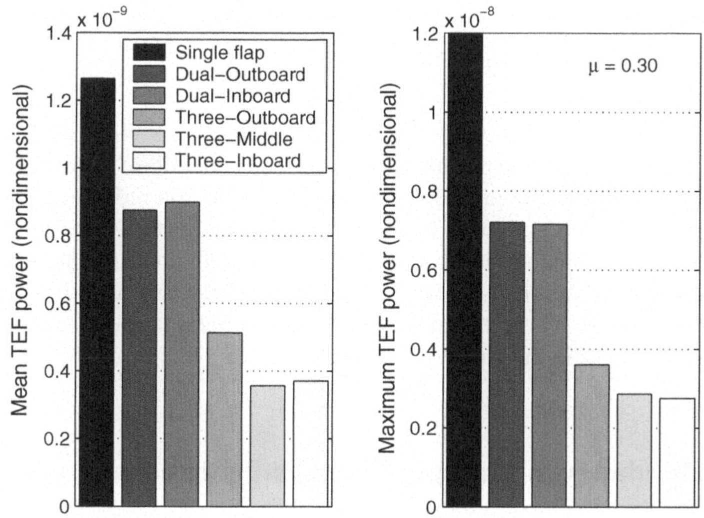

Fig. 4.17 Mean and maximum flap control power; $\mu = 0.30$

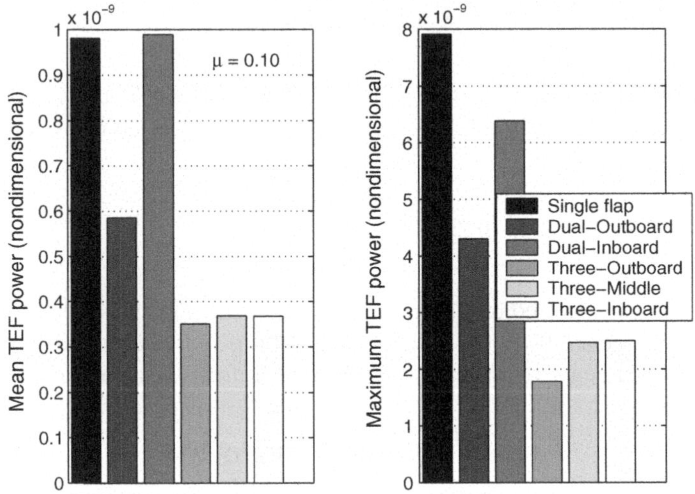

Fig. 4.18 Mean and maximum flap control power; $\mu = 0.10$

requires considerably more power than the outboard flap. However, each flap in the multiple-flap configurations requires almost the same power (around one-third of the power required by a single flap) which is desirable in terms of workload sharing between the flaps. Smart material-based actuators used to move the control surfaces are restricted in their output. It is therefore desirable to move several small control surfaces to achieve vibration reduction than moving a single large control surface.

4.8 Trailing-Edge Flap Hinge Loads

Hinge load is an another important parameter for designing trailing-edge flaps. The mechanical design of the flap mechanism needs to consider these loads which should be as low as possible while providing the required vibration reduction. Figure 4.19 shows the maximum hinge loads encountered by single-, dual-, and multiple-flap configurations at advance ratio of $\mu = 0.30$. The maximum hinge moment in the multiple-flap configurations is about 25 % less than that encountered in the single-flap configuration. However, the maximum hinge moment in the case of dual flaps is almost the same as in the single-flap case.

4.9 Redundancy in Multiple-Flap Configurations

A key advantage of multiple trailing-edge flaps compared to the single flap lies in their redundancy. Thus even if some of the flaps are not functional, the vibration control system for the helicopter can still yield considerable vibration reduction. In

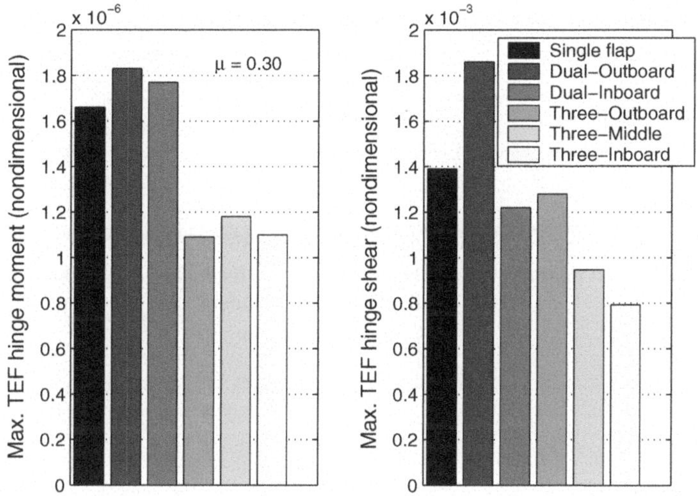

Fig. 4.19 Maximum flap hinge loads; $\mu = 0.30$

this chapter, it is assumed that the inoperable flap is locked in zero position along with the corresponding flaps in other blades. To determine the effect of redundancy with three trailing-edge flaps on hub vibration suppression, three cases are considered. In each case, the control law obtained using three flaps is used. In case(1), only the outboard flap is actuated, while the other two flaps are locked at zero position. Vibration reduction of 47 and 36 is achieved at high- and low-speed flight conditions, respectively. In case(2), only the middle flap is actuated which gives 36 and 18 % reduction in hub vibration at high- and low-speed flight conditions, respectively. In case(3), only the inboard flap is actuated which yields 42 and 20 % reduction in hub vibration at high- and low-speed flight, respectively. This indicates that outboard flap and middle flap are the most and least effective in reducing hub vibration, respectively. The outboard flap is the most effective control surface due to high dynamic pressure conditions near the blade tip. The middle flap is least effective since it is located very near to the node of second flap bending mode (0.79R). However, all the flaps are less effective at low-speed flight due to overall lower dynamic pressure conditions. These results are summarized in Table 4.3. It is clear that considerable vibration reduction of 36–47 % at high forward speed and 18–36 % at low forward speed can be obtained with only one flap being operational. Therefore, multiple flaps provide redundancy and have the features of a fault tolerant system when performing the vibration reduction function. Table 4.4 shows the performance of the multiple-flap configuration in the event of failure of one flap.

Table 4.3 Vibration reduction obtained with only one-flap operational

Flight condition	All flaps on	Only outboard flap on	Only middle flap on	Only inboard flap on
$\mu = 0.30$	76	47	36	42
$\mu = 0.10$	64	36	18	20

Table 4.4 Vibration reduction obtained with two flaps operational

Flight condition	All flaps on	Inboard flap failure	Middle flap failure	Outboard flap failure
$\mu = 0.30$	76	65	68	63
$\mu = 0.10$	64	50	52	36

4.10 Summary

The advantages of multiple active trailing-edge flaps for helicopter vibration control in forward flight are investigated. Uniform weighting of the trailing-edge flap control effort typically leads to complete utilization of only one trailing-edge flap. The other flaps are under-utilized leading to inferior performance in terms of hub vibration reduction. This drawback is solved by *differentially weighting* each trailing-edge flap so that each flap is used to complete authority without significant additional computational effort. Among the various configurations considered, there is a direct correlation between the reduction in hub vibration and the extent to which the contribution of the *second flap bending mode* (toward blade motion) is affected due to trailing-edge flap motion. The effect of trailing-edge flap motion on the contribution of first blade torsion mode does not greatly affect the extent of hub vibration reduction. *Dual- and multiple-flap configurations are capable of altering the contribution of second flap bending mode (towards blade motion) more effectively than the single trailing-edge flap configuration.* This is achieved by placing one or more flaps on either side of the node of the second flap bending mode of the rotor blade and actuating them in out-of-phase manner. The actuation power requirements of multiple trailing-edge flaps are lower than single large trailing-edge flap. Multiple-flap configuration also has the advantage of redundancy over single-flap configuration. It is possible to achieve reasonable levels of vibration control even when one or more flaps fail during operation.

References

1. Hariharan, N., Leishman, J.G.: Unsteady aerodynamics of a flapped airfoil in subsonic flow by indicial concepts. J. Aircr. **33**(5), 855–868 (1996)
2. Theodorsen, T.: General theory of aerodynamic instability and the mechanism of utter. NACA Report 496 (1935)
3. Theodorsen, T., Garrick, I.E.: Nonstationary flow about a wing-aileron-tab combination including aerodynamic balance. NACA Report 736 (1942)
4. Johnson, W.: Wake model for helicopter rotor in high speed flight. NASA CR-177507, November (1988)
5. Viswamurthy, S.R.: Using the complete authority of multiple active trailing-edge flaps for helicopter vibration control. J. Vib. Control **14**(8), 1175–1199 (2008)

Chapter 5
Active Flap Controller Evaluation

The performance of a global and a local controller is studied in this chapter. It is well-known that the global controller yields optimal control input in a single step for a linear system. For a nonlinear system, however, the global controller gives only a suboptimal solution, since the transfer matrices ($\mathbf{T_i}$) are determined only once. Section 5.1 provides a comparison of the performance of a global controller with a local controller. Section 5.2 presents the control law sensitivity studies. Section 5.3 explains the noise in hub loads measurements. Section 5.4 gives the summary of this chapter.

Sensitivity is an important aspect of the controller as it describes the robustness of the optimal solution and indicates the penalties that may be incurred for wavering from the optimal solution. It is desirable to have an optimal trajectory of state variables which is insensitive to small perturbations in control input and system parameters. This sensitivity is related to the adjoint variables which can be derived from Pontryagin's maximum principle [1]. Several other works have studied the performance sensitivity of optimal control systems to system parameter variations [2, 3]. In this chapter, it is assumed that a decrease in sensitivity of the system to parameter variation is desirable and that a certain amount of increase in the cost function can be allowed to achieve this reduction in sensitivity.

5.1 Performance of Global Controller

The helicopter aeroelastic problem is inherently nonlinear in nature. However, the extent of nonlinear relationship between the flap control inputs and blade response is not known a priori. The choice of the controller (local or global) depends on the extent of this nonlinearity. The local controller used in this chapter attempts to minimize a quadratic cost functional containing the 4/rev hub loads with constraints posed on the peak deflection of all flaps. The optimization problem posed above is a constrained minimization problem. The unknowns to be determined are the flap harmonics. The

R. Ganguli et al., *Smart Helicopter Rotors*, Advances in Industrial Control,
DOI 10.1007/978-3-319-24768-7_5

method of feasible directions (MFD) is used to determine a feasible direction of descent. The gradient of the cost function with respect to the design variables is calculated using the forward finite difference method. The optimum step size along this direction is then determined using a constrained one-dimensional search like the golden section method. The advantage of the local controller is that it is applicable to systems that have strong nonlinear properties. However, the implementation of a local controller is computationally expensive since gradient calculations need to be made repeatedly at every design point using a computationally expensive aeroelastic simulation code.

To compare the performances of the global and local controllers, only the single-flap configuration is considered as the local controller becomes computationally prohibitive as the number of flaps increases. This is because there are more number of unknown variables and constraints in a multiple-flap case and numerically computing the gradient at any point in the feasible domain becomes cumbersome and time consuming. The single trailing-edge flap is located at 70 % radial location and has a span of 12 % (Fig. 5.1). Both local and global controllers are used to determine the optimal flap control inputs. Figures 5.2 and 5.3 show the optimal flap motion history obtained at high-speed and low-speed flight, respectively. Though the solution obtained using the global and local controllers are different, it is seen that they are almost in phase throughout the rotor azimuth. This is because the phase of the trailing-edge flap motion with respect to rotor azimuth is critical in determining the extent of vibration reduction. The differences between the two controllers occur mostly at the retreating side of the blade ($180° < \psi < 360°$) where dynamic pressures are less. This effect is more pronounced at low-speed flight than at high-speed flight.

In high-speed forward flight, both controllers yielded about 73 % reduction in hub vibration. In low-speed flight, however, the local controller gave 60 % reduction as compared to 51 % reduction by the global controller. Though the global controller gives a suboptimal solution to the vibration control problem, it is worth noting here

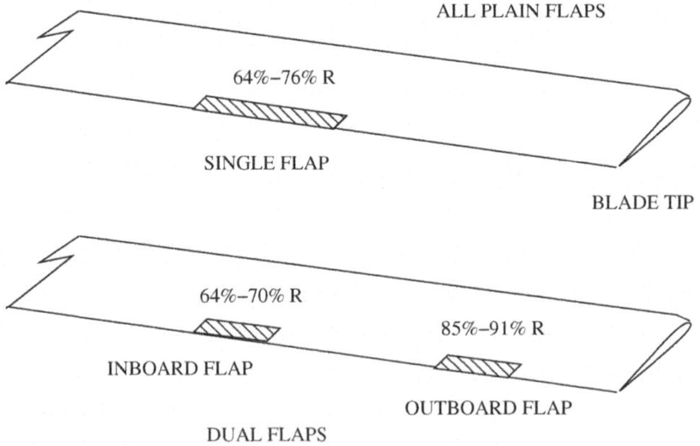

Fig. 5.1 Single- and dual-flap configurations considered in this chapter

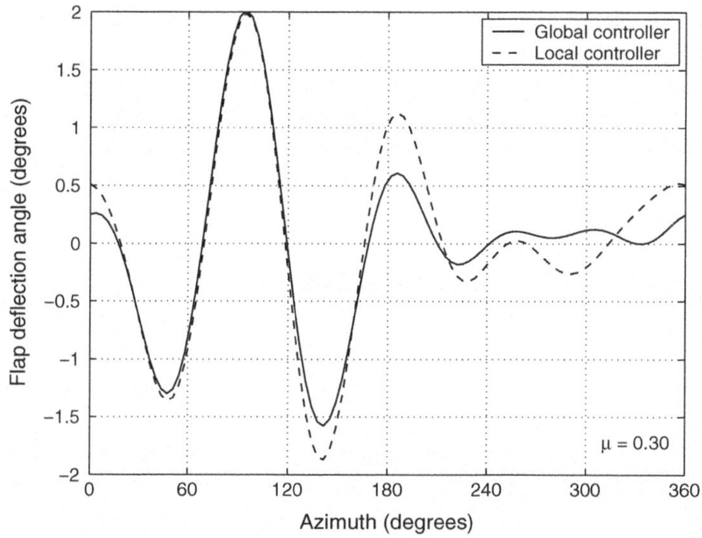

Fig. 5.2 Comparison of local and global controllers, $\mu = 0.30$

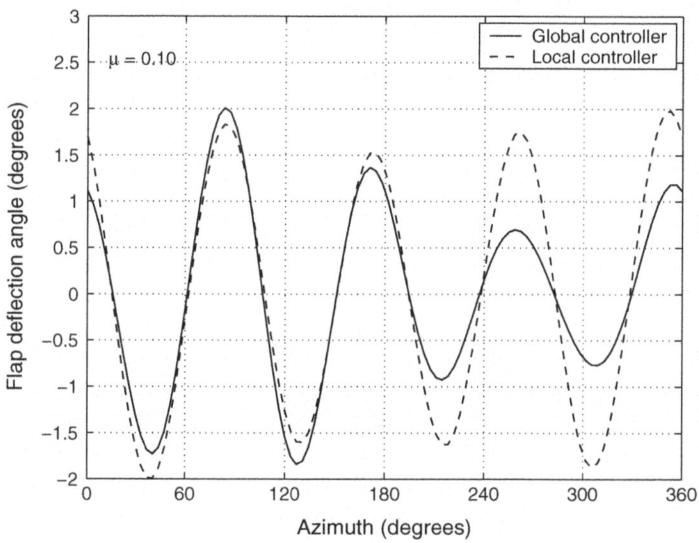

Fig. 5.3 Comparison of local and global controllers, $\mu = 0.10$

that it requires much lesser computational effort than the local controller. For instance, at low-speed flight, the global controller takes about 22 min as compared to 160 min taken by the local controller (on a 256 MB RAM computer) to give the optimal solution [4].

Though the overall helicopter aeroelastic problem is quite nonlinear, it is seen that the relationship between trailing-edge flap input and blade response is only mildly nonlinear within the limits, ($\delta_{sat} = 2°$). Hence, the simple and efficient global controller is used for further studies in this chapter. The global controller when applied to the dual-flap configuration leads to 75 and 62 % reduction in hub vibration levels at high- and low-speed flight conditions, respectively.

5.2 Control Law Sensitivity Studies

In practice, the actual control input for vibration control is not the flap motion in itself but the electric field or voltage applied to the smart material-based actuators that actuate the trailing-edge flaps. It is often not possible to achieve the desired flap motion exactly, due to damping, friction, freeplay, nonlinearity, hysteresis, and other factors in the actuator-flap mechanism. This section discusses the sensitivity and robustness of the harmonic control law for helicopter vibration reduction.

5.2.1 Error in Flap Phasing

The extent of hub vibration reduction achieved depends on the phasing between flap motion and rotor azimuth. During the practical implementation of the trailing-edge flap concept, the actual flap motion could lead or lag the commanded flap motion due to the unmodeled nonlinearity, hysteresis in actuator-flap system, and/or actuator-flap dynamics. Figure 5.4 shows the degradation in the performance of the vibration control system due to an error in the phasing of trailing-edge flap motion. The degradation is fairly symmetric about the optimum phasing angle for both single- and dual-flap configurations at both high- and low-speed flight regimes. In high-speed forward flight, the degradation in performance is similar for both single- and dual-flap configurations. It is seen that an error of 5° in phasing yields about 67 % reduction in vibration which is about 6 % lower than the optimum solution. However, if the actual flap motion leads or lags the optimum/commanded flap motion by 10° in phase, the performance of the flap system comes down sharply, resulting in about 51 % reduction in hub vibration. In low-speed flight, dual-flap configuration performs better than the single-flap configuration. For the dual-flap configuration, the vibration reduction drops from 63 to 45 % due to an error of 10° in phasing. For the same error, the vibration reduction decreases from 52 to 36 % for the single-flap configuration.

5.2.2 Error in Flap Control Inputs

In this section, the performance degradation of the trailing-edge flap system due to perturbation in the optimal flap control input ($\mathbf{u_i}$) is studied. It is desirable to have a

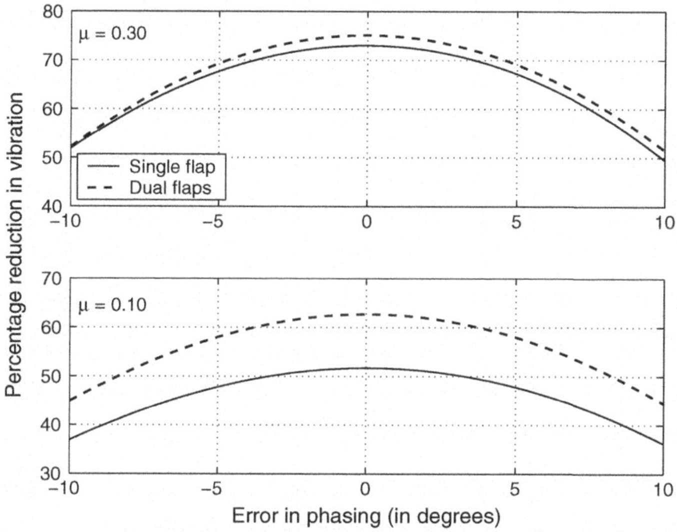

Fig. 5.4 Performance of the controller with error in flap phasing

robust control law whose performance does not degrade sharply due to perturbation in the optimal flap control input angles, $\mathbf{u_i}$. Figure 5.5 shows the minimum vibration reduction that could be achieved (worst case scenario) when every component of the optimum control input is allowed a maximum perturbation limit of Δ. For instance, for $\Delta = 0.2°$, at low-speed flight conditions, hub vibration levels can be reduced at the least by 39 and 53 % using single- and dual-flap concepts, respectively. It is readily seen that the dual-flap configuration is more robust to errors in optimal flap control inputs than the single-flap configuration. Furthermore, it is apparent that the drop in performance is more severe in high-speed flight as compared to low-speed flight condition. For instance, in the dual-flap configuration, for $\Delta = 0.3°$, the performance reduced from 75 to 46 % (a drop of 29 %) under high-speed flight and from 63 to 47 % (a drop of 17 %) under low-speed flight conditions.

5.2.3 Sensitivity to Plant Parameters

It is of interest to study the sensitivity of the global controller to system/plant parameters. The elastic rotor blade stiffness and mass are considered as plant parameters. The global controller is used to determine the optimal flap control inputs. Figure 5.6 shows the optimum flap motion history following the stiffness change for the single-flap configuration at high-speed flight conditions. The flap motion is almost identical in phasing after the stiffness change. However, only 67 and 59 % reductions in hub vibration were achieved for stiffness reductions of 5 and 10 %, respectively, compared to 73 % for the baseline rotor. Figure 5.7 shows the optimum flap motion history

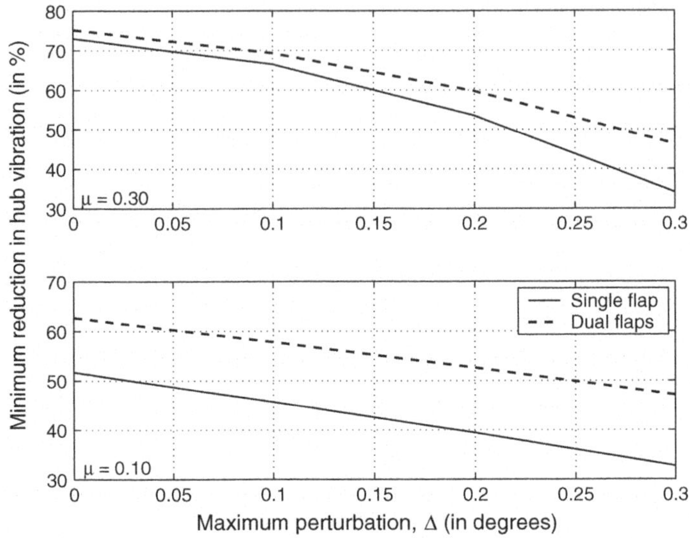

Fig. 5.5 Minimum vibration reduction achievable versus maximum control input perturbation

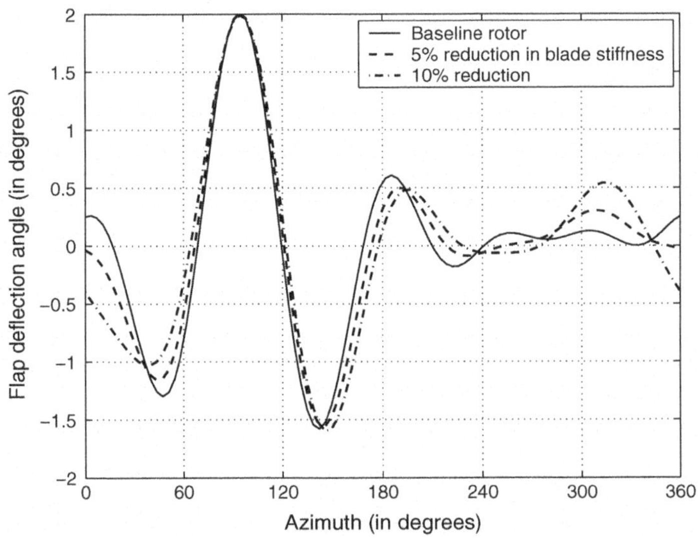

Fig. 5.6 Optimal flap control input due to stiffness perturbations, $\mu = 0.30$

following the stiffness change for the single-flap configuration at low-speed flight conditions. The optimum flap motion is identical for all the three cases considered. In this case, the hub vibration was reduced by 54 and 56 % from baseline values, for stiffness reductions of 5 and 10 %, respectively, compared to 52 % for the baseline rotor. In another study, the mass of the rotor blade was increased by up to 1 %.

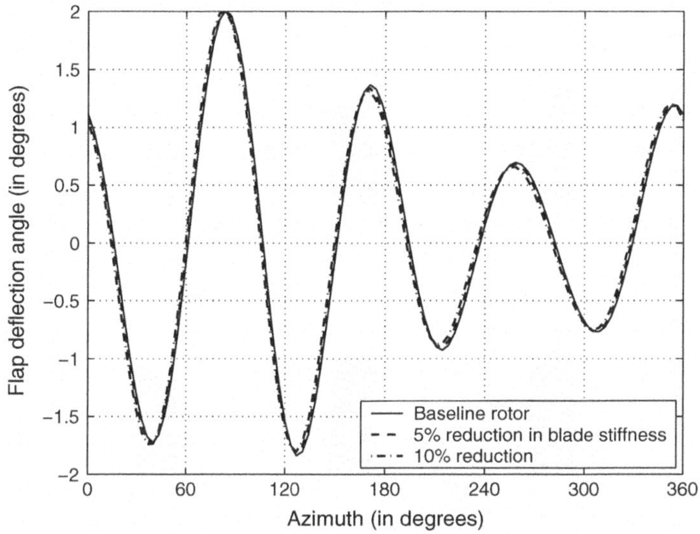

Fig. 5.7 Optimal flap control input due to stiffness perturbations, $\mu = 0.10$

The optimal flap control input determined using the global controller was found to be very insensitive to such small changes in rotor blade mass. The controller is therefore quite insensitive to gross changes in rotor plant parameters.

5.3 Noise in Hub Loads Measurements

In reality, we cannot measure the vibratory hub loads exactly in wind tunnel or flight tests. Hence, it is important to look into the performance of the control strategy when the measured hub load data contains noise. Typically, the global controller can be developed using tests where the hub loads contain noise. In this chapter, zero-mean Gaussian noise was added to the hub loads in the time-domain to simulate noisy data. Noise levels of 5 and 10 % were considered to check the performance of the controller. The gradient transfer matrix ($\mathbf{T_i}$) is now numerically calculated in the presence of noise and subsequently used by the global controller to obtain the optimal flap control inputs. This is in a way equivalent to developing the control law from flight tests with noisy data. Both single- and dual-flap configurations were considered. To get a statistical measure of the performance of the control law, the above operation is repeated with several sets of noisy data. Figures 5.8 and 5.9 show the performance of the global controller at different noise levels at high-speed and low-speed flight conditions, respectively. The performance of the controller degrades sharply in low-speed flight compared to high-speed flight. Both single- and dual-flap configurations show promising results in terms of mean percentage reduction in

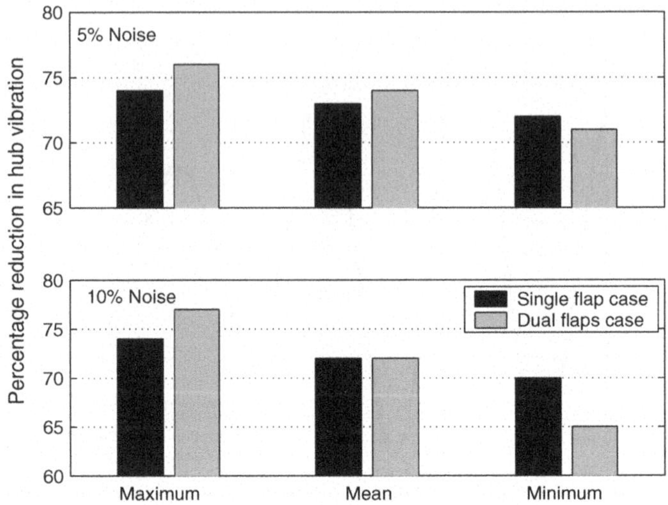

Fig. 5.8 Performance of controller with noisy hubload data, $\mu = 0.30$

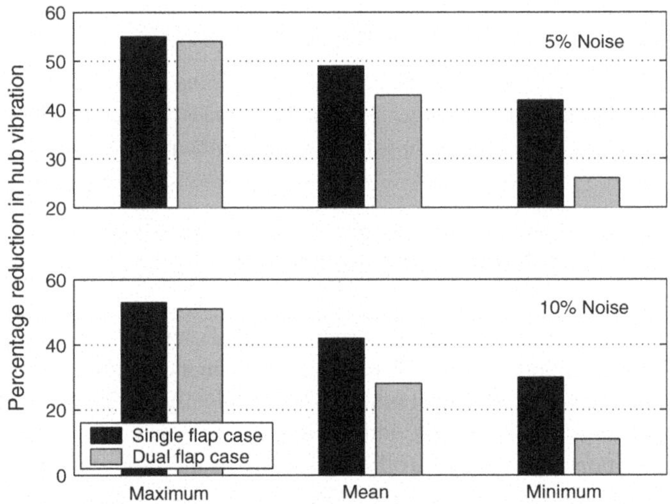

Fig. 5.9 Performance of controller with noisy hubload data, $\mu = 0.10$

vibration in high-speed flight. In low-speed flight, the single-flap case shows an average vibration reduction of 49 and 42 % with 5 and 10 % noise, respectively. The dual-flap case shows a mean vibration reduction of 43 and 28 % with 5 and 10 % noise, respectively. In the single-flap case, the worst event shows a vibration reduction of 30 %, a reduction of about 22 % from the ideal situation. In the dual-flap case, the worst case shows a vibration reduction of 11 %, a reduction of about 52 % from the ideal situation.

5.4 Summary

In this chapter, we investigate the performance and sensitivity of the control law used for helicopter vibration control using single/multiple trailing-edge flaps. Studies show that the global controller performs as well as the local controller in high-speed forward flight within the limits considered in this chapter ($\delta_{sat} = 2°$). However, in low-speed flight, global controller gives a slightly suboptimal solution to the vibration control problem. However, the global controller requires much lesser computational effort as compared to the local controller. While helicopter rotor vibration is nonlinearly related to blade structural properties and other such design variables, the relationship to trailing-edge flap control angles is only mildly nonlinear for angles in the range of $\pm 2°$ which is typical for flaps actuated with smart materials. Single and dual flaps perform equally well at high-speed forward flight, yielding about 73 and 75 % reduction in hub vibration. However, at low-speed flight, dual-flap configuration performs better than single-flap (63 % against 52 % reduction in vibration). The phasing of the trailing-edge flap(s) is/are closely linked to the performance of the vibration control system. Numerical studies show that the performance is quite good upto an error of $\pm 5°$ in phasing from the optimal phase. Also, the degradation in performance due to error in phasing is more severe in high-speed flight than in low-speed flight. The control law is found to be robust even after small perturbations (upto $\pm 0.3°$) are applied to the flap harmonics. Dual-flap configuration is found to be more robust to errors in optimal flap control harmonics than single-flap configuration. The effect of system/plant parameters on the optimal flap motion is also studied. The elastic rotor blade stiffness and mass are considered as plant parameters. It is found that the optimal control inputs are, in general, fairly insensitive to small changes in structural stiffness and mass of the rotor blade. The global controller used in this chapter performs well in the presence of zero-mean Gaussian noise in the measured hub load data. The performance of both single and dual flaps are found to be good in the presence of two different noise levels (5 and 10 %). Results indicate that dual-flap configuration is more sensitive to noise in hubload data than single-flap configuration.

References

1. Pontryagin, L.S.: The Mathematical Theory of Optimal Processes. Wiley-Interscience, New York (1962)
2. Keys, P.: The solution and sensitivity of a general optimal control problem. Appl. Math. Model. **4**(4), 287–294 (1980)
3. Pagurek, B.: Sensitivity of the performance of optimal control systems to plant parameter variations. IEEE Trans. Autom. Control. **10**, 178–180 (1965)
4. Viswamurthy, S.R., Ganguli, R.: Performance sensitivity of helicopter global and local optimal harmonic vibration controller. Comput. Math. Appl. **56**(10), 2468–2480 (2008)

Chapter 6
Trailing-Edge Flap Placement

The optimal location of flaps for multiple-flap configuration shown in Chap. 4 was obtained through parametric studies and the objective was to minimize only the hub vibration levels. These trailing-edge flaps are typically operated using piezoceramic actuators placed inside the blade section. Typically, the electrical power required to drive these actuators is transmitted from the fixed frame to the rotating frame through slip rings. It is therefore desirable to keep the trailing-edge flap power requirements to a minimum. In this chapter, we aim to determine the optimal locations of dual trailing-edge flaps to achieve minimum hub vibration levels in a helicopter while incurring low penalty in terms of required trailing-edge flap control power. The reduced hub vibration levels and required flap control power are the two objectives considered in this chapter and the flap locations along the blade are the design variables. A formal multiobjective optimization approach is followed. Response surface methodology is used to develop an inexpensive surrogate model of the high-fidelity and expensive helicopter aeroelastic analysis. The rest of the chapter is organized as follows. Section 6.1 provides a brief introduction to surrogate models. The optimization problem is posed in Sect. 6.2. Section 6.3 provides a brief explanation of the response surface methodology. Numerical results and discussions are provided in Sect. 6.4. Section 6.5 gives the summary of this chapter.

6.1 Surrogate Models

Integration of high-fidelity, computationally intensive analyses in design optimization problems lead to enormous computational challenges. One approach to reduce the computational cost is to develop inexpensive "surrogates" or "metamodels" of the high-fidelity analyses which can replace the expensive models during the design optimization stage. Simpson et al. [1] gave an excellent survey of metamodeling applications in engineering design. Metamodeling techniques have also been widely used in

R. Ganguli et al., *Smart Helicopter Rotors*, Advances in Industrial Control,
DOI 10.1007/978-3-319-24768-7_6

structural and multidisciplinary design optimization. Response surface methodology (RSM) and Kriging and Radial basis function method (RBF) are among the most popular metamodeling techniques. In RSM, approximations known as *response surfaces* are created by sampled numerical experiments over the feasible design space. Typically, lower order polynomial response surfaces are used to approximate the objectives and constraints. The unknown regression coefficients in the response surfaces are determined by the method of least-square error. Once the response surfaces are obtained, the optimum design variables can be determined easily, since evaluating polynomial functions involves very low computational cost. Response surface approximations have also been used in reliability-based design optimization problems. Kriging is a regression technique originally used in geostatistics to approximate or interpolate data. Kriging models are capable of accurately approximating a highly nonlinear system by using fewer samples than RSM. However, the computational effort required to fit a Kriging model is much greater than RSM, and the interpretation of Kriging model parameters is not straightforward [2]. Comparative studies of various metamodeling methods show that RSM is ideal for small-scale, mildly nonlinear problems.

Response surfaces are known to be global approximations in nature as against Taylor series approximations that are local in nature [3]. The low computational cost of evaluating response surfaces also allows the use of global search strategies like evolutionary algorithms and exhaustive search methods. However, one disadvantage of response surface methods is that the computational cost to obtain them grows rapidly as the number of design variables increase. This is the so-called "curse of dimensionality."

6.2 Optimization Problem

The objective of the current chapter is to determine the optimal placement of dual trailing-edge flaps to achieve maximum attenuation in hub vibration levels. However, it is desirable to achieve this objective with minimum penalty in terms of power consumed by the active-flap system. The initial dual-flap configuration and move limits are shown in Fig. 6.1. The move limits are the lower and upper bounds of the design variables. The blade tip region is not included to avoid 3D flow and blade tip shape effects. A saturation limit of $\pm 2^\circ$ is placed on each trailing-edge flap. The vibration cost functional due to optimal control input for a given flap configuration is denoted by J_v^*. The mean actuation power for the complete flap system for a N_b-bladed rotor with N_f flaps in each rotor blade can be written as

$$P_t = \frac{N_b}{2\pi} \sum_{i=1}^{N_f} \int_0^{2\pi} \max(-M_{hi} \frac{d\delta_i}{d\psi}, 0)\, d\psi. \tag{6.1}$$

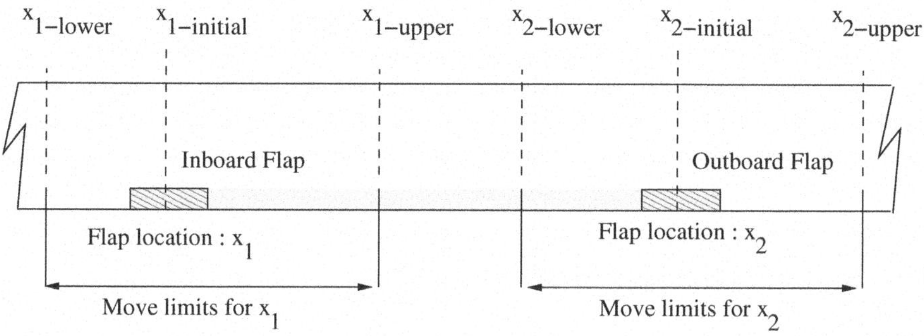

Fig. 6.1 Initial configuration of dual flaps

The values of J_v^* and P_t are normalized with respect to their corresponding values at the initial configuration (Fig. 6.1). These normalized values are denoted by F_v and F_p, respectively. The optimization problem is posed as

$$\begin{aligned} &\text{Minimize} && \{F_v\ ,\ F_p\}, \\ &\text{Subject to:} && x_{1-lower} \leq x_1 \leq x_{1-upper} \\ & && x_{2-lower} \leq x_2 \leq x_{2-upper}. \end{aligned} \tag{6.2}$$

It is unlikely that both objectives would be minimized by the same choice of design variables. This problem falls under the domain of multiobjective optimization [4]. The nature of the trade-off between the two objectives (F_v and F_p) depends on the shape of the Pareto curve/surface. A solution is said to be *Pareto optimal* if it is impossible to minimize any one objective without allowing an increase in one or more of the other objectives. A standard method of finding a Pareto point involves the minimization of a scalar objective function obtained by combining the various objectives. Typically, the scalar objective is a positively weighted convex sum of the individual objectives. However, this method has a few drawbacks. First, a uniform distribution of weighting parameters seldom yields a uniform distribution of points on the Pareto surface. Often, most points are found clustered in certain parts of the Pareto surface. Furthermore, only convex parts of the Pareto surface can be obtained using the method of minimizing convex combinations of the individual objectives.

Another drawback of the weighted function method is the issue of computational cost. Optimization has to be performed for different sets of weighting parameters to obtain a fair idea of the Pareto surface which is computationally expensive when function evaluations involve complex problems such as helicopter aeroelasticity. For instance, each evaluation of F_v or F_p involves calling the helicopter aeroelastic code 17 times (16 for the evaluation of transfer matrix $\mathbf{T}$ and 1 for determining the hub loads due to optimal control input, $\mathbf{Z}^*$). Each call to the aeroelastic code takes about 170 s of CPU time on a 256 MB RAM Pentium-4 computer. This translates to 48 min of CPU time for one function evaluation of F_v or F_p. This prohibits the use of conventional optimization techniques to determine the optimal location of trailing-edge flaps. Response surface approximations provide a solution by decoupling the optimization problem from the aeroelastic analysis problem. Once the approximate model is constructed, the optimum solution can be determined easily since the response surfaces are typically polynomial expressions.

6.3 Response Surface Methodology

RSM provides an approximate relationship between the response of a system y and the n inputs to the system $\mathbf{x}$ (independent variables). Typically smooth analytical functions such as polynomials are used for approximations. A lower order smooth polynomial response surface removes the spurious minima/maxima while capturing the global trend of the response thereby leading to a robust optimal design. For instance, a second-order polynomial response surface has the following form:

$$f(\mathbf{x}) = A_0 + \sum_{i=1}^{n} A_i x_i + \sum_{i=1}^{n} \sum_{j=1}^{i} A_{ij} x_i x_j. \tag{6.3}$$

By appropriate choice of the regression coefficients (A_0, A_i and A_{ij}) in the above equation, the function f can be made to approximately represent the actual response y of the system around a point $\mathbf{x_0}$ in the design space. One method of determining the unknown regression coefficients is by comparing the above equation with the second-order Taylor series expansion of the actual system response y around $\mathbf{x_0}$. However, this method requires the gradient and Hessian information of system response y at $\mathbf{x_0}$. The determination of gradient and Hessian in problems involving large computer programs like helicopter aeroelastic analysis is a cumbersome and expensive process involving the method of finite differences. Also, in such cases, the choice of step-size is critical in determining the gradient and Hessian accurately.

The unknown regression coefficients in Eq. 6.3 can also be determined using the method of curve-fitting. A second-order response surface as shown in Eq. 6.3 has $[n(n+1)/2+n+1]$ regression coefficients. For instance, if $n = 2$, the second-order response surface with six regression coefficients can be written as:

$$f(x_1, x_2) = A_0 + A_1 x_1 + A_2 x_2 + A_{11} x_1^2 + A_{12} x_1 x_2 + A_{22} x_2^2. \tag{6.4}$$

To determine the six regression coefficients uniquely, we need to evaluate the actual system response at six design points (called data points), at the least. A better fit can be constructed by evaluating the system response at more than six data points and using the method of least-square error to determine the coefficients. However, the computational cost to evaluate the system response at every additional data point limits the total number of response observations. In this chapter, a second-order response surface is used to approximate the functions F_v and F_p in terms of the

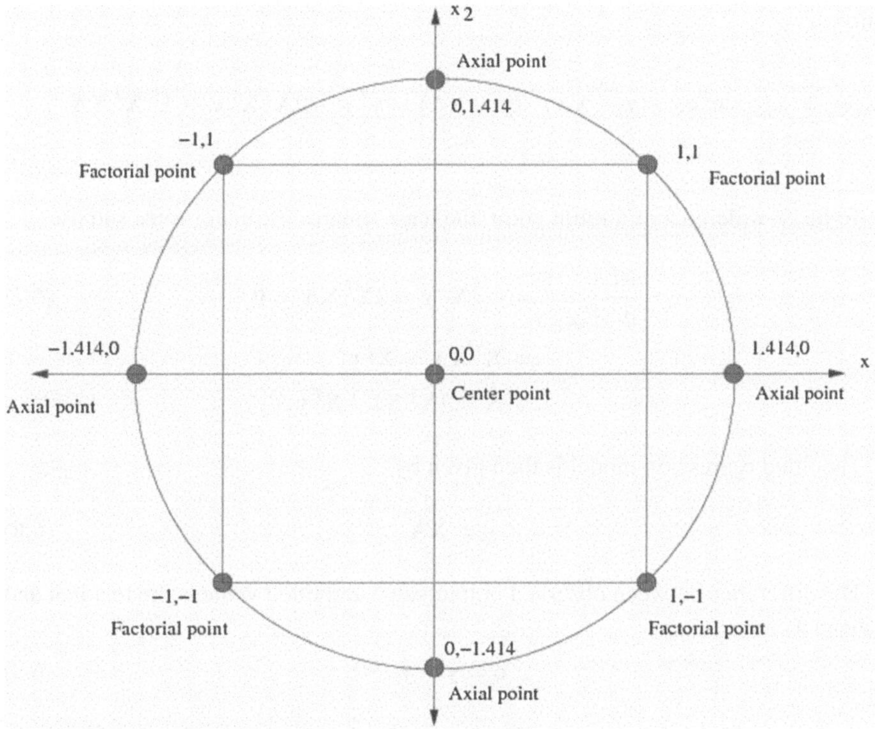

Fig. 6.2 Central composite design for two design variables

independent variables x_1 and x_2 (flap positions). Central composite design (CCD) is used for the optimal selection of data points to obtain a high-fidelity response surface approximation. This design consists of a center point, encompassed by 4 axial and 4 factorial data points. Each independent variable is scaled such that each of the factorial points lies at the vertex of a unit square with its center at the central data point. Also, each axial point lies at a distance $\alpha = 1.414$ from the center (Fig. 6.2). This choice of axial points ensures that any two design points which are equally distant from the center have equally good prediction in response values. Once the system response is observed at the nine data points, the regression coefficients can be obtained using the method of least-square error as follows:

$$\mathbf{y} = \mathbf{X}\mathbf{A} + \mathbf{e}, \tag{6.5}$$

where $\mathbf{y}$ is the vector of observations, $\mathbf{X}$ is a matrix of the levels of the independent variables, $\mathbf{A}$ is a vector of regression coefficients, and $\mathbf{e}$ is a vector of error terms. For a second-order response surface with six regression coefficients and nine data points from CCD, $\mathbf{y}$ and $\mathbf{e}$ are 9×1 vector, $\mathbf{X}$ is a 9×6 matrix and $\mathbf{A}$ is a 6×1 vector. The regression parameters can be obtained by minimizing the least-squares error as follows:

$$L = \sum_{i=1}^{9} e_i^2 = \mathbf{e}^T\mathbf{e} = (\mathbf{y} - \mathbf{XA})^T (\mathbf{y} - \mathbf{XA}) = \mathbf{y}^T\mathbf{y} - 2\mathbf{A}^T\mathbf{X}^T\mathbf{y} + \mathbf{A}^T\mathbf{X}^T\mathbf{XA}. \tag{6.6}$$

To be considered a minimum point, the least-squares estimator must satisfy

$$\left.\frac{\partial L}{\partial \mathbf{A}}\right|_{\mathbf{A}=\hat{\mathbf{A}}} = -2\mathbf{X}^T\mathbf{y} + 2\mathbf{X}^T\mathbf{X}\hat{\mathbf{A}} = \mathbf{0} \tag{6.7}$$

$$\Rightarrow \mathbf{X}^T\mathbf{X}\hat{\mathbf{A}} = \mathbf{X}^T\mathbf{y}$$

$$\Rightarrow \hat{\mathbf{A}} = \left(\mathbf{X}^T\mathbf{X}\right)^{-1}\mathbf{X}^T\mathbf{y}.$$

The fitted regression model is then given by

$$\hat{\mathbf{y}} = \mathbf{X}\hat{\mathbf{A}}. \tag{6.8}$$

The difference between observed output value and fitted value is the residual and is given as

$$\mathbf{e} = \mathbf{y} - \hat{\mathbf{y}}. \tag{6.9}$$

Formal methods of estimating the error in the response surfaces exist. However, engineering judgement can often be used to account for the accuracy of the fitted model.

6.4 Numerical Results

Numerical results are obtained for the four-bladed, uniform hingeless rotor whose properties are given in Chap. 3 (Table 3.1). The trailing-edge flap properties are the same as considered in Chap. 4 (Table 4.1). Results are obtained at the cruise speed corresponding to an advance ratio of $\mu = 0.30$ [5]. Plain trailing-edge flaps are considered in this chapter where the flap is integrated into the airfoil cross-section as shown in Fig. 6.3. Each flap spans 6 % of the rotor blade span. The numerical values for initial configuration and move limits for variables x_1 and x_2 as defined in Fig. 6.1 are given in Table 6.1. The inboard and outboard flaps are constrained to be placed between 59 to 71 and 77 to 89 % blade span, respectively. The lower limit on the inboard flap prevents it from operating in comparatively low dynamic pressure region. The outboard flap is prevented from being placed too near the blade tip since the flow there is 3D in nature and is dominated by strong tip-vortices whose effect on the flap cannot be modeled using 2D aerodynamic models. The midpoint of each flap can therefore move one flap span length (6 %R) to the left or right.

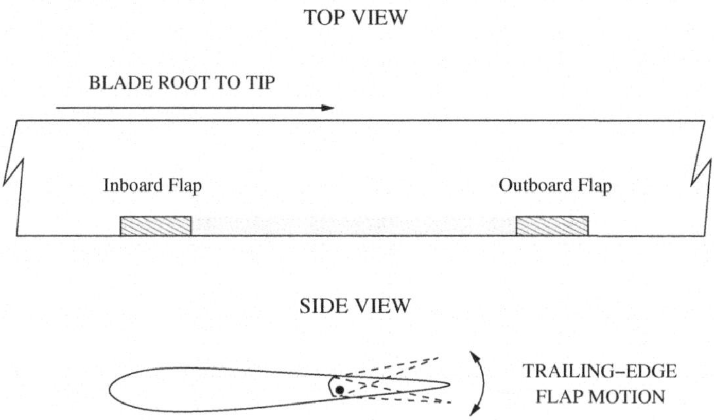

Fig. 6.3 Rotor with two trailing-edge flaps

Table 6.1 Initial configuration and move limits

	Physical variable (R)	Coded value
$x_{1-lower}$	0.59	−1.414
$x_{1-initial}$	0.65	0
$x_{1-upper}$	0.71	1.414
$x_{2-lower}$	0.77	−1.414
$x_{2-initial}$	0.83	0
$x_{2-upper}$	0.89	1.414

6.4.1 Response Surface Generation

The two objective functions F_v and F_p are evaluated at the data points corresponding to the CCD using the helicopter aeroelastic analysis. The response surfaces obtained by minimizing the error square are

$$\begin{aligned} F_v &= 1.000 + 0.0039x_1 - 0.0657x_2 + 0.0199x_1^2 + 0.0099x_1x_2 - 0.1039x_2^2, \\ F_p &= 1.000 - 0.0239x_1 + 0.1161x_2 - 0.0048x_1^2 + 0.0211x_1x_2 - 0.0351x_2^2. \end{aligned} \tag{6.10}$$

Tables 6.2 and 6.3 show the comparison of response surface prediction and aeroelastic analysis prediction at the nine data points for F_v and F_p, respectively. It can be seen that the response surfaces approximate the actual system output (objective functions F_v and F_p) fairly well (0–4 % error). The mean error of the response surfaces predicting F_v and F_p are 0.001 and 0.04 %, respectively. The standard deviation in both cases are 1.15 and 2.33 % respectively. Note that for each CCD point, the optimal controller is designed to minimize the vibration levels as discussed in

Table 6.2 Comparison of response surface prediction and aeroelastic analysis prediction of F_v at data points

S. No.	x_1 coded value	x_2 coded value	x_1 physical variable (R)	x_2 physical variable (R)	F_v RSM prediction	F_v analysis prediction	% error
1	0	0	0.65	0.83	1.0000	1.0000	0.00
2	0	1.414	0.65	0.89	0.6992	0.7097	−1.48
3	0	−1.414	0.65	0.77	0.8851	0.8947	−1.07
4	1.414	0	0.71	0.83	1.0454	1.0584	−1.22
5	−1.414	0	0.59	0.83	1.0343	1.0415	−0.69
6	−1	−1	0.61	0.79	0.9878	0.9801	0.79
7	−1	1	0.61	0.87	0.8364	0.8281	1.01
8	1	−1	0.69	0.79	0.9758	0.9640	1.22
9	1	1	0.69	0.87	0.8641	0.8517	1.46

Table 6.3 Comparison of response surface prediction and aeroelastic analysis prediction of F_p at data points

S. No.	x_1 coded value	x_2 coded value	x_1 physical variable (R)	x_2 physical variable (R)	F_p RSM prediction	F_p analysis prediction	% error
1	0	0	0.65	0.83	1.0000	1.0000	0.00
2	0	1.414	0.65	0.89	1.0940	1.0799	1.31
3	0	−1.414	0.65	0.77	0.7656	0.7907	−3.17
4	1.414	0	0.71	0.83	0.9567	0.9404	1.73
5	−1.414	0	0.59	0.83	1.0243	1.0515	−2.59
6	−1	−1	0.61	0.79	0.8890	0.8542	4.07
7	−1	1	0.61	0.87	1.0791	1.0721	0.65
8	1	−1	0.69	0.79	0.7990	0.7950	0.50
9	1	1	0.69	0.87	1.0735	1.0973	−2.17

Sect. 4.3 in Chap. 4. As mentioned earlier, each such evaluation takes about 48 min of CPU time and obtaining the response surfaces takes about 7 h of CPU time. However, this is far less than if conventional optimization algorithms were used.

6.4.2 *Analysis of Response Surfaces*

The next step is to determine the stationary point of the response surfaces shown in Eq. 6.10. The stationary point of a response surface is obtained by setting the gradient to zero. A second-order response surface has only one stationary point which could fall inside or outside the move limits. The stationary point can either be a global maximum, global minimum, or a saddle point. The Hessian matrix containing the second-order derivatives of the objective function with respect to the design variables gives the information needed to determine the nature of the stationary point. For a second-order response surface, the Hessian is a constant square matrix and its eigenvalues are related to the nature of the stationary points. If all eigenvalues are positive, then the Hessian is a positive definite and the stationary point is a global minimum. If all eigenvalues are negative, the Hessian is negative definite and the stationary point is a global maximum. If the eigenvalues are of different signs, the Hessian is indefinite and the stationary point is a saddle point. If the stationary point is a global minimum and if it falls inside the move limits, then it is the best design point for that objective. Otherwise, a constrained optimization has to be performed to determine the best design point which satisfies the move limit constraints. Typically, this is done using an exhaustive search of the feasible domain since the objective function can now be evaluated very quickly using the response surface.

Best Design Point for Minimum Vibration

The stationary point for the F_v response surface is found to be $x_1 = 0.65R$ and $x_2 = 0.82R$. The eigenvalues of the Hessian are $\lambda_1 = -0.208$ and $\lambda_2 = 0.040$. Since the eigenvalues are of different signs, the Hessian is indefinite and the stationary point is a saddle point. Figure 6.4 shows the variation of F_v with respect to variables x_1 and x_2 in coded form. It is clear that the stationary point is a saddle point. The best design point for obtaining minimum vibration using exhaustive search is found to be $x_1 = 0.63R$ and $x_2 = 0.89R$. At this point, the move limit for x_2 becomes active. At this design point, the response surfaces predict $F_v = 0.695$ and $F_p = 1.090$. Numerical results were also obtained by running the aeroelastic analysis at this optimal point predicted by the response surfaces. The aeroelastic analysis predicted $F_v = 0.712$ and $F_p = 1.071$. This is about 29 % reduction in F_v value from the initial design point. The response surface is therefore used as a guide to a better design.

From Fig. 6.4 it is apparent that the objective function J_v is more sensitive to changes in x_2 than x_1. This is also evident from the coefficients in Eq. 6.10. The coefficients relating F_v to x_2 are at least one order of magnitude higher than those

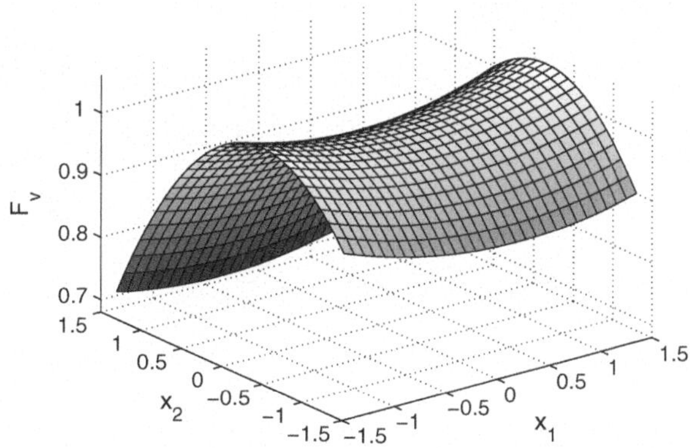

Fig. 6.4 Variation in F_v with respect to trailing-edge flap locations

relating F_v to x_1. It is therefore possible to further reduce F_v by moving the outboard flap further toward the blade tip. However, the inadequacies of the aerodynamic model used to predict the aerodynamic loads acting on the flap in vortex-dominated 3D flow prevents us from placing the flap too close to the tip. Furthermore, the blade tip can have sweepback or planform taper to alleviate compressibility effects and so it is best not to consider flaps close to the tip for a generic design study.

Best Design Point for Minimum Flap Power

Now the second-order response surface representing F_p is considered. The stationary point for this surface is found to be $x_1 = 0.80R$ and $x_2 = 0.94R$ which is outside the move limits. Also, the eigenvalues of the Hessian are $\lambda_1 = -0.077$ and $\lambda_2 = -0.003$. Since both eigenvalues are negative, the stationary point is a global maximum. Figure 6.5 shows the variation of F_p with respect to x_1 and x_2. The variables shown in the figure are in coded form. The best design point for minimum trailing-edge flap power inside the move limits is found using an exhaustive search. This point is found to be $x_1 = 0.71R$ and $x_2 = 0.77R$. The value of F_p as predicted by the response surface is 0.680. The aeroelastic analysis predicts a value of $F_p = 0.734$ at this design point. Also, from Fig. 6.5 it is clear that the trailing-edge flap power is dominated by the position of the outboard flap.

6.4.3 *Identification of Best Design Point*

Figures 6.4 and 6.5 indicate that the optimum flap locations to achieve minimum vibration levels and minimum trailing-edge flap power are different. For minimum vibration, $(x_1, x_2) = (0.63R, 0.89R)$ and for minimum flap power, $(x_1, x_2) = (0.71R, 0.77R)$. To minimize vibration, the inboard flap is moved further inboard from the initial design (0.65R–0.63R). However, for minimizing flap power, the inboard flap is moved towards the blade tip (0.65R–0.71R). The outboard flap has

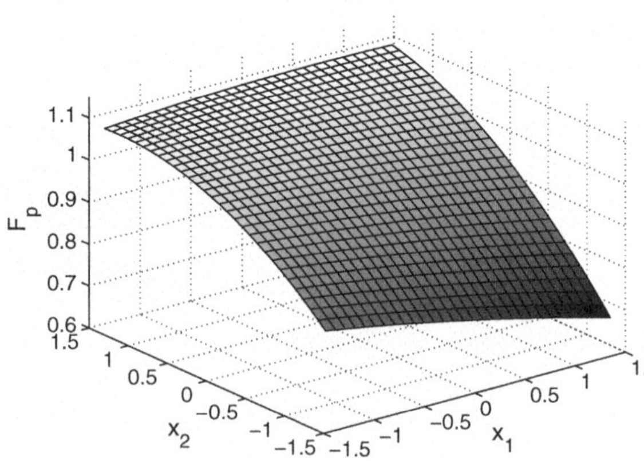

Fig. 6.5 Variation in F_p with respect to trailing-edge flap locations

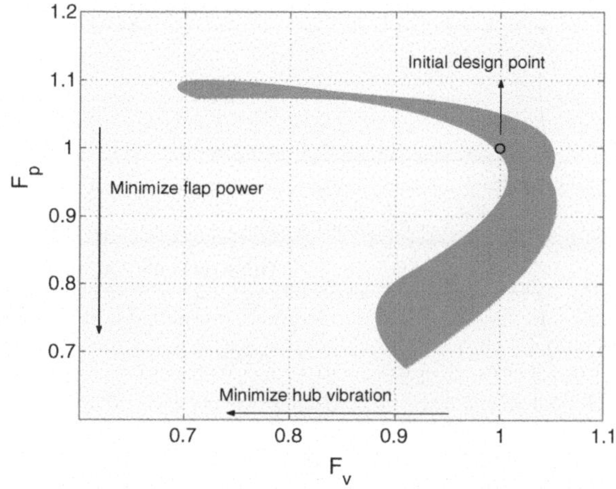

Fig. 6.6 F_p versus F_v for various design points within the move limits

to be placed at its upper move limit near the blade tip (0.89R) to achieve minimum vibration while to minimize flap power, it has to be placed at its lower move limit (0.77R). These two requirements are of conflicting nature. Therefore, it is of interest to determine a "compromise design" which results in lower vibration levels and trailing-edge flap power compared to the initial design. Figure 6.6 is obtained by combining Figs. 6.4 and 6.5 and by moving from the design space (x_1, x_2) to the criterion space (F_v, F_p). Figure 6.6 shows the vibration level and flap power resulting from various design choices within the move limits. As mentioned earlier, the design point $F_v = F_p = 1.0$ corresponds to the initial design. The initial design point and the points on the Pareto surface are shown in Fig. 6.7. It is interesting to note that the Pareto surface is divided into two distinct parts with no feasible design point lying in the "compromise region". Thus, the Pareto surface is disjoint. The design point

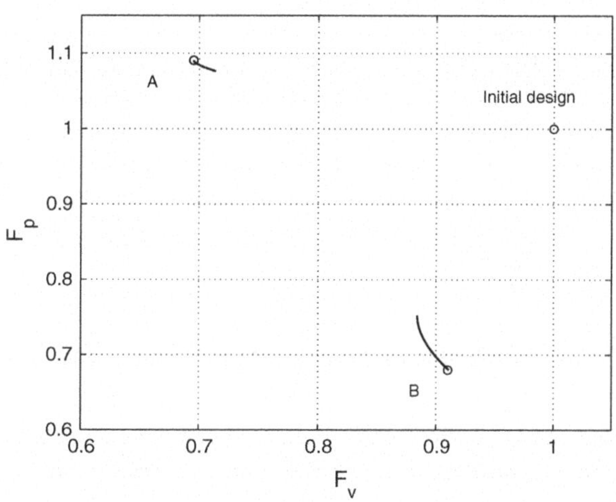

Fig. 6.7 Initial design point and Pareto surface

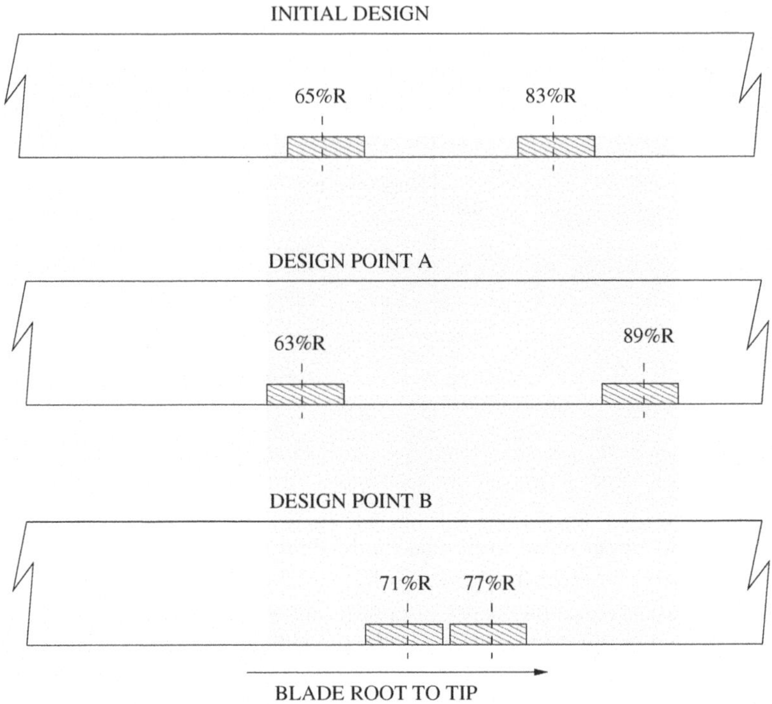

Fig. 6.8 Initial design and design points A and B

A corresponds to $(x_1, x_2) = (0.63R, 0.89R)$. This is the design point obtained in Sect. 6.4.2 by minimization of vibration levels at the hub without consideration to flap power consumption. Design point A and other Pareto points in its vicinity (Fig. 6.7) yield about 30 % reduction in hub vibration compared to the initial design but at the cost of increasing flap power by about 7–10 % compared to the initial design. Design point A is optimal if there exists no constraint on the available trailing-edge flap power. Design point B shown in Fig. 6.7 corresponds to $(x_1, x_2) = (0.71R, 0.77R)$ as obtained in Sect. 6.4.2. According to response surface predictions, design point B and other Pareto points in its vicinity require about 68–75 % of the flap power consumed by the initial design and yet yield about 9–13 % further reduction in hub vibration compared to the initial design. The flap positions for the design points A and B and the initial design are shown schematically in Fig. 6.8.

6.4.4 Performance at Various Forward Speeds

The above studies were conducted at cruise speed corresponding to an advance ratio (μ) of 0.30. Figure 6.9 shows the performance of the initial and Pareto optimal

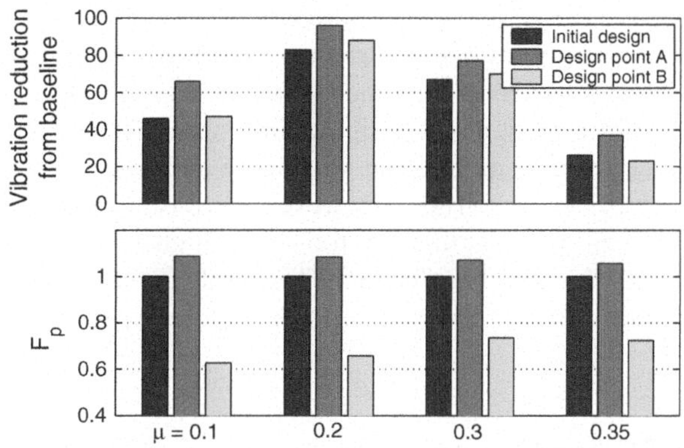

Fig. 6.9 Performance of initial design and Pareto optimal designs at different forward speeds

design configurations (A and B) at different forward speeds. The vibration reductions shown are from the baseline conditions (zero flap motion). It is seen that the design configuration A yields better vibration reduction at all forward speeds but at the cost of higher flap power. However, if sufficient power for the actuators is available, then design point A is the best design for vibration reduction. In comparison, design point B gives modest reductions in vibration but requires much lesser flap power. It is seen that configuration B requires around 27–37 % lower flap power compared to the initial design. Design configuration B is an attractive solution when the available flap power is constrained due to other practical reasons. The application of optimization to the trailing-edge flap placement problem has resulted in the discovery of this novel design where substantial vibration reduction is possible with less flap power

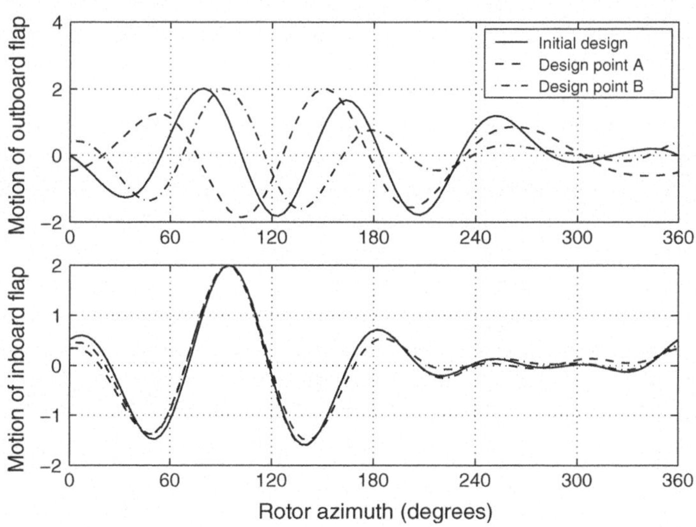

Fig. 6.10 Time history of flap motion for various design points, $\mu = 0.30$

requirements. Since flap power is provided through piezoelectric actuators which require cumbersome electronics, it is advantageous to have a low power requirement. Figure 6.10 shows the time history of flap motion (both outboard and inboard) for the initial design and design points A and B. The motion of inboard flap in all the three cases is quite close with large deflection in the advancing side of the rotor disk ($0° < \psi < 180°$). However, there is considerable difference in the motion of outboard flap and this flap too shows higher deflection in the advancing side of the rotor. This could be due to the higher dynamic pressure on the advancing side of the rotor disk.

6.5 Summary

An optimal dual trailing-edge flap configuration that consumes low flap power to achieve considerable reduction in hub vibration is obtained. The development of optimal controller for flap motion is computationally expensive and involves many function calls to the aeroelastic analysis, making the use of conventional optimization methods difficult. The second-order polynomial response surfaces are created by solving the aeroelastic problem at limited number of data points obtained using the CCD. The design variables are the spanwise location of the two flaps and the objectives are hub vibration level and required trailing-edge flap power. The response surface prediction and aeroelastic prediction differ by at most 2 % for vibration level and 4 % for flap control power. The use of response surface method significantly reduces the computational effort required to determine the optimal flap locations compared to if conventional optimization methods were used. The analytical form of the polynomial response surfaces are ideally suited for very quick function evaluations and allow the use of global exhaustive search methods. Minimum hub vibration level is obtained when the inboard and outboard flaps are located at 63 and 89 % spanwise locations, respectively. This configuration leads to 77 % reduction in hub vibration from baseline condition (zero flap motion) at cruise speed. However, this configuration requires about 7 % more trailing-edge flap power compared to the initial design. This configuration yields better reduction in vibration than the initial design at other forward speeds too. By adjusting the flap spanwise locations, the power consumed by the dual-flap system can be reduced by about 27 % from the initial design at cruise speed. This configuration corresponds to placing the inboard and outboard flaps at 71 and 77 % spanwise locations, respectively and yields about 70 % reduction in hub vibration from baseline condition. This is marginally better than the initial design configuration which gave 67 % reduction in hub vibration. The Pareto surface between the two objectives, hub vibration level and flap control power, is disjoint. One region is dominated by configurations yielding substantial reduction in vibration but at the cost of higher flap actuation power. The other set contains configurations that require less flap control power yet leading to marginally better reductions in vibration compared to the initial configuration.

References

1. Simpson, T.W., Peplinski, J.D., Koch, P.N., Allen, J.K.: Metamodels for computer-based engineering design: survey and recommendations. Eng. Comput. **17**(2), 129–150 (2001)
2. Rajagopal, S., Ganguli, R.: Conceptual design of long endurance UAV using Kriging approximation and multi-objective genetic algorithm. Aeronaut. J **112**(1137), 653–662 (2008)
3. Engineering Optimization: A Modern Approach. Universities Press, Madison (2012)
4. Deb, K.: Multi-Objective Optimization: Search Methodologies, pp. 403–449. Springer, Heidelberg (2014)
5. Viswamurthy, S.R., Ganguli, R.: Optimal placement of trailing-edge flaps for helicopter vibration reduction using response surface methods. Eng. Optim. **39**(2), 185–202 (2007)

Chapter 7
Piezoceramic Actuator Hysteresis

In this chapter, we aim to investigate the effect of piezoelectric actuator hysteresis on helicopter vibration control using trailing-edge flaps. The material and mechanical hysteresis in the piezoelectric actuator is modeled using the classical Preisach model. Experimental data from literature is used to estimate the Preisach distribution function. Section 7.1 provides a brief introduction to the classical Preisach model and its numerical implementation. A modified flap control algorithm is explained in Sect. 7.2. Section 7.3 deals with a comparative study on the performance of an ideal-linear actuator and a real-hysteretic actuator. Hysteresis compensation algorithms and their performance are discussed in Sect. 7.4. Section 7.5 gives the summary of this chapter.

7.1 Classical Preisach Model

The Preisach model was originally proposed by the German physicist Preisach as a model for magnetic hysteresis [1]. In a pioneering work, Krasnoselskii separated Preisach's model from its physical meaning and represented it in a pure mathematical form [2]. This model can therefore be used for the mathematical description of hysteresis of any physical nature. The Preisach model has several appealing features including its ability to model complex hysteresis types, a well defined identification algorithm, and a convenient numerical simulation form. This hysteresis model has received considerable attention in the past and has been widely used to describe hysteresis in piezoceramic actuators, shape memory alloys and magnetostrictive actuators.

The piezo actuator along with the amplification mechanism can be considered as a hysteresis transducer (Fig. 7.1), which can be characterized by an input $u(\psi)$ (excitation voltage) and an output $\delta(\psi)$ (flap deflection) [3, 4]. Hysteresis is a type of nonlinearity in which the input–output relationship is a multibranch nonlinearity for which a branch-to-branch transition occurs after each input extremum (Fig. 7.2).

R. Ganguli et al., *Smart Helicopter Rotors*, Advances in Industrial Control,
DOI 10.1007/978-3-319-24768-7_7

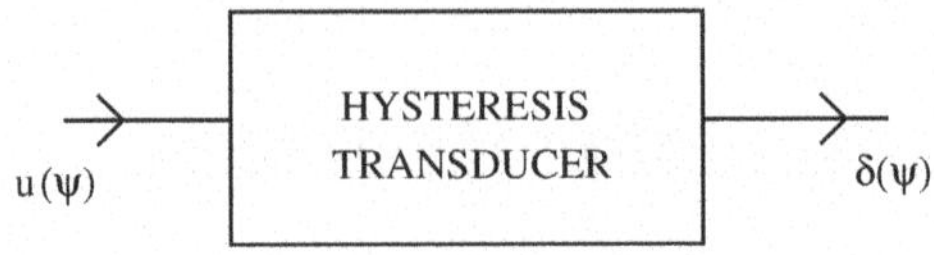

Fig. 7.1 Schematic representation of a hysteresis transducer system

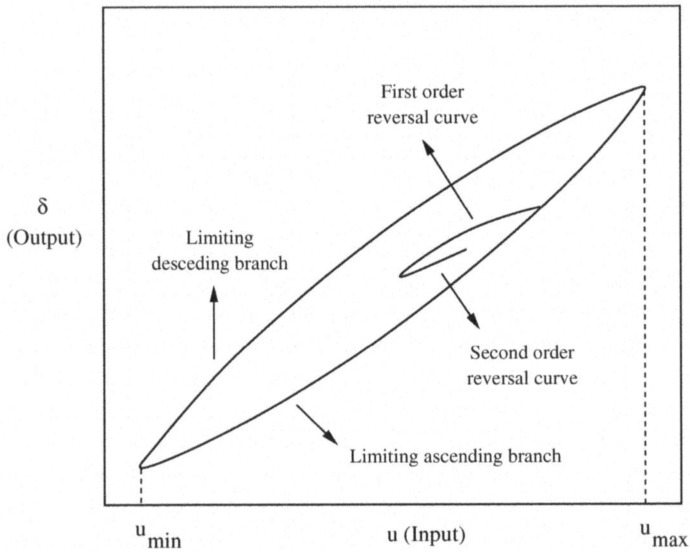

Fig. 7.2 Hysteresis multi-branch nonlinearity

As mentioned before, the classical Preisach model is a 'static' hysteresis model. The term 'static' means that the branches of hysteresis nonlinearities are determined by the past extremum values of the input, while the speed of input variation between the extremum points has no influence on branching.

The Preisach model can be described by considering an infinite set of elemental hysteresis operators $\hat{\gamma}_{\alpha\beta}$ and a weighting function $\mu(\alpha, \beta)$. Figure 7.3 shows the elemental hysteresis operator (sometimes called *hysteron*) in the input–output plane. The output of each elemental hysteresis operator traces a rectangular loop on the input–output diagram switching from -1 to $+1$ when the input increases above the threshold α. The output switches from $+1$ to -1 when the input decreases below the value of β. This operator can be considered as a two-position relay that has different up and down switch points. The Preisach model can be represented in the following mathematical form:

$$\delta(\psi) = \int\int_{\alpha \geq \beta} \mu(\alpha, \beta)\, \hat{\gamma}_{\alpha\beta}[u(\psi)]\; d\alpha\, d\beta. \tag{7.1}$$

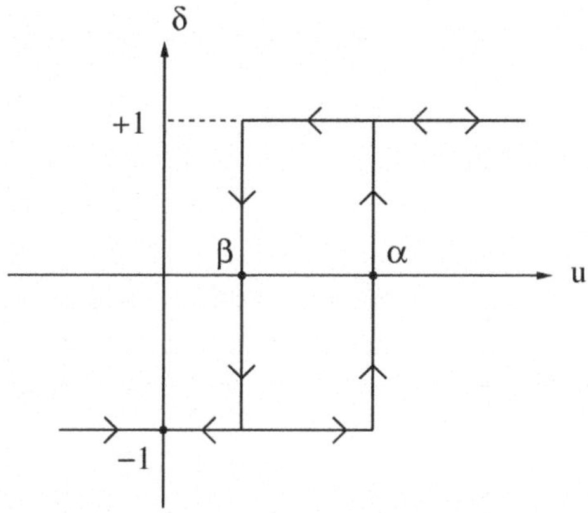

Fig. 7.3 Elemental hysteresis operator

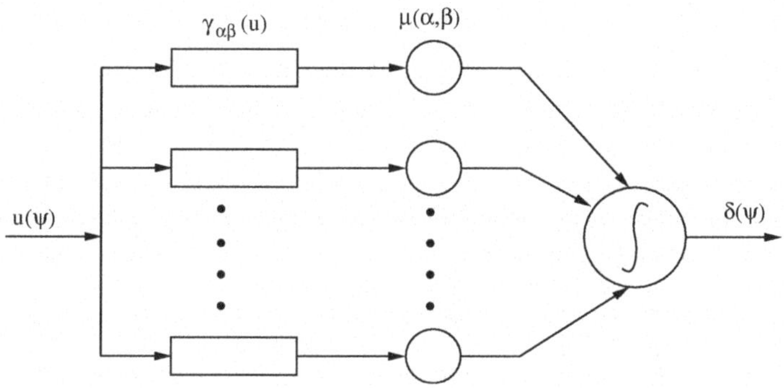

Fig. 7.4 Block diagram interpretation of the classical Preisach model

The Preisach model can be visualized as a continuous analog of a system of two-position relays connected in parallel (Fig. 7.4). The input $u(\psi)$ is applied to each of the relays. Their individual outputs are multiplied by $\mu(\alpha, \beta)$ and then integrated over all appropriate values of α and β to obtain the output $\delta(\psi)$. Thus the Preisach model is described without any reference to the actual physical nature of hysteresis. The weighting function, $\mu(\alpha, \beta)$ is characteristic of the hysteresis transducer shown in Fig. 7.1 and is sometimes referred to as the Preisach distribution function. If the weighting function is known, then Eq. 7.1 can be solved directly by integration. If the function $\mu(\alpha, \beta)$ is not known explicitly, then experimental data is needed to estimate the weighting function through the geometric interpretation and numerical implementation [5].

The Preisach model can be geometrically interpreted as follows. This interpretation is based on the fact that there is a one-to-one correspondence between operators $\hat{\gamma}_{\alpha\beta}$ and points (α, β) on the half plane $\alpha \geq \beta$. At any instant of time ψ, the limiting triangle T_0 can be divided into two sets (Fig. 7.5): $S^+(\psi)$ consisting of points (α, β) for which $\hat{\gamma}_{\alpha\beta}[u(\psi)] = 1$ and $S^-(\psi)$ consisting of points (α, β) for which

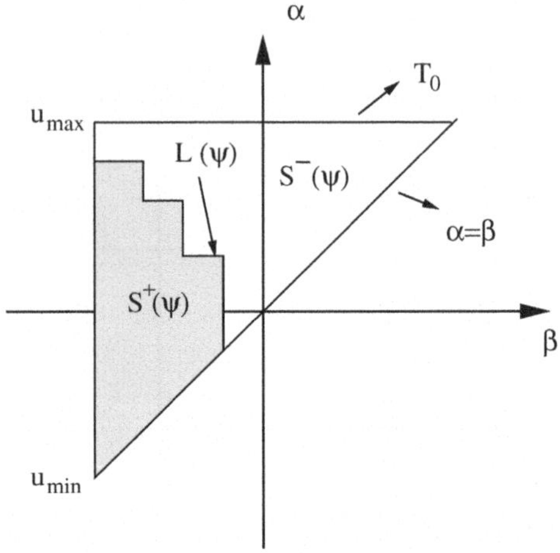

Fig. 7.5 Geometric interpretation of Preisach model

$\hat{\gamma}_{\alpha\beta}[u(\psi)] = -1$. It can be shown that the interface $L(\psi)$ between $S^+(\psi)$ and $S^-(\psi)$ is a staircase line whose vertices have α and β coordinates coinciding with local maxima and minima of input at previous instants of time (Fig. 7.6). The final link of $L(\psi)$ is attached to the line $\alpha = \beta$ and moves when the input changes. This link is horizontal and moves up when the input increases, and it is vertical and moves from right to left when the input decreases. Using this geometric interpretation, the Preisach model can be written in the following form

$$\delta(\psi) = \int\int_{S^+(\psi)} \mu(\alpha, \beta)\, d\alpha\, d\beta - \int\int_{S^-(\psi)} \mu(\alpha, \beta)\, d\alpha\, d\beta. \tag{7.2}$$

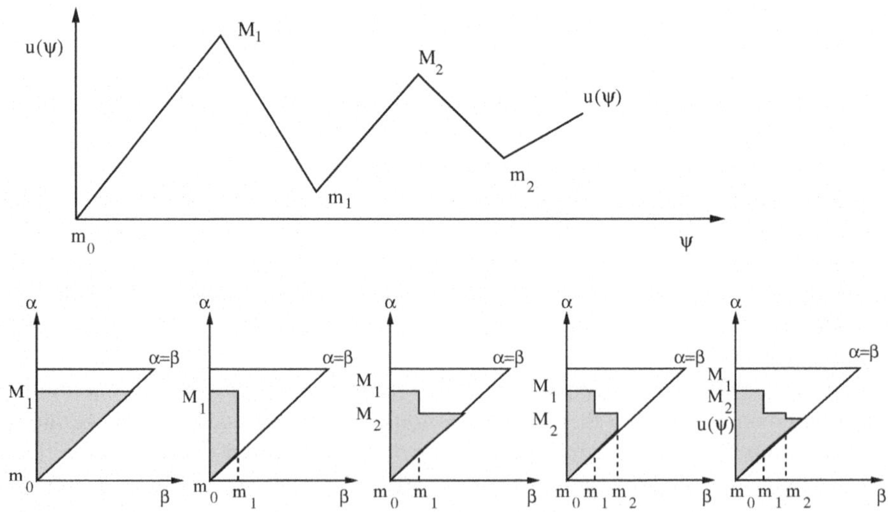

Fig. 7.6 α–β plane diagrams corresponding to different input extrema

From Eq. 7.2 it follows that an instantaneous value of output depends on the shape of the interface $L(\psi)$, which in turn, is determined by the extremum values of input at previous instants of time. Consequently, the past extremum input values shape the interface $L(\psi)$, and in this way they leave their mark upon the future. However, not all extremum input values are accumulated by the model. Some of them can be wiped out. It can be shown using the rules of motion of the final link of $L(\psi)$ that the following property is valid.

Property A: Wiping-out property Each local maximum wipes out the vertices whose α coordinates are below this maximum, and each local minimum wipes out the vertices whose β coordinates are above this minimum.

It is clear that the wiping out of vertices is equivalent to the erasing of the history associated with these vertices. Thus, subsequent variations of input might erase some previous history. This property has been found to be in agreement with some experimental facts. Another important property of the Preisach model is stated as follows.

Property B: Congruency property All hysteresis loops corresponding to the same extremum values of input are congruent. To determine the weighting function $\mu(\alpha, \beta)$, the set of first-order reversal curves should be experimentally found. This can be done by bringing first the input to such a value that outputs of all operators $\hat{\gamma}_{\alpha\beta}$ are equal to -1. If we now gradually increase the input value, then we will follow along a limiting ascending branch (Fig. 7.7). This branch is called 'limiting' because there is no branch below it. The notation f_α is normally used to denote the output value on this branch corresponding to the input $u = \alpha$. The first order reversal curves are attached to the limiting ascending branch (Fig. 7.7). Each of these curves is formed when the above monotone increase of input is followed by the subsequent monotone decrease. The notation $f_{\alpha\beta}$ is normally used for the output value on the reversal curve which is attached to the limiting ascending branch at the point f_α. This output value corresponds to the input $u = \beta$. The function $F(\alpha, \beta)$ (called the *Everett surface*) is then defined as

$$F(\alpha, \beta) = \frac{1}{2} \left(f_\alpha - f_{\alpha\beta} \right) . \tag{7.3}$$

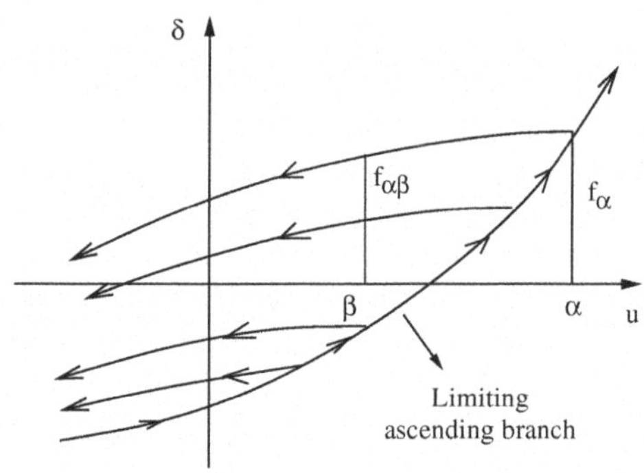

Fig. 7.7 First order reversal curves

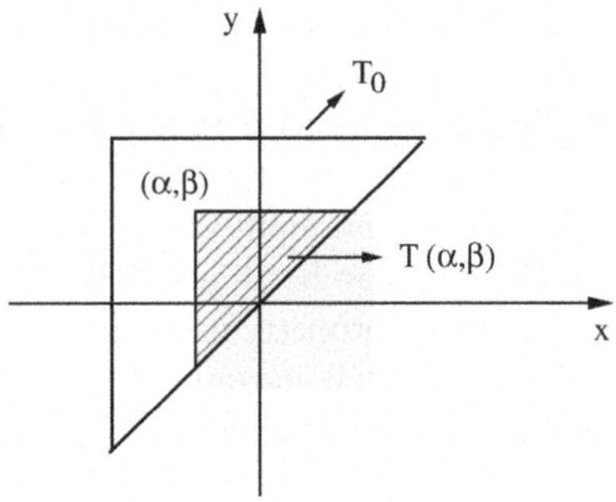

Fig. 7.8 Graphical representation of Eq. 7.4

Using the geometric interpretation of the Preisach model it can be shown that

$$F(\alpha,\beta)=\int\int_{T(\alpha,\beta)}\mu(x,y)dxdy=\int_{\beta}^{\alpha}\left(\int_{\beta}^{y}\mu(x,y)dx\right)dy, \tag{7.4}$$

where $T(\alpha,\beta)$ is the triangle formed by the intersection of the lines $x=\beta$, $y=\alpha$ and $y=x$ (Fig. 7.8). From Eq. 7.4 we get

$$\mu(\alpha,\beta)=-\frac{\partial^2F(\alpha,\beta)}{\partial\alpha\partial\beta}. \tag{7.5}$$

The experimental data provided by the set of first-order reversal curves allows the determination of the weighting function $\mu(\alpha,\beta)$ through Eq. 7.5. Numerical implementation of the Preisach model using Eqs. 7.1 and 7.5 has two main disadvantages. Firstly, differentiation of experimentally obtained function $F(\alpha,\beta)$ is required in order find $\mu(\alpha,\beta)$. This might amplify any error which is inherently present in $F(\alpha,\beta)$. Secondly, numerical evaluation of double integrals is needed to compute the output $\delta(\psi)$. The above disadvantages are overcome by using the following explicit formula

$$\delta(\psi)=-F(\alpha_0,\beta_1)+2\sum_{k=1}^{n}\left(F(\alpha_k,\beta_k)-F(\alpha_k,\beta_{k+1})\right), \tag{7.6}$$

where α_k and β_k are decreasing and increasing sequences of α and β coordinates of interface vertices, respectively, and n is the number of horizontal links of interface $L(\psi)$. In this chapter, experimental data obtained by Hall and Prechtl is used to estimate the weighting function. Another important requirement for the numerical implementation of the classical Preisach model is the knowledge of the initial state of the hysteresis transducer. The initial state of the transducer along with the initial input to the transducer determines the initial output. The following output values obviously depend on the initial output value. In this chapter, it is assumed that the initial output of the system either corresponds to the value on the limiting ascending branch (when initial input rate is positive) or the limiting descending branch (when initial input rate is negative). Such an assumption allows us to easily fix the initial state of the hysteresis transducer.

7.2 Modified Control Algorithm

The flap control algorithm used in this chapter is similar to the algorithm discussed in Chap. 4 (Sect. 4.3). However, the control input in this chapter is the voltage applied to the piezostack and not the flap deflection angle itself. As mentioned before, for a 4-bladed rotor, the control input to the actuator is typically 2/rev, 3/rev, 4/rev or 5/rev or a combination of all the above harmonics (multicyclic input) to reduce the vibratory hub loads. *In this chapter, a DC bias is also given to the stack actuator. This DC bias is adjusted so as to obtain a flap motion with zero steady component.* The total voltage applied to the piezostack driving the jth flap is then given by

$$u^j(\psi) = u^j_{st} + \sum_{N=2}^{5} \left[u^j_{Nc} \cos(N\psi) + u^j_{Ns} \sin(N\psi) \right], \quad j = 1 \text{ to } N_f. \tag{7.7}$$

At the ith control step, the objective function for optimal control is given by

$$J\left(\mathbf{Z}_i, \mathbf{u}_i^1, \mathbf{u}_i^2, \ldots, \mathbf{u}_i^{N_f}\right) = \mathbf{Z}_i^T \mathbf{W}_Z \mathbf{Z}_i + \sum_{j=1}^{N_f} \beta^j \mathbf{u}_i^{jT} \mathbf{u}_i^j. \tag{7.8}$$

Here $\mathbf{Z_i}$ is the vibratory hub load vector at the ith control step containing the 4/rev sine and cosine harmonics (three hub shears and three hub moments), $\mathbf{u_i^j}$ is the control input vector containing cosine and sine higher harmonics of the jth flap at the ith control step and N_f is the number of flaps on each rotor blade. Since u^j_{st} is not an independent parameter, it is not included in the vector of control input harmonics $\mathbf{u}_i^j$. The diagonal matrix $\mathbf{W}_Z$ is a weighting matrix for hub vibratory loads. The first term in the above equation is a scalar quantity that is purely related to the magnitude of the hub vibratory loads and is a measure of hub vibration. The non-dimensional weighting parameters β^j can be changed to establish the relative importance of hub loads versus flap motions in the objective function. When $\beta^j = 0$, the controller tries to minimize hub loads without regard to the motion of the jth trailing-edge flap. As β^j increases towards unity, the controller will try to reduce the motion of the jth trailing-edge flap to zero.

$$\mathbf{Z}_i = \left[F_{xH}^{4p} \ F_{yH}^{4p} \ F_{zH}^{4p} \ M_{xH}^{4p} \ M_{yH}^{4p} \ M_{zH}^{4p} \right]^T,$$

$$\mathbf{u}_i^j = \left[u_{2c}^j \ u_{2s}^j \ u_{3c}^j \ u_{3s}^j \ u_{4c}^j \ u_{4s}^j \ u_{5c}^j \ u_{5s}^j \right]^T, \quad j = 1 \text{ to } N_f,$$

$$\mathbf{W}_Z = \left(1 - \sum_{j=1}^{N_f} \beta^j \right) \begin{bmatrix} \alpha \, \mathbf{I}_{3\times 3} & \mathbf{0}_{3\times 3} \\ \mathbf{0}_{3\times 3} & (1-\alpha)\, \mathbf{I}_{3\times 3} \end{bmatrix},$$

$$\sum_{j=1}^{N_f} \beta^j \leq 1. \tag{7.9}$$

The subscript i in Eqs. 7.8 and 7.9 refers to the ith control step, reflecting the discrete-time nature of the control. The time interval between each control step must be sufficient to allow the system to return to the steady state so that the vibration level can be measured accurately. Here, the "feedback form of the global controller" is implemented since the system is only moderately nonlinear. This controller is a compromise between the global and the local controller. The feedback form of the global controller can save computation time without losing accuracy for a moderately nonlinear system. The transfer matrix is assumed to be constant over the entire range of control inputs just like in the global controller. However, this controller is a closed-loop form where the control input during each control step is determined by feedback of the measured vibration levels of the previous control step. Linearizing the system about the current control inputs using Taylor series expansion gives

$$\mathbf{Z}_i = \mathbf{Z}_{i-1} + \sum_{j=1}^{N_f} \mathbf{T}_0^j \left(\mathbf{u}_i^j - \mathbf{u}_{i-1}^j \right), \tag{7.10}$$

where $\mathbf{T}_0^j$ is the transfer matrix that relates the system response to the motion of the jth trailing-edge flap. This transfer matrix is assumed to be constant over the entire range of the control input and hence is calculated only once by perturbing the control harmonics individually around zero control inputs. Equation 7.10 is substituted in Eq. 7.8 and the optimal controller is obtained by applying the following optimality criteria

$$\frac{\partial J}{\partial \mathbf{u}_i^j} = 0, \quad j = 1 \text{ to } N_f. \tag{7.11}$$

In this chapter, a single trailing-edge flap configuration is considered as a first step ($N_f = 1$). The flap spans 12 % of the blade and is located at 70 % blade radial location. Numerical studies are conducted at both high ($\mu = 0.30$) and low speed ($\mu = 0.15$) flight conditions.

7.3 Performance of Linear and Hysteretic Actuators

To obtain a realistic model of hysteresis, experimental data from previous studies is used. Figure 7.9 shows the quasistatic response of a trailing-edge flap actuator on a bench top test obtained by Hall and Prechtl [6]. The hysteresis nonlinearity is due to both material effects and friction in the mechanical device. The stacks used in this research were made of lead magnesium niobate-lead titanate (PMN-PT) composition. The experimental data shown in Fig. 7.9 is used to determine the unknown Everett surface $F(\alpha, \beta)$ required by the Preisach model in Eq. 7.6. If the entire flap actuation mechanism is assumed to have no hysteresis, then it is reasonable to presume that the relationship between piezostack voltage input and flap deflection angle is linear. This presumed linear relationship is also shown in Fig. 7.9. When a pure sinusoidal voltage input is given to such a linear actuator with no hysteresis, the output (flap deflection) is also a pure sinusoid with zero phase shift and at the same frequency. However, a sinusoidal input given to an actuator with hysteretic behavior, yields an output (flap deflection) which has a phase shift at the input frequency and a non-zero response at other frequencies.

Two cases are considered:

Case (1): Ideal-linear actuator

Case (2): Real actuator, hysteresis addressed using Preisach model

The linear actuator is termed ideal because hysteresis which occurs in the real actuator leads to an unwanted nonlinearity which complicates the controller design

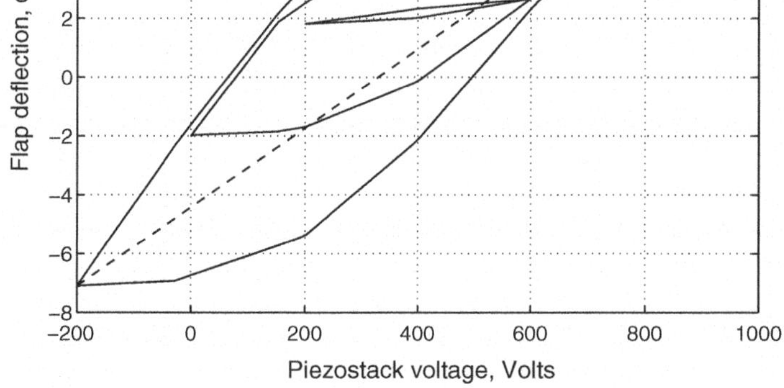

Fig. 7.9 Trailing-edge flap actuator hysteresis from Hall and Prechtl [6] and the linear approximation

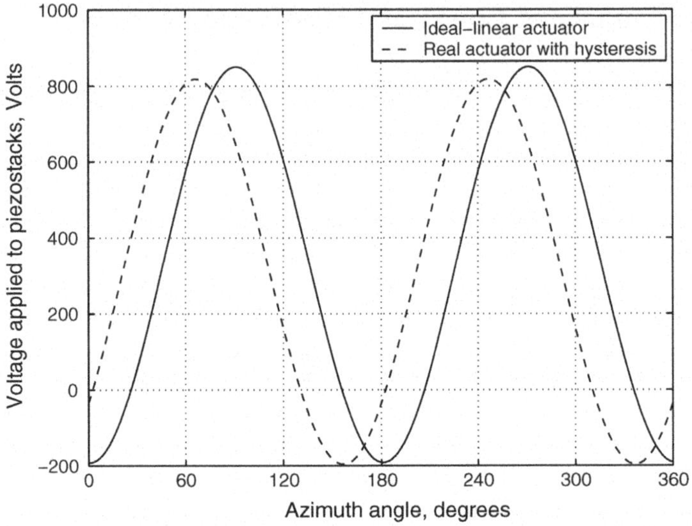

Fig. 7.10 Optimal 2/rev control input, $\mu = 0.30$

and adversely affects performance. The helicopter aeroelastic model is found to be almost linear with respect to the actuator control input. Hence, for the ideal-linear actuator, the global controller is used to determine the optimal control input. For the real actuator with hysteresis, the "feedback form of the global controller" is used to determine the optimal control input, since the inclusion of the actuator hysteresis model made the overall system nonlinear. Initial studies are conducted by applying the higher harmonics of the control input individually. This provides a general understanding of the influence of the control input on the various components of the fixed frame hub vibratory loads.

Vibration Reduction at High-Speed Flight ($\mu = 0.30$)

2/rev Control Input

Figure 7.10 shows the time history of the optimal 2/rev control input (voltage input to the piezostack) for the two cases mentioned above. The optimal control input for the two cases are different, particularly in terms of phasing with respect to the rotor azimuth. The resulting flap motion for the two cases are shown in Fig. 7.11. As expected, the flap motion in the ideal actuator case contains only the 2/rev component. In addition to the dominant 2/rev component, the flap motion in the real-hysteretic actuator contains non-negligible 4/rev and 6/rev harmonics (Fig. 7.11). This spillover can be attributed to the hysteresis in the actuator. Figure 7.12 shows the flap deflection angle as a function of piezostack voltage for the two cases. Hub vibration is reduced by 63 and 51 % from the baseline value for the ideal and real actuator, respectively. This difference in performance of the two actuators is due to the disparity in the flap motion between the two cases. In the case of ideal actuator, the longitudinal,

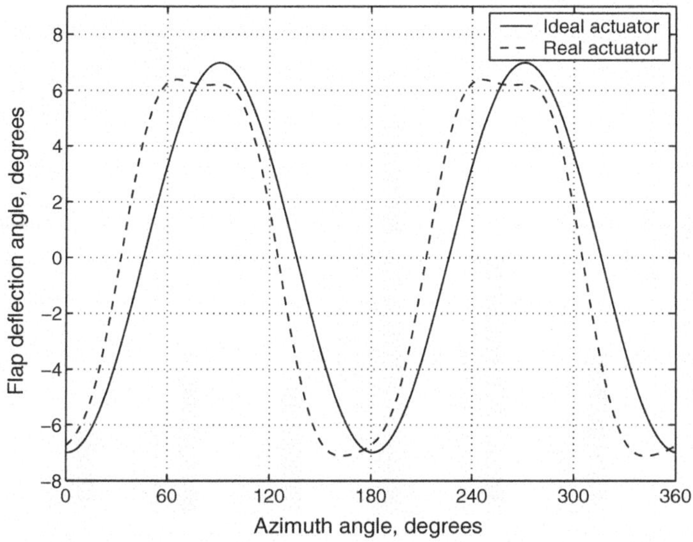

Fig. 7.11 Flap motion history due to 2/rev control input, $\mu = 0.30$

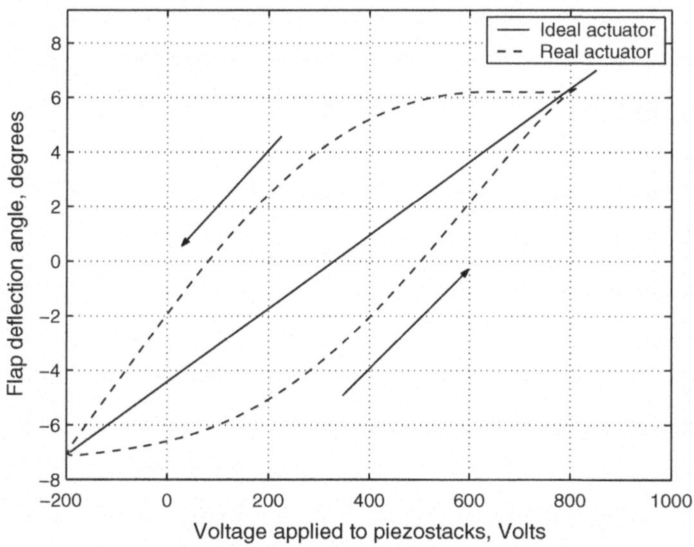

Fig. 7.12 Flap motion history versus piezostack voltage for 2/rev control input, $\mu = 0.30$

lateral and vertical hub shear forces are reduced by 53, 35 and 39 %, respectively. The rolling, pitching and yawing moments are reduced by about 45, 43 and 49 %, respectively (Fig. 7.13). For the real actuator, the longitudinal, lateral and vertical hub shear forces are reduced by 58, 27 and 28 %, respectively. The rolling, pitching and yawing moments are reduced by about 23, 28 and 51 %, respectively.

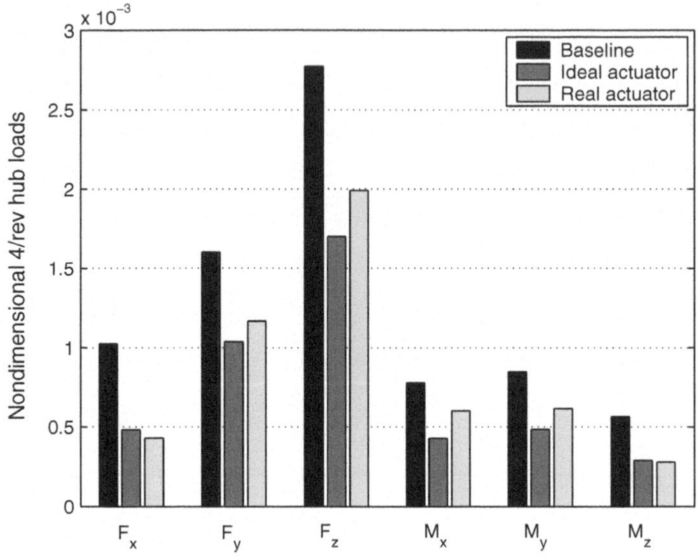

Fig. 7.13 Comparison of hub vibratory loads for 2/rev control input, $\mu = 0.30$

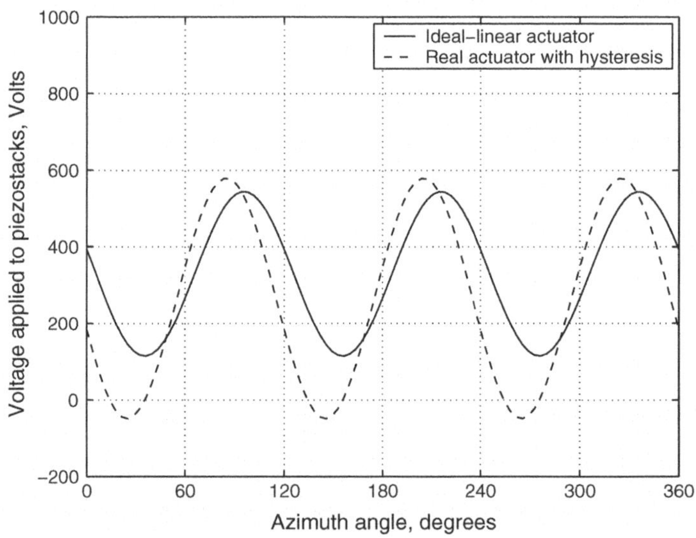

Fig. 7.14 Optimal 3/rev control input, $\mu = 0.30$

3/rev Control Input

3/rev control input is known to be very effective in vibration control of 4-bladed rotors at high forward speeds. Figure 7.14 shows the time history of the optimal 3/rev control input (voltage input to the piezostack) for the two cases. There is a notable difference in magnitude and phase of the optimal control input obtained for

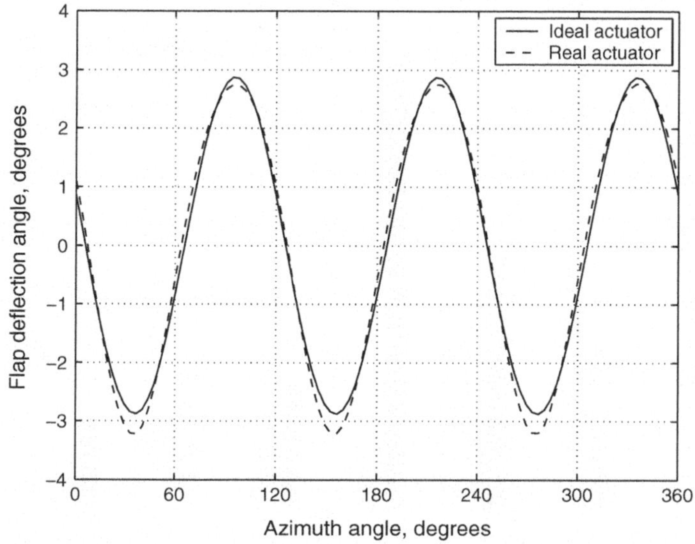

Fig. 7.15 Flap motion history due to 3/rev control input, $\mu = 0.30$

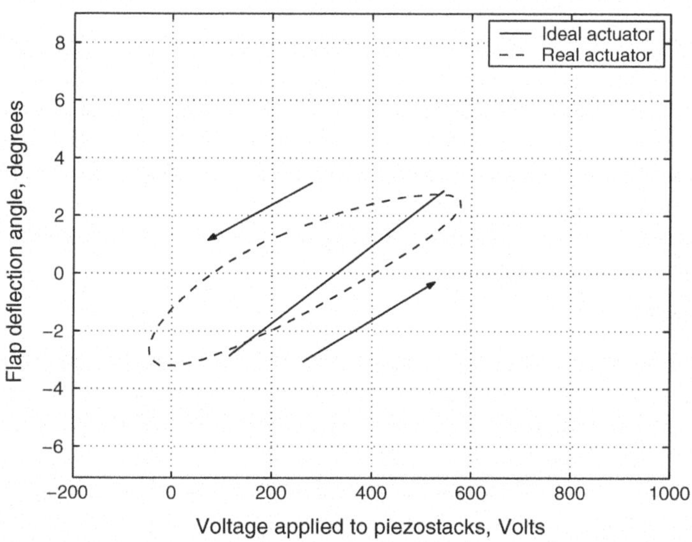

Fig. 7.16 Flap motion history versus piezostack voltage for 3/rev control input, $\mu = 0.30$

the ideal and real actuator. It is seen that the flap motion history is almost identical for the ideal and real actuator, though the 3/rev control input required to achieve this flap motion is different in the two cases (Fig. 7.15). In other words, the controller is able to adjust the control input for the real actuator to obtain a flap motion similar to the ideal actuator. Figure 7.16 shows the flap deflection angle as a function of piezostack voltage for the two cases.

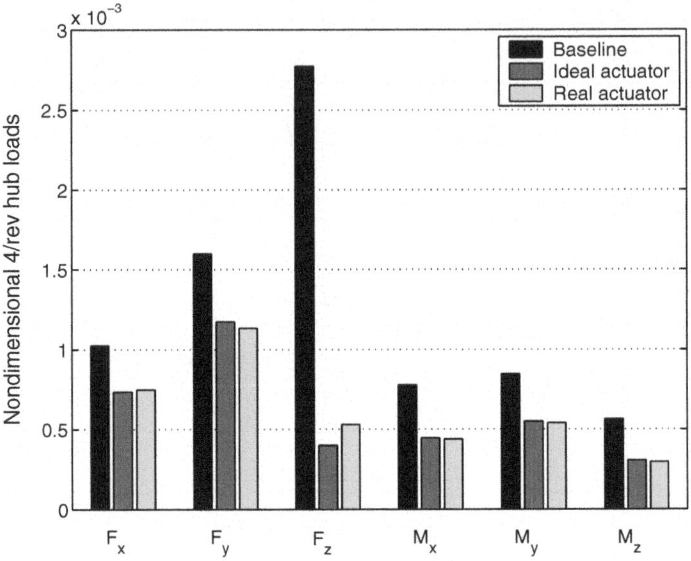

Fig. 7.17 Comparison of hub vibratory loads for 3/rev control input, $\mu = 0.30$

An overall hub vibration reduction of 79 % is obtained in both the ideal and real actuators. For the ideal actuator, longitudinal and lateral hub shear are reduced by about 28 % each. However, the vertical shear is reduced by about 86 %. The rolling, pitching and yawing moments are reduced by 43, 35 and 46 %, respectively. Similar reductions in hub loads are achieved for the real actuator (Fig. 7.17). The identical performance of the two cases is not surprising since the optimal flap motion achieved in both cases are nearly same.

4/rev Control Input

The optimal 4/rev piezostack voltage input for the ideal and real actuators are shown in Fig. 7.18. Again, there is a notable difference in magnitude and phase of the optimal 4/rev control input obtained for the ideal and real actuator. The resulting flap motion in both cases are again found to be nearly identical (Fig. 7.19). Overall vibration reduction of 70 % is obtained in both cases. For the ideal actuator, longitudinal and lateral hub shear are reduced by about 15 % each (Fig. 7.20). However, the vertical shear is reduced by about 92 %. The rolling, pitching and yawing moments are reduced by 15, 9 and 25 %, respectively. Similar reductions in hub loads are achieved for the real actuator. Figure 7.21 shows the variation in flap angle versus piezostack voltage for the two cases.

5/rev Control Input

Figure 7.22 shows the optimal 5/rev control input for the ideal and real actuators. The resulting flap motion in both cases are identical, just as seen with 3/rev and 4/rev control input. The 5/rev control input required to achieve this identical flap motion

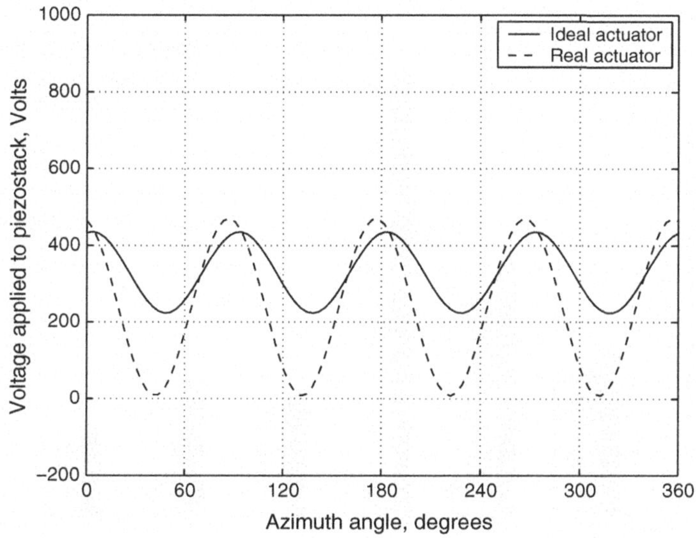

Fig. 7.18 Optimal 4/rev control input, $\mu = 0.30$

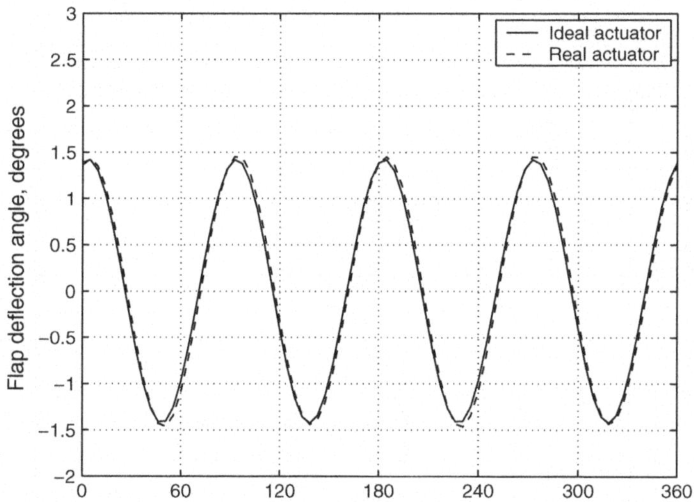

Fig. 7.19 Flap motion history due to 4/rev control input, $\mu = 0.30$

is different in the two cases (Fig. 7.23). We notice that if the actuator response is modeled fairly well, the controller is able to adjust the magnitude and phase of the control input so as to get the precise optimal motion of the trailing edge flap to obtain maximum vibration reduction. Overall vibration reduction of 66 % is obtained for both the ideal and real actuators. As expected, the performance of 5/rev control input is marginally lower than 3/rev and 4/rev control inputs. For the ideal actuator, the

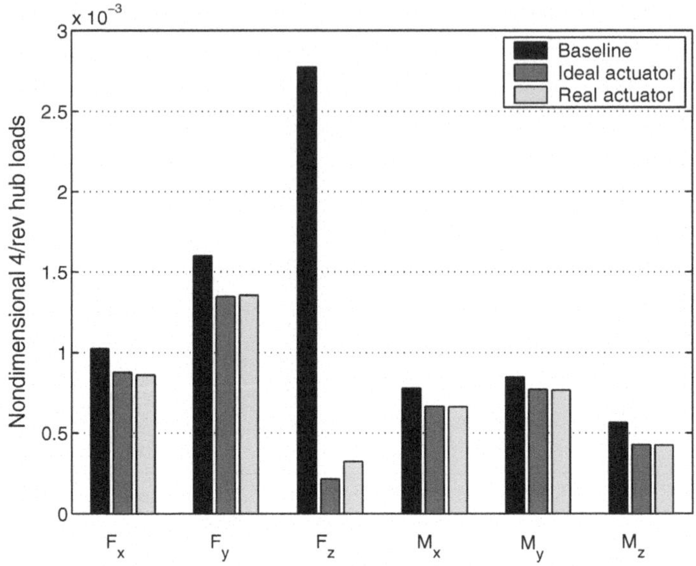

Fig. 7.20 Comparison of hub vibratory loads for 4/rev control input, $\mu = 0.30$

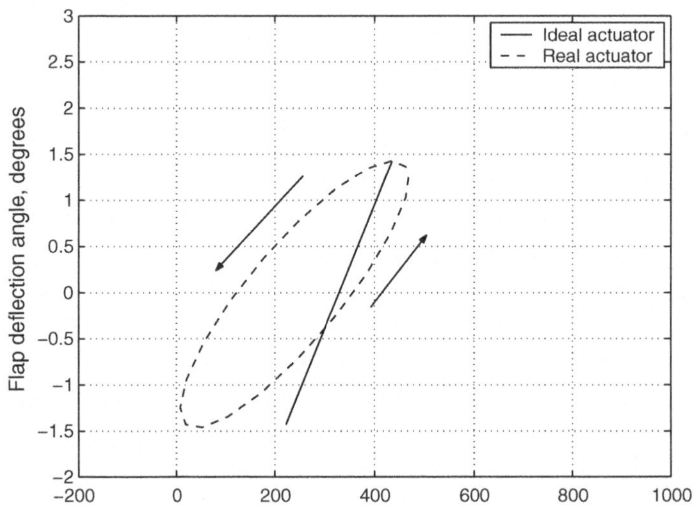

Fig. 7.21 Flap motion history versus piezostack voltage for 4/rev control input, $\mu = 0.30$

longitudinal and vertical hub shear forces are reduced by 41 and 96 % respectively. However, the lateral shear force increased marginally by about 2 %. Both pitching and yawing moments are reduced by about 24 %. The rolling moment increased by 5 % from the baseline value. The reduction in 4/rev hub loads is nearly identical for the case of real actuator (Fig. 7.24).

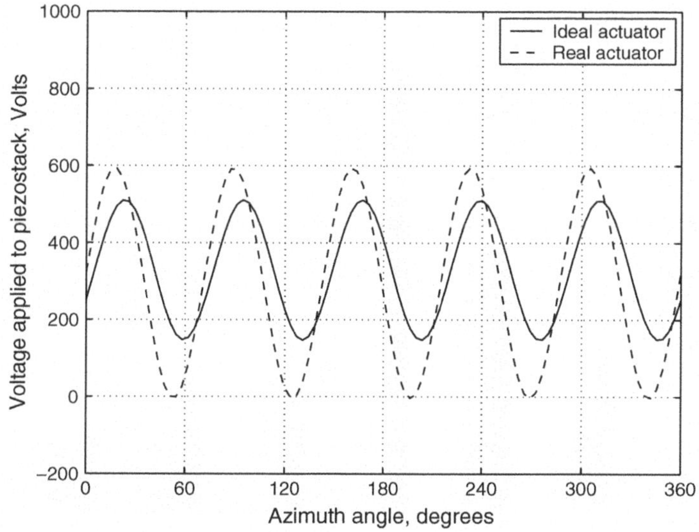

Fig. 7.22 Optimal 5/rev control input, $\mu = 0.30$

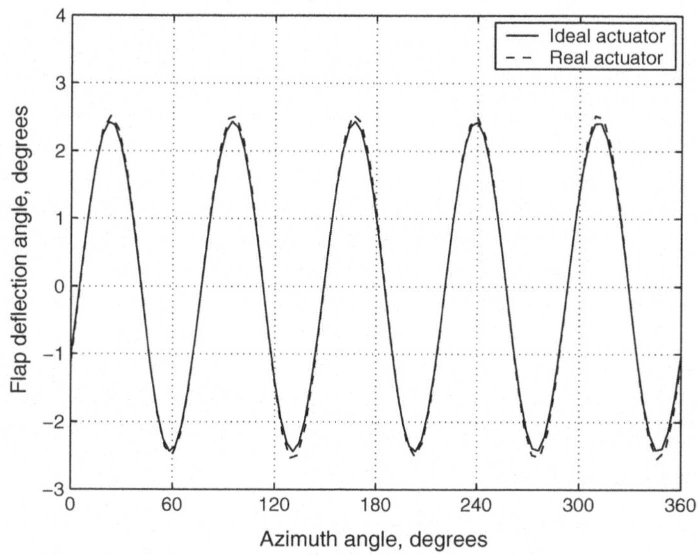

Fig. 7.23 Flap motion history due to 5/rev control input, $\mu = 0.30$

In the three cases with individual 3/rev, 4/rev and 5/rev piezostack control input, the flap motion obtained using the "feedback form of global controller" for a real actuator has nearly the same magnitude and phasing as the flap motion obtained using the global controller for an ideal-linear actuator. This is the reason why similar reduction in hub vibratory loads were obtained for the ideal and real actuator using

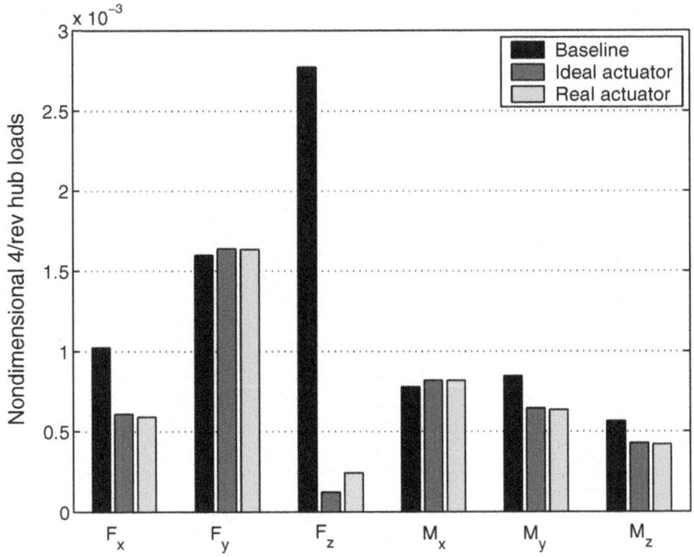

Fig. 7.24 Comparison of hub vibratory loads for 5/rev control input, $\mu = 0.30$

a single frequency control input. This is because, for the real actuator, the system response is only mildly nonlinear with respect to individual voltage control inputs. Hence, the "feedback form of global controller" used works quite well. However, in the case of individual 2/rev control input, the real actuator performs marginally below the ideal actuator. This is due to the non-zero 4/rev and 6/rev harmonics of the flap motion resulting from hysteresis in the actuator.

Figures 7.14, 7.18 and 7.22 indicate that both linear and hysteretic actuators are not used to their complete potential for 3/rev, 4/rev, and 5/rev control input at a single frequency. That is, the constraints on the minimum and maximum voltage that could be applied to the piezostack actuator (from Fig. 7.9) are not active at the optima. However, in the case of 2/rev control input, both actuators are used to their full potential (Fig. 7.10).

Multicyclic Control Input

A multicyclic control algorithm is used where 2/rev, 3/rev, 4/rev and 5/rev harmonics of the control input are given simultaneously (Eq. 7.7). Figure 7.25 shows the optimal control input for the two cases. Note that the optimal control input which needs to be given to the piezostack is different for the real actuator compared to the ideal actuator. However, in both cases, the actuators are used to complete authority in contrast to 3/rev, 4/rev and 5/rev single frequency control input cases. The optimal flap motion history for the multicyclic control input is different for the two cases (Fig. 7.26). Hub vibration is reduced by 90 and 81 % from the baseline value for the ideal and real actuator, respectively. This is because, for the real actuator, the system response is fairly nonlinear with respect to multicyclic control input. This shows that neglecting

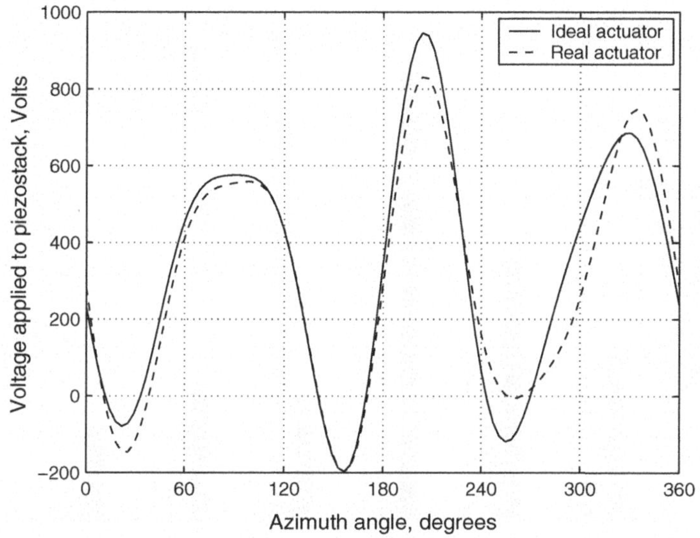

Fig. 7.25 Optimal multicyclic control input, $\mu = 0.30$

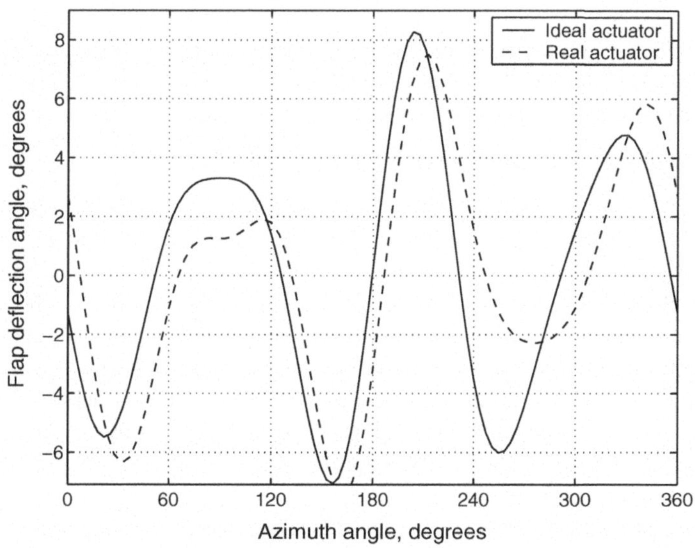

Fig. 7.26 Flap motion history due to multicyclic control input, $\mu = 0.30$

hysteresis in the actuator leads to over-prediction in vibration reduction compared to an actuator with hysteresis modeled.

Figure 7.27 shows the reduction in the individual 4/rev hub loads. For the ideal actuator case, the longitudinal, lateral and vertical hub shear forces are reduced by about 49, 50 and 95 %, respectively. The rolling, pitching and yawing moments are

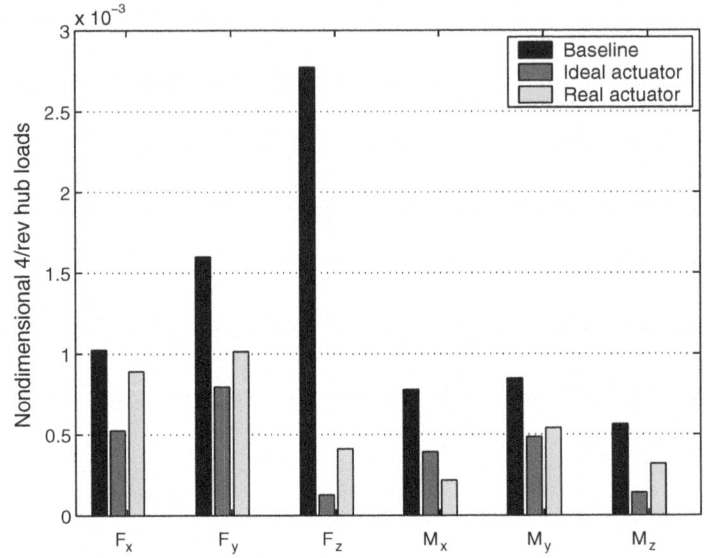

Fig. 7.27 Comparison of hub vibratory loads for multicyclic control input, $\mu = 0.30$

reduced by 50, 43 and 75 %, respectively. In the case of real hysteretic actuator, a reduction of 13, 37 and 85 % in 4/rev hub shear forces are achieved. The 4/rev hub moments are reduced by about 72, 36 and 44 % from baseline levels.

Vibration Reduction at Low-Speed Flight ($\mu = 0.15$)

The numerical results presented above were obtained at a high forward speed corresponding to an advance ratio of 0.30. Studies are also performed at a low forward speed of advance ratio 0.15. The primary source of vibration in these two flight conditions are different. In high speed flight, the main source of vibration is the increased magnitude and periodicity of the aerodynamic loads. In low speed flight, the rotor wake remains close to the blades and results in higher wake induced loading. For the ideal actuator, overall vibration reduction of 90, 98 and 91 % is achieved by using individual 3/rev, 4/rev and 5/rev control input, respectively. For the real actuator, hub vibration is reduced by 88, 84 and 91 % from baseline level through 3/rev, 4/rev and 5/rev control input, respectively.

In low speed flight regime, the vertical hub shear force dominates the vibratory loading. From Eq. 2.55 in Chap. 2 we infer that 4/rev actuator control input is likely to be most effective for vibration suppression in low-speed flight conditions. This is confirmed by the numerical results presented above. The optimal 4/rev control input for the ideal and hysteretic actuators are shown in Fig. 7.28. Once again, both actuators are required to operate well below complete authority. The resulting flap motion is shown in Fig. 7.29. The dominant 4/rev vertical hub shear force is reduced by about 99 and 63 % from baseline level for the ideal and hysteretic actuators, respectively (Fig. 7.30). Other vibratory hub loads are almost unchanged from their corresponding baseline values.

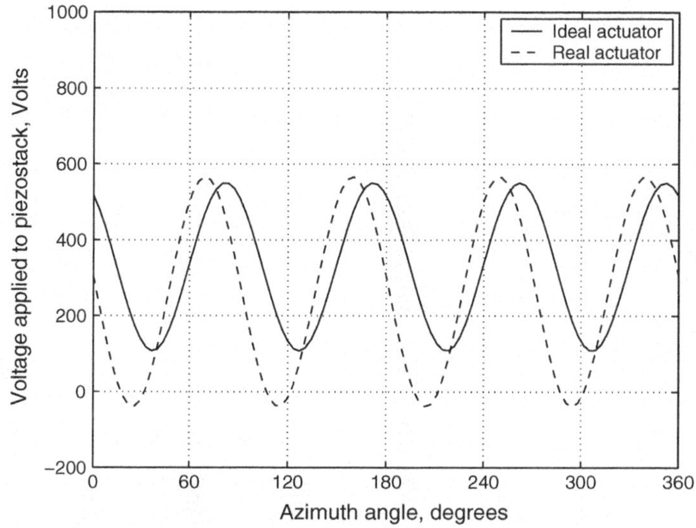

Fig. 7.28 Optimal 4/rev control input, $\mu = 0.15$

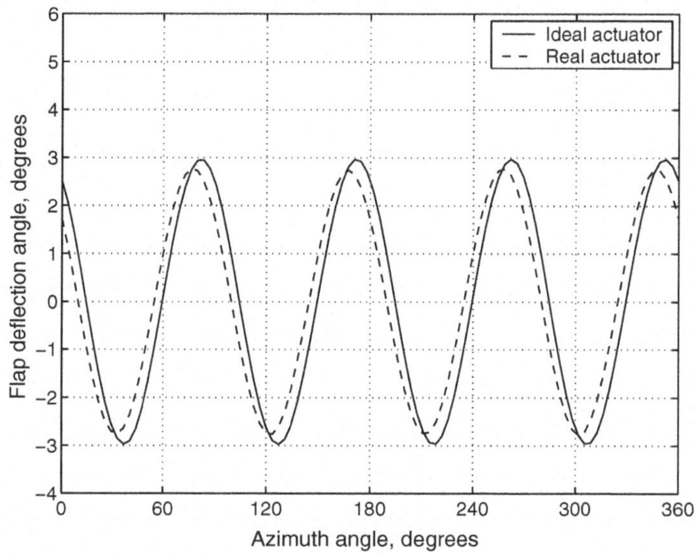

Fig. 7.29 Flap motion history due to 4/rev control input, $\mu = 0.15$

In low speed flight, multicyclic control input gave 99 and 86 % reduction in hub vibration for the ideal and real actuator, respectively. This indicates that not modeling hysteresis leads to results showing much larger reduction in vibration compared to when hysteresis is modeled. Figure 7.31 shows the optimal multicyclic control input obtained for the two cases. The flap motion for the two cases are quite different

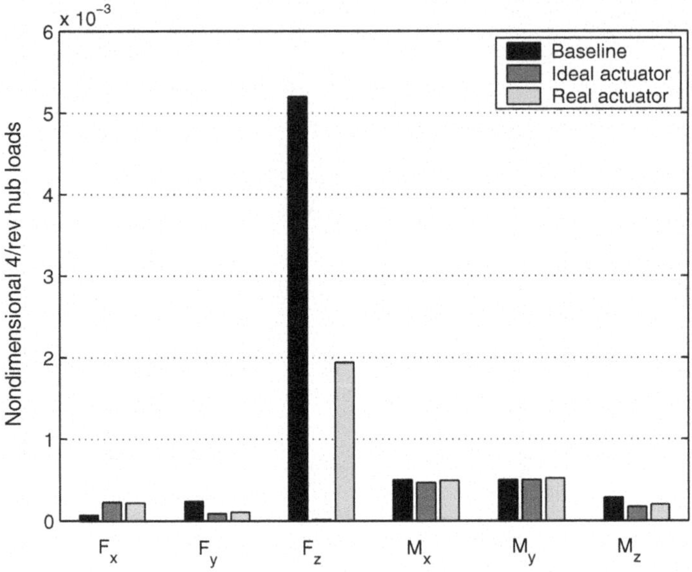

Fig. 7.30 Comparison of hub vibratory loads for 4/rev control input, $\mu = 0.15$

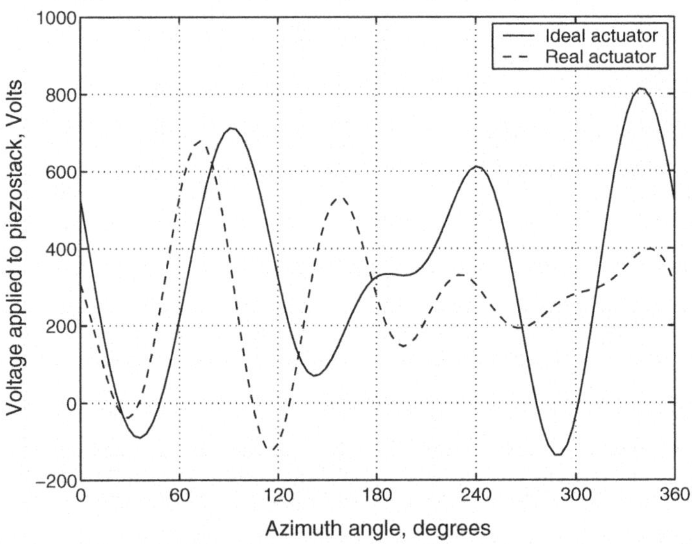

Fig. 7.31 Optimal multicyclic control input, $\mu = 0.15$

(Fig. 7.32). The reductions in individual hub vibratory loads are presented in Fig. 7.33. It is clear that the ideal-linear actuator performs better than the real actuator with hysteresis modeled.

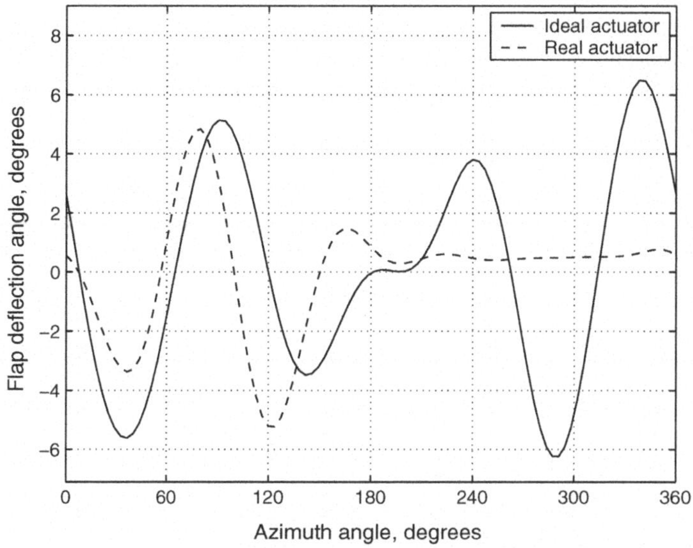

Fig. 7.32 Flap motion history due to multicyclic control input, $\mu = 0.15$

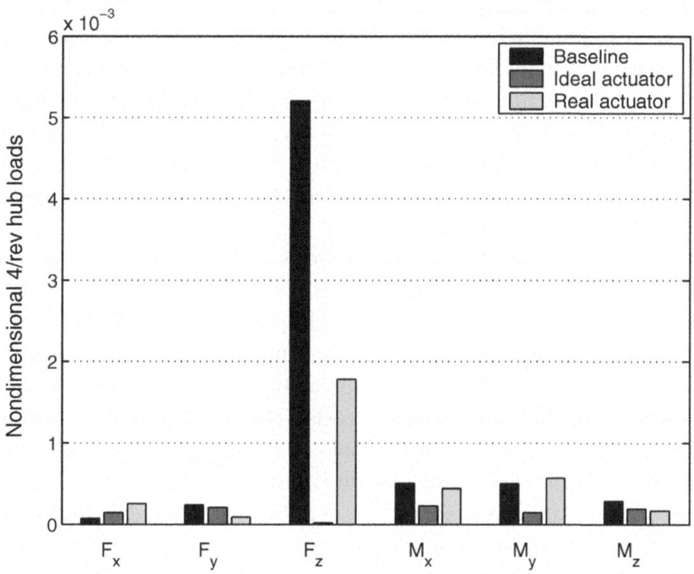

Fig. 7.33 Comparison of hub vibratory loads for multicyclic control input, $\mu = 0.15$

Table 7.1 summarizes the numerical results obtained in terms of hub vibration reduction for the ideal and real actuator at high and low forward speeds.

Table 7.1 Percentage reduction in hub vibration for all cases

Type of control input	Ideal-linear actuator	Real actuator with hysteresis
	High speed flight ($\mu = 0.30$)	
2/rev	63	51
3/rev	79	79
4/rev	70	70
5/rev	66	66
Multicyclic	90	81
	Low speed flight ($\mu = 0.15$)	
3/rev	90	88
4/rev	98	84
5/rev	91	91
Multicyclic	99	86

7.4 Hysteresis Compensation Algorithm

The previous section was based on deriving the control law for a flap-actuator system in the presence of actuator hysteresis. Another approach is to develop a hysteresis compensator based on an inverse hysteresis operator. Such a compensator can be used as a part of an open-loop filter to produce a linear relationship between the input and output. The hysteresis in the actuator is primarily due to material nonlinearity and to a lesser extent due to losses in the amplification mechanism. In this section, we aim to study the effect of inaccurate compensation of piezostack actuator hysteresis on the performance of trailing-edge flap system.

An "ideal" compensator when placed in series with a hysteretic system will make the overall system "exactly" linear thereby allowing us to use linear control theory techniques to control the actuator. However, in practice it is difficult to develop an exact compensation scheme. So the overall system cannot be made exactly linear but approximately linear. In this chapter, an approximate inverse of the Preisach operator is used to compensate for actuator hysteresis. This nonlinear inversion algorithm called the *closest match algorithm* was proposed by Tan et al. [7]. In this method, the Preisach plane (α–β plane) is discretized into a finite number of grids. The algorithm aims to determine the input whose output matches the desired output most closely among all possible inputs. Since the Preisach plane is discretized, the input can only take values from a finite set $U = \{u_l, 1 \leq l \leq L\}$ with each $u_l, 1 \leq l \leq L$, representing an input level. We define,

$$\Delta u = \frac{u_{max} - u_{min}}{L - 1}. \tag{7.12}$$

Then, $u_l = u_{min} + (l-1)\Delta u$. Given the initial memory curve $\phi^{(0)}$ (from which the initial input $u^{(0)}$ and the output $\delta^{(0)}$ can be derived) and a desired output δ^*, the aim of the inverse operator is to determine the input $u^* \in U$, such that

$$|\Gamma(u^*; \phi^{(0)}) - \delta^*| = \min_{u \in U} |\Gamma(u; \phi^{(0)}) - \delta^*|. \tag{7.13}$$

The algorithm should also return the resulting memory curve ϕ^*. The algorithm is as follows

- Step 0. Set n $= 0$. Compare $\delta^{(0)}$ and δ^*: if $\delta^{(0)} = \delta^*$, let $u^* = u^{(0)}$, $\phi^* = \phi^{(0)}$, go to Step 3; if $\delta^{(0)} < \delta^*$, go to Step 1.1; otherwise go to Step 2.1;
- Step 1.
  - Step 1.1: If $u^n = u_L$, let $u^* = u^n$, $\phi^* = \phi^n$, go to Step 3; otherwise $u^{(n+1)} = u^{(n)} + \Delta u$, $\tilde{\phi} = \phi^{(n)}$ [backup the memory curve], $n = n + 1$, go to Step 1.2;
  - Step 1.2: Evaluate $\delta^{(n)} = \Gamma(u^n; \phi^{(n-1)})$ and (at the same time) update the memory curve to $\phi^{(n)}$. Compare δ^n with δ^*: if $\delta^{(n)} = \delta^*$, let $u^* = u^n$, $\phi^* = \phi^n$, go to Step 3; if $\delta^{(n)} < \delta^*$, go to Step 1.1; otherwise go to Step 1.3;
  - Step 1.3: If $|\delta^{(n)} - \delta^*| \le |\delta^{(n-1)} - \delta^*|$, let $u^* = u^n$, $\phi^* = \phi^n$, go to Step 3; otherwise $u^* = u^{(n-1)}$, $\phi^* = \tilde{\phi}$ [restore the memory curve], go to Step 3;
- Step 2.
  - Step 2.1: If $u^n = u_1$, let $u^* = u^n$, $\phi^* = \phi^n$, go to Step 3; otherwise $u^{(n+1)} = u^{(n)} - \Delta u$, $\tilde{\phi} = \phi^{(n)}$ [backup the memory curve], $n = n + 1$, go to Step 2.2;
  - Step 2.2: Evaluate $\delta^{(n)} = \Gamma(u^n; \phi^{(n-1)})$ and (at the same time) update the memory curve to $\phi^{(n)}$. Compare δ^n with δ^*: if $\delta^{(n)} = \delta^*$, let $u^* = u^n$, $\phi^* = \phi^n$, go to Step 3; if $\delta^{(n)} > \delta^*$, go to Step 2.1; otherwise go to Step 2.3;
  - Step 2.3: If $|\delta^{(n)} - \delta^*| \le |\delta^{(n-1)} - \delta^*|$, let $u^* = u^n$, $\phi^* = \phi^n$, go to Step 3; otherwise $u^* = u^{(n-1)}$, $\phi^* = \tilde{\phi}$ [restore the memory curve], go to Step 3;
- Step 3. Exit.

The above algorithm yields the best input u^* in at most L iterations.

7.4.1 Numerical Results

The hysteresis in the X-frame actuator developed by Hall and Prechtl is shown in Fig. 7.34. Figure 7.34a shows the hysteresis only due the piezostack material. Figure 7.34b shows the combined effect of both material and mechanical hysteresis of the actuator, which is significantly greater.

To study the effect of material and mechanical hysteresis on the performance of helicopter vibration control system, three different cases are considered (Fig. 7.35). They represent different levels of modeling and compensation of X-frame actuator hysteresis. In case (1), the compensator is derived based on the complete actuator nonlinearity (Fig. 7.34b). This model is the most complete and more accurate than

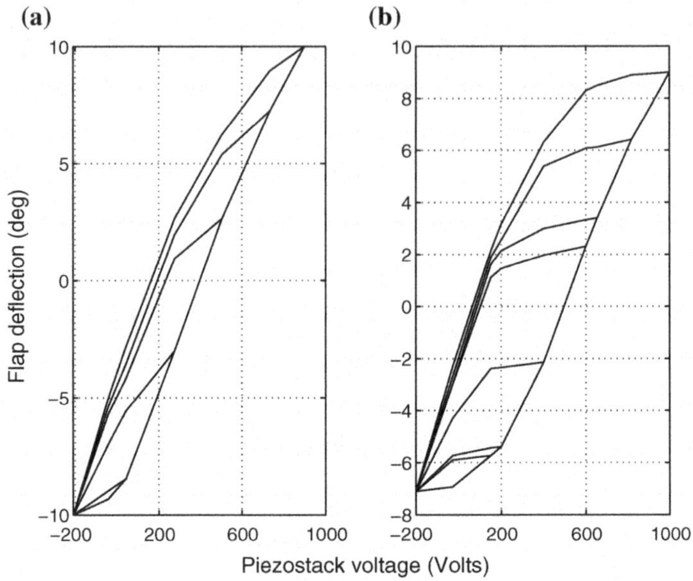

Fig. 7.34 Trailing edge flap actuator hysteresis from Hall and Prechtl [6]. **a** Only material hysteresis. **b** Both material and mechanical hysteresis

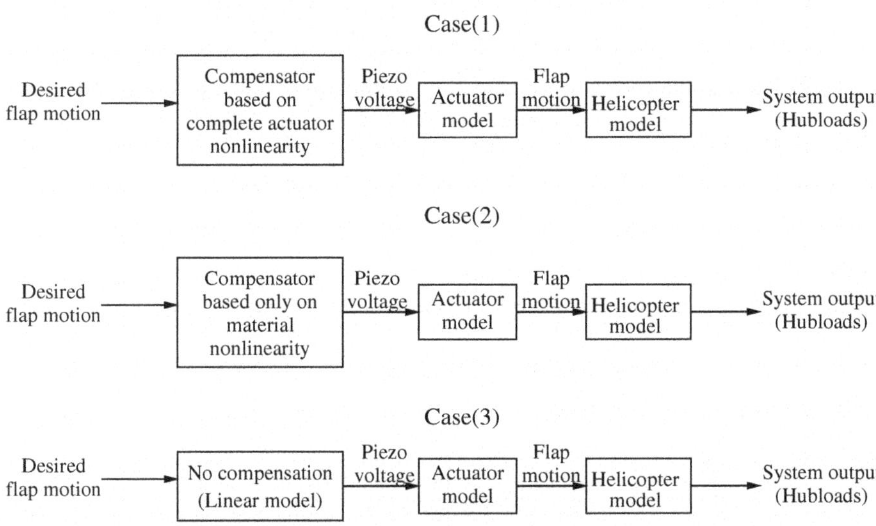

Fig. 7.35 Three different hysteresis compensation levels considered in this study

the other two cases. In case (2), the compensator is derived based only on the material nonlinearity in the actuator and the effect of mechanical hysteresis in the actuator is ignored (Fig. 7.34a). This model is expected to be less accurate than case (1) but better than the case of no compensation (Case (3)). In case (3), no compensation

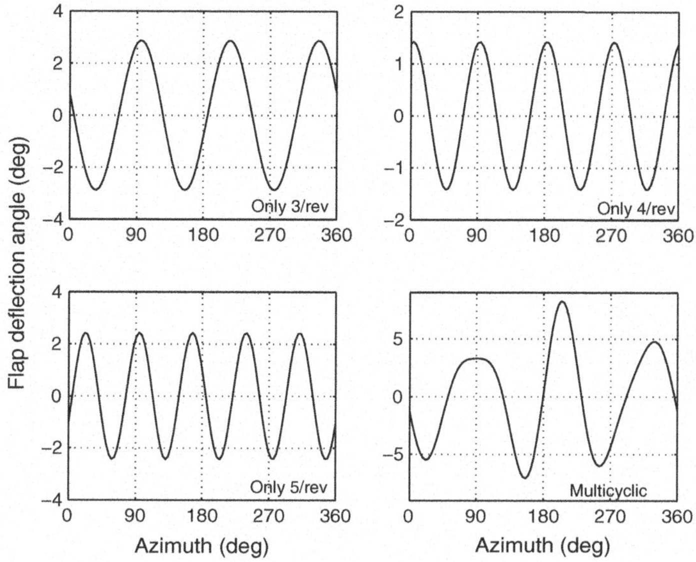

Fig. 7.36 Desired flap motion for independent 3/rev, 4/rev and 5/rev control input and multicyclic control input

is made for the actuator hysteresis and hence this is the least accurate of the three models. However, it is simple to implement since it does not involve inversion of the Preisach model.

Firstly, the control algorithm described in Sect. 4.3 of Chap. 4 is applied to the helicopter aeroelastic model to determine the optimal/desirable flap control input for maximum reduction in hub vibration level. Figure 7.36 shows the desirable flap motion when 3/rev (δ_{3c}, δ_{3s}), 4/rev (δ_{4c}, δ_{4s}) and 5/rev (δ_{5c}, δ_{5s}) control input are applied independently and when applied together (multicyclic input) at an advance ratio of $\mu = 0.30$. This provides a general understanding of the influence of various components of the control input on the fixed frame hub vibratory loads. In practice, the desired flap motion obtained above can be fine-tuned using wind-tunnel and flight test data.

3/rev Control Input

Initially, the desired 3/rev flap motion from Fig. 7.36 is given as input to the three different compensators mentioned above. Figure 7.37 shows the piezostack voltage obtained from the compensators in all three cases. The voltage history is appreciably different in all three cases. The actuator response to the voltage inputs obtained for the three cases are shown in Fig. 7.38 along with the desired flap motion history. As expected, the result from case (1) is almost indistinguishable from the desired flap motion, since the compensator in this case was derived based on the complete actuator nonlinearity. The helicopter model predicts a reduction of 79 % in hub vibration levels due to this flap motion. The actuator response in cases (2) and (3) are quite different

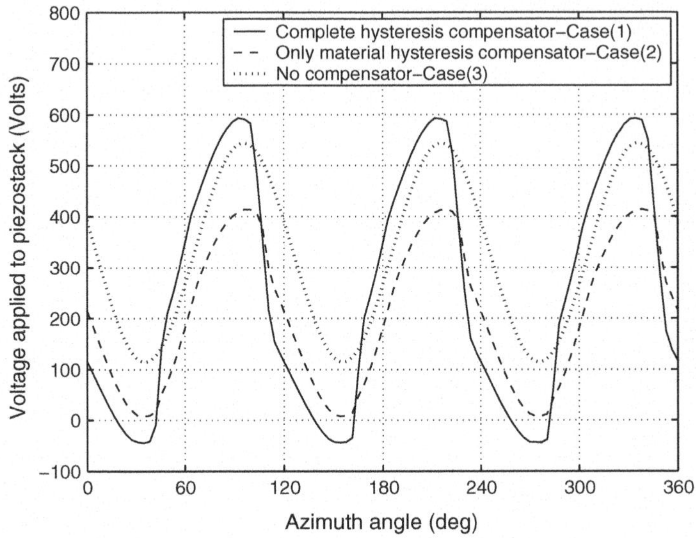

Fig. 7.37 Piezostack voltage computed from the compensators (3/rev control input)

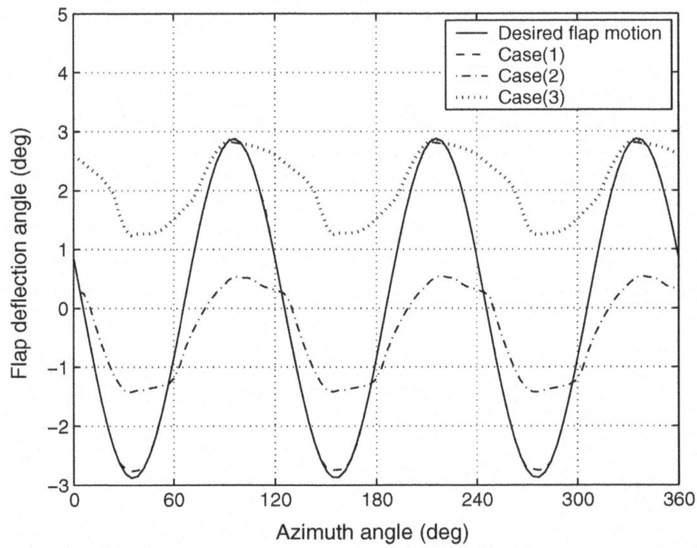

Fig. 7.38 Flap motion history for the three cases (3/rev control input)

from the desired flap motion due to the use of inaccurate compensation methods. The helicopter aeroelastic model predicted 42 and 36 % reduction in hub vibration in case (2) and case (3), respectively. Figure 7.39 shows the reduction in 4/rev component of individual hub forces and moments for the three cases of compensation. The

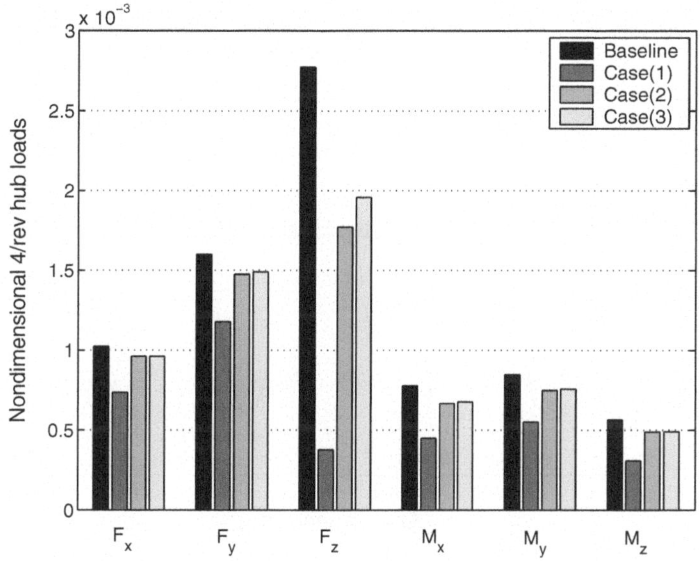

Fig. 7.39 Reduction in hub vibratory loads from baseline values (3/rev control input)

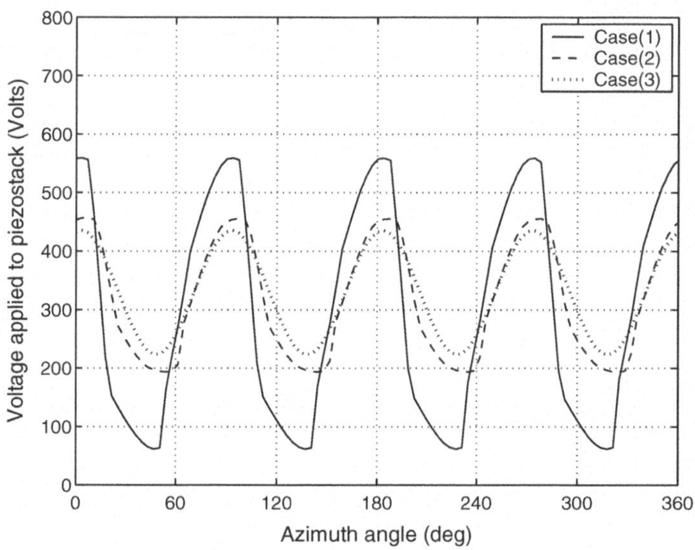

Fig. 7.40 Piezostack voltage computed from the compensators (4/rev control input)

dominant vertical hub force (F_z) shows much less reduction for case (2) and case (3) where inaccurate compensation is used.

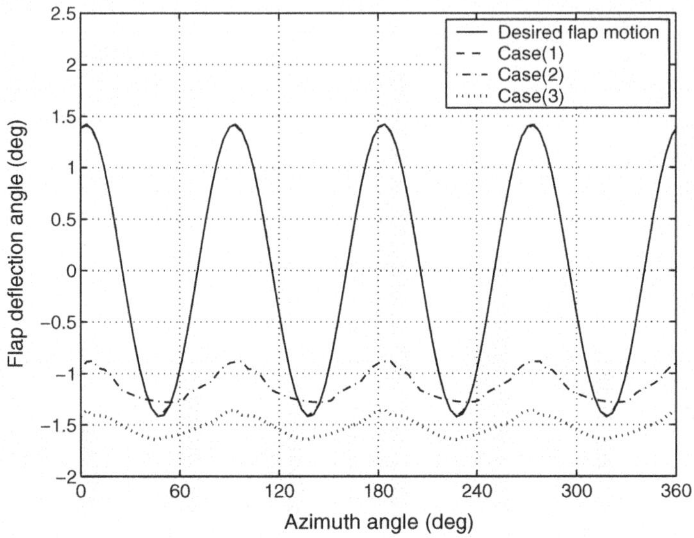

Fig. 7.41 Flap motion history for the three cases (4/rev control input)

4/rev Control Input

Figure 7.40 shows the time history of the voltage input to the piezostack for all three cases. Again, there is a notable difference in magnitude and phase of the voltage input obtained for the three cases. The flap motion obtained in case (1) is almost indistinguishable from the desired flap motion, since the compensator in this case was based on complete hysteresis inversion (Fig. 7.41). The helicopter aeroelastic model predicts about 70 % reduction in 4/rev hubloads due to this flap motion. However, the flap motion in the other two cases are entirely different from the desired flap motion, thereby leading to very poor performance in terms of helicopter hub vibration reduction (Fig. 7.42), especially in the dominant F_z component. In case (2), the overall hub vibration reduced by only 14 % and in case (3), it was even lower at 8 %.

5/rev Control Input

The voltage input and flap motion for 5/rev control input are shown in Figs. 7.43 and 7.44, respectively. Again, it is seen that the flap motion due complete hysteresis compensation is identical to the desired flap motion. In this case, the hub vibration is reduced by about 66 % from baseline conditions. Hub vibration levels reduced by about 24 and 18 % from baseline conditions in the case of material hysteresis compensation and no compensation case, respectively. This is due to the large difference between the actual flap motion in these two cases and the desired flap motion. Figure 7.45 presents the reductions in the individual hub vibratory forces and moments.

Multicyclic Control Input

In this section, 2/rev, 3/rev, 4/rev and 5/rev harmonics of the control input are given simultaneously. Figure 7.46 shows the piezostack voltage obtained from the com-

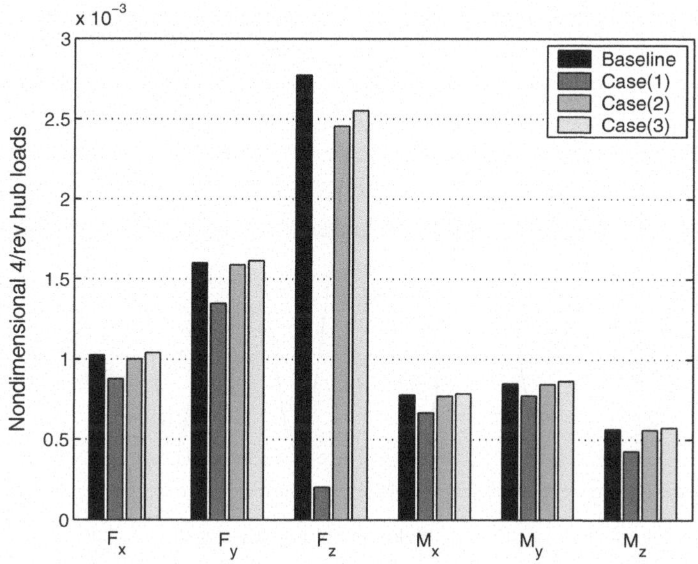

Fig. 7.42 Reduction in hub vibratory loads from baseline values (4/rev control input)

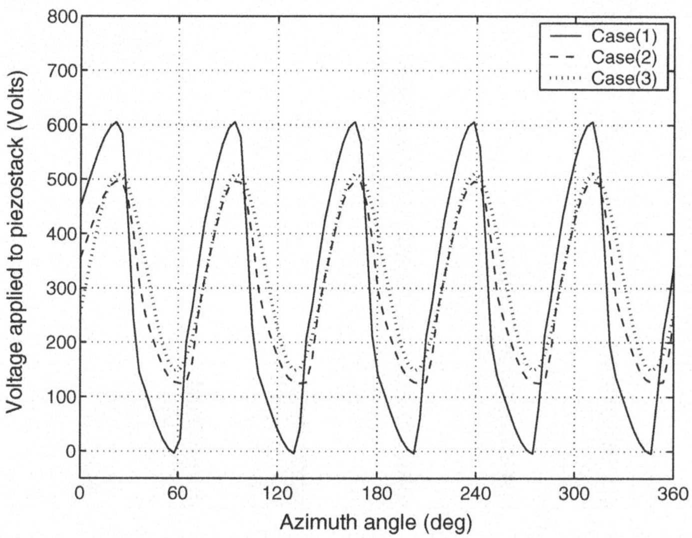

Fig. 7.43 Piezostack voltage computed from the compensators (5/rev control input)

pensators in all three cases. The actuator response for the three cases along with the desired flap motion is shown in Fig. 7.47. In case (1), where the compensator is based on complete actuator hysteresis inversion, the hub vibration is reduced by about 90 % from baseline condition. Compensation of only material hysteresis leads

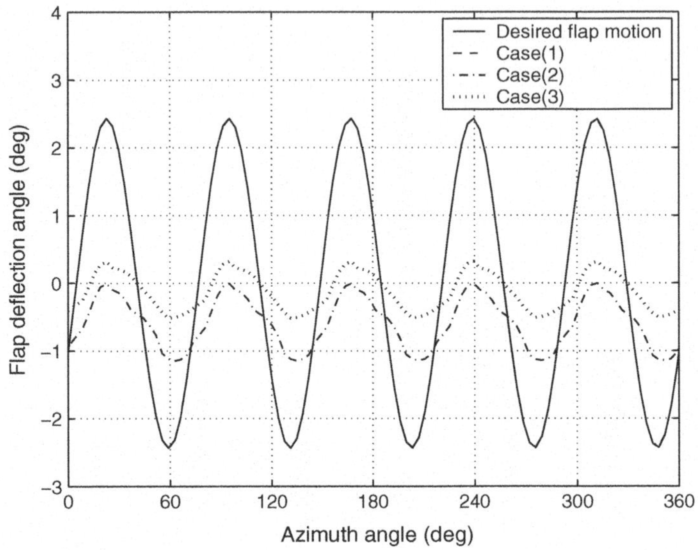

Fig. 7.44 Flap motion history for the three cases (5/rev control input)

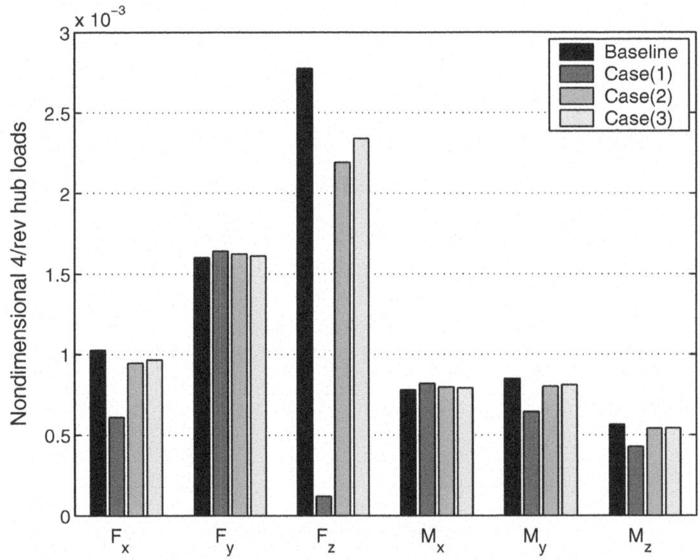

Fig. 7.45 Reduction in hub vibratory loads from baseline values (5/rev control input)

to degradation in the performance of the trailing-edge flap system, yielding only 64 % reduction in hub vibration. However, when hysteresis in the actuator is not compensated, the hub vibration is reduced by 69 % from baseline (Fig. 7.48). The performance in case (3) is better than case (2) because the flap motion in case (3) is

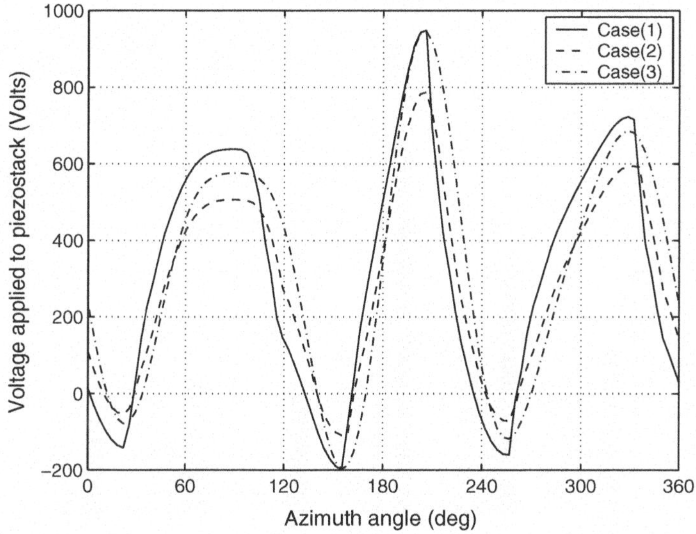

Fig. 7.46 Piezostack voltage computed from the compensators (multicyclic control input)

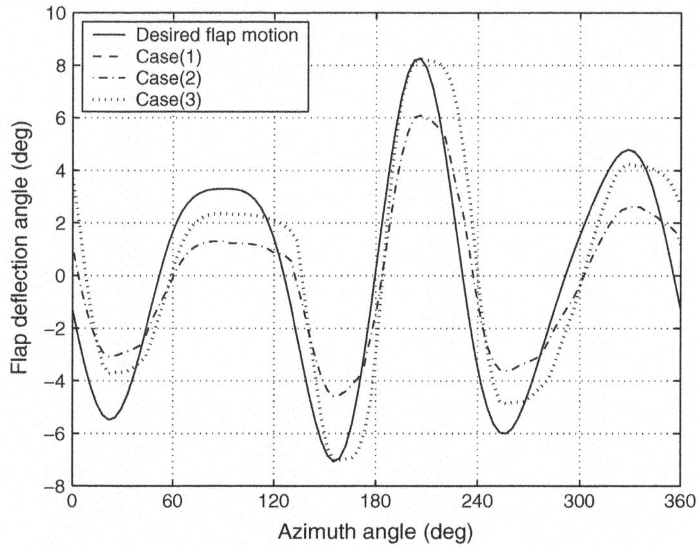

Fig. 7.47 Flap motion history for the three cases (multicyclic control input)

closer to the desired flap motion than the flap motion in case (2). It also shows that addressing only material hysteresis is not enough for actuators which have significant mechanical hysteresis.

Figure 7.49 summarizes the numerical results obtained in terms of percentage reduction in hub vibration for all three cases. It is evident that partial or no com-

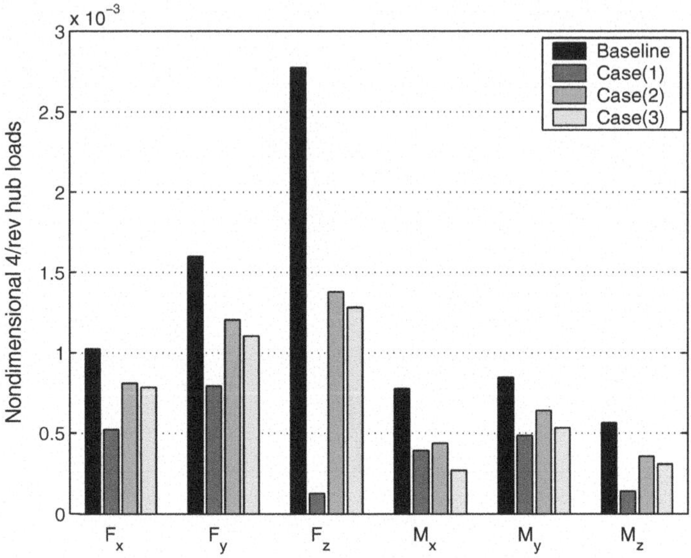

Fig. 7.48 Reduction in hub vibratory loads from baseline values (multicyclic control input)

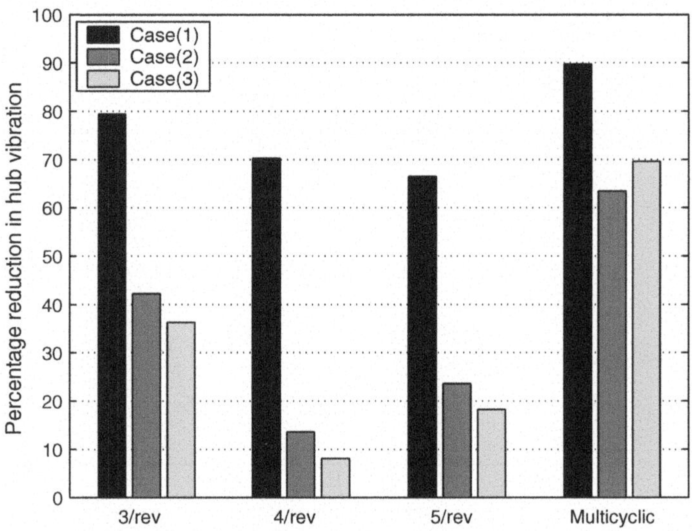

Fig. 7.49 Percentage reduction in hub vibration for the three cases

pensation of the actuator hysteresis leads to deterioration in the performance of the vibration control system. However, it should be noted here that the compensator in case (2) performed poorly due to significant mechanical hysteresis in the X-frame actuator compared to the piezostack material hysteresis (Fig. 7.34). The hysteresis in the amplification mechanism due to friction and freeplay can be minimized by reducing the number of joints or by using compliant mechanisms. However, since it is difficult to completely eliminate mechanical hysteresis, both material and mechanical hysteresis must be addressed when using piezostack actuators for vibration control.

7.5 Summary

The effect of trailing-edge flap actuator hysteresis on helicopter vibration control is studied. Two cases are considered: (1) Ideal-linear actuator and (2) Real actuator with hysteresis nonlinearity. The global controller is used to determine the optimal control input for the ideal-linear actuator. The "feedback form of the global controller" is used in the case of the real actuator with hysteresis. The controller performance is compared for the above two cases. The hysteresis in the real actuator is modeled using the classical Preisach model. The weighting function which is characteristic of the hysteresis transducer is estimated from experimental data. Active vibration control is simulated for steady-state high and low forward speed conditions. In the case of real actuator, the addition of actuator hysteresis made the overall system more nonlinear. The ideal-linear actuator with multicyclic control input gives 90 % reduction in hub vibration at high speed flight. The real actuator with multicyclic control input resulted in only 81 % in hub vibration reduction. This loss in controller performance (about 9 %) in the real actuator is due to the presence of actuator hysteresis. The optimal voltage control input and the resulting flap motion for the two cases are different. For an ideal-linear actuator in low speed flight, vibration levels are reduced by about 99 % with multicyclic control input. However, the real actuator gave only 86 % reduction in vibration levels. Again this performance penalty (a drop of 13 %) can be attributed to the presence of hysteretic nonlinearity in the flap actuator. The effect of inaccurate compensation of piezostack actuator hysteresis on the performance of vibration control system is also studied. The hysteresis in the trailing-edge flap actuator is due to both material property of piezostacks and mechanical linkage losses. Compensator based on complete actuator hysteresis gave upto 90 % reduction in hub vibration at high speed flight. Compensator based only on material nonlinearity of the piezostack does not yield the desired flap motion accurately, resulting in only 64 % reduction in hub vibration. In comparison, the hub vibration reduced by 69 % when actuator hysteresis was not compensated. Numerical results demonstrate the importance of complete compensation of both material and mechanical hysteresis present in the trailing-edge flap actuator system.

References

1. Hughes, D., Wen, J.T.: Preisach modeling of piezoceramic and shape memory alloy hysteresis. Smart Mater. Struct. **6**(3), 287–300 (1997)
2. Krasnoselskii, M.A., Pokrovskii, A.V.: Systems with Hysteresis. Springer, Berlin (1983)

3. Viswamurthy, S.R., Ganguli, R.: Effect of piezoelectric nonlinearity on helicopter vibration control using trailing edge flaps. AIAA J. Guid. Control Dyn. **29**(5), 1201–1209 (2006)
4. Viswamurthy, S.R., Ganguli, R.: Modeling and analysis of piezoceramic actuator hysteresis for helicopter vibration control. Sens. Actuators: Phys. **35**(2), 801–810 (2007)
5. Mayergoyz, I.D.: Mathematical models of hysteresis. IEEE Trans. Magn. **22**(5), 603–608 (1986)
6. Hall, S.R., Prechtl, E.F.: Preliminary testing of a Mach-scaled active rotor blade with a trailing-edge servo flap. In: Wereley, N.M. (ed.) Proceedings of SPIE, vol. 3668, pp. 14–21 (1999)
7. Tan, X., Venkataraman, R., Krishnaprasad, P.S.: Control of hysteresis: theory and experimental results. In: Rao, V.S. (ed.) Proceedings of SPIE, vol. 4326, pp. 101–112 (2001)

Chapter 8
Active Rotating Beams

From this chapter onwards, we will address the active twist rotor using piezoelectric actuation. The helicopter rotor blades can be considered as a closer approximation to rotating beams due to their high aspect ratios. However, the structural dynamic and aerodynamic behavior are highly coupled for helicopter rotor blades. To obtain an idea about the physics behind the active twist concept and to validate the formulation and analysis developed in the previous chapter, aerodynamic loads are not included in this chapter and the effect of time varying tip loads on the dynamic behavior of rotating rectangular section beams is studied. Introduction to the theory of rotating beams is given in Sect. 8.1. Section 8.2 describes the model selected for the dynamic analysis of rotating beam. In Sect. 8.3, the bending actuation mechanism is investigated. Section 8.4 presents the analysis utilizing torsion actuation and discusses its effect on beam response. Section 8.5 discusses the importance of nonlinear terms by applying a tip sinusoidal load in the framework of torsion actuation. A similar study, in the framework of bending actuation is presented in Sect. 8.6. Section 8.7 highlights the effect of torsion actuation on a pretwisted beam. Lastly, important findings are summarized in Sect. 8.8.

8.1 Background

The area of "modeling beams with piezoelectric actuators" [1, 2] has received considerable attention in the past years due to the fact that many important structural members can be modeled as beams. Because of the practical importance of such structures, work has been done in the dynamic analysis of rotating beams [3, 4]. However, only a few studies have been done in the dynamic analysis of rotating smart beams which typically address vibration control and shape control using active materials.

However, none of these studies considered shear-based piezoceramic actuation to obtain the response of pretwisted rotating beams [5]. The approach in this chapter is to demonstrate the feasibility of shear actuation in rotating beams and progress

R. Ganguli et al., *Smart Helicopter Rotors*, Advances in Industrial Control,
DOI 10.1007/978-3-319-24768-7_8

further to refine the structural and aerodynamic model to achieve vibration reduction in helicopter rotor.

The governing nonlinear equations of motion for the elastic bending and torsion of twisted nonuniform rotor blades made of isotropic material were derived by Hodges and Dowell using the Hamilton's principle and have become the underlying basis for comprehensive aeroelastic analysis developed and used by several researchers. Geometrical nonlinearities, which result in moderate blade deflections, play an important role in the aeroelastic stability analysis of rotor blades. Generally, moderate deflection type theories based on an ordering scheme are quite adequate for hingeless rotor blades. However, large deflection theories can be important for the bearingless rotors [6]. The governing equations for rotating smart beams were derived in Chap. 2. Here, numerical results are obtained to analyze the influence of smart actuation, centrifugal stiffening, nonlinear effects, and pretwist on deformations. The possibility of using the piezoceramic shear actuation mechanism (d_{15} effect) for torsion and the bending actuation (d_{31} effect) for bending of rotating beams is explored. The idea is to bend or twist the beam with a certain deflection. It should be noted that small amount of deflections can be used to reduce airloads and evolve an active control strategy for rotating beam type structures.

Assessment of modeling techniques for rotor blade applications have been given in [7, 8]. In this chapter, we consider a solid rectangular cross-section beam. Such sections are not widely used in the rotorcraft industry. However, a notable exception is flexbeams used for bearingless rotors.

8.2 Model Details

The model used for numerical results is a cantilevered aluminum beam with surface mounted piezoceramic actuators as shown in Fig. 8.1. This model is similar to an experimental beam devised by Park et al. [9] allowing validation of the results with closed form solutions for a nonrotating bending beam. The cross-section of the beam is shown in Fig. 8.1a.

The cross-section can undergo twisting motion by an angle of $\theta_0 + \hat{\phi}$; where θ_0 is the applied control pitch and $\hat{\phi}$ is elastic twist (see Fig. 2.12). Thus, it can move from the undeformed zy axes to the deformed $\eta\zeta$ axes about the axial coordinate x or ξ. The beam has a height of 1.28 mm, breadth of 50.8 mm, actuator thickness of 0.245 mm, and adhesive thickness of 0.10 mm. The length of the beam is 406.4 mm. The adhesive (resin) layer has been considered here to incorporate the shear lag effects. The results are obtained for a potential of 400 V in case of bending actuation and for 200 V in case of torsion actuation. The mechanical properties considered for simulation are listed in Table 8.1.

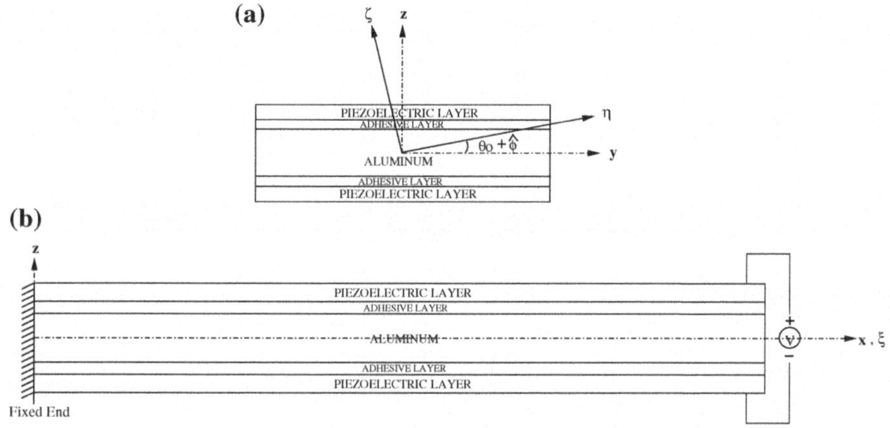

Fig. 8.1 **a** Cross-sectional view of the beam. **b** Side view of the beam

Table 8.1 Mechanical properties considered for simulation

	PZT-5h	Adhesive	Aluminum	Units
E	63	3.068	68.94	10^3 MPa
ν	0.3	0.3485	0.3	–
ρ	7600	1000	2700	kg/m^3
d_{15}	730	–	–	10^{-09} mm/V
d_{31}	254	–	–	10^{-09} mm/V

8.3 Bending Actuation

For obtaining the bending actuation equal but opposite potentials are given to the top and the bottom piezo layers (bending motor concept) [10], which results into the expansion of one surface and contraction of other. As it can be seen from Fig. 8.2, the poling direction is parallel to the direction of the potential. To bend the beam, a voltage of equal magnitude but opposite sign is applied to the top and bottom of actuators. Thus, the beam bends to produce transverse bending deflection. For the nonrotating condition, a tip bending deflection of about 14 mm is obtained. This is slightly more than 3 % of the beam length. The bending deflection obtained here matches with the closed form solution at the nonrotating level. Response of a rotating beam over a wide range of rotational speeds (0–100 Hz) is investigated. The transverse tip bending deflection (nondimensionalized with respect to the radius R of the blade) and its variation with rotation speed is shown in Fig. 8.3. There is no other load on the beam except a bending moment due to bending actuation and the centrifugal loading due to rotation. It can be observed from Fig. 8.3 that there is a rapid fall in the bending deflection with increase in rotating speed. On studying the Eqs. 2.28–2.32, it is noticed that the centrifugal stiffening effect ($\int_0^1 F_A H'^T H' ds$ in the stiffness matrix

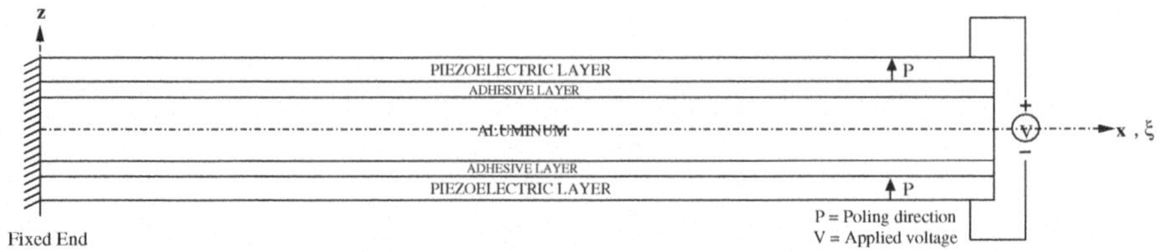

Fig. 8.2 Schematic of bending actuation (poling direction is parallel to applied voltage)

Fig. 8.3 Transverse tip bending deflection for beam with bending actuation and increasing rotation speed

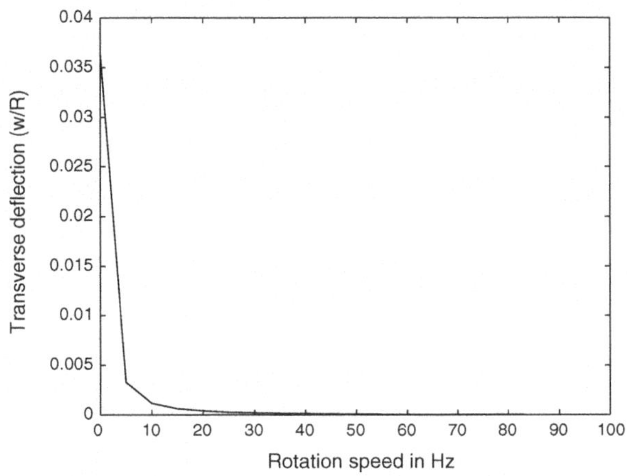

K) counters the piezoceramic actuation force ($\int_0^1 A_3 H''^T ds$ in the linear force vector $\mathbf{F_L}$) and decreases the deflection which reduces to almost zero at higher rotation speeds. It therefore appears that bending actuation is useful for nonrotating and slowly rotating structures.

8.4 Shear or Torsional Actuation

Conventional smart structure applications induce strain in the structure using the d_{31} and d_{33} piezoelectric constants, applying the electric field parallel to the poling direction. The shear-based constant, d_{15} also induces strain but requires application of electric fields in the direction perpendicular to that of poling as shown in Fig. 8.4. These shear-based actuators were proposed by Sun and Zhang [11] in a study in which a sandwich structure was constructed using the shear mode of actuation. The transverse shear deformation in the core was utilized to give rise to the desired transverse deflection of the sandwich. A comparative study between the sandwich structure and surface mounted actuator systems was demonstrated using finite element analysis. It was shown that for short and thick actuators the sandwiched structure had a better performance over surface mounted, *whereas for long and thin actuators the surface mounted systems performed better than the sandwiched systems.*

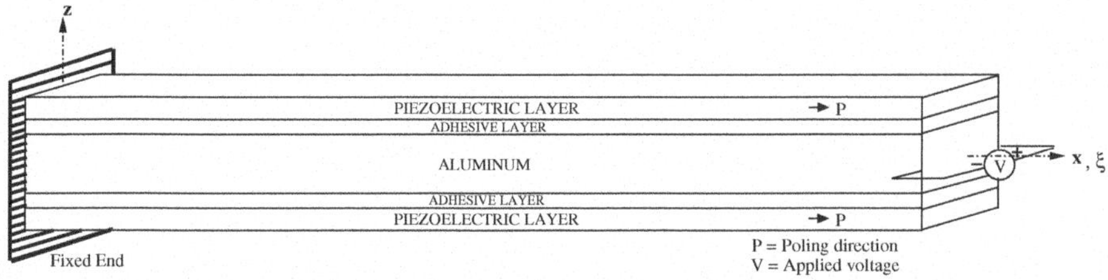

Fig. 8.4 Schematic of shear actuation (poling direction is perpendicular to applied voltage)

Later, in 1999, Benjeddou et al. [12] compared the extension actuation mechanism with the shear actuation mechanism. For both cases the stress and vibration level were compared. Parametric studies were done by varying the structure/actuator stiffness, thickness ratios, actuator position, and length. It was shown that shear actuator performance was less dependent on the structure's stiffness. It was also noted that for stiffer structures the shear actuation mechanism gave better results than the extension mechanism. The vibration modes were found to be equivalent in both cases. It was shown that surface mounted actuators, acting through the e_{31} piezoelectric constant, induce boundary concentrated forces, and moments in the structure. Whereas, sandwich shear actuators, acting through the e_{15} piezoelectric constant, induce distributed moments in the structure. Extension actuators were found to be efficient when placed near the clamped end of the beam for long spans of actuators. However, shear actuators were found to be effective for very short lengths and over a larger range of positions.

Glazounov et al. [13] proposed a piezoelectric actuator generating torsional displacement from the piezoelectric d_{15} shear response. A torsional actuator (tube) which generated high angular displacement from shear strain was developed. The tube consisted of an even number of piezoelectric segments adhesively bonded together. These segments were poled in the length direction and the electric field was applied in the direction perpendicular to the poling, and hence shear strain was induced. The study was a proof of concept type of study, which stated that this kind of torsional tube actuator could be used to generate twist. Thus, shear mode actuators have been suggested for obtaining large torque and angular displacement.

Recently, a comparison of stresses in a sandwich laminate with those in the corresponding surface mounted extension actuator configuration revealed that the stresses within the shear actuators are significantly smaller than those in extension actuators. Thus, it is advantageous to use shear actuators since large stresses can be detrimental to the brittle piezoceramic material [14].

Also, a coupled finite element procedure was derived using a two-noded beam element to model two different kinds of piezoelectric actuation in a smart sandwich beam. Active vibration control of smart sandwich beams was performed. Two constitutive models are incorporated so that the transverse electric field couples the longitudinal strain and the shear strain fields. It was concluded in the study that shear actuator performs better in controlling the first three bending modes than the

Fig. 8.5 Torsion deflection at tip for beam with torsion actuation and increasing rotating speed

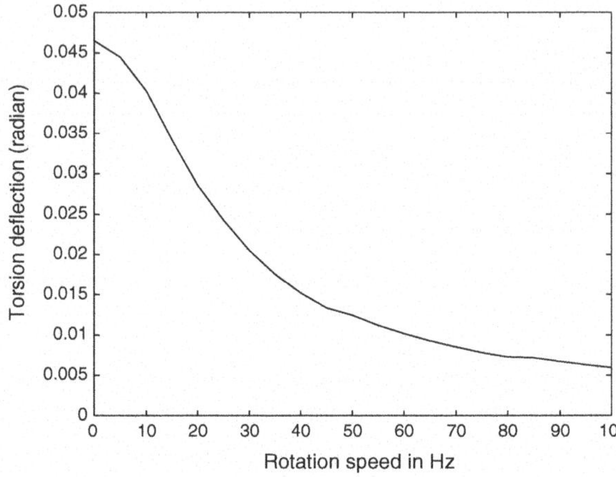

extension-bending actuator [15]. Note, however that most of the studies use shear actuation in sandwich beams to obtain bending deformation, whereas we use it for torsion deformation. The principal torsional/shear actuation term in the linear load vector is $\int_0^1 A_4 H_{\hat{\phi}}^{'T} ds$ (Eq. 2.31). Among the stiffening terms; $\int_0^1 GJH_{\hat{\phi}}^{'T} H_{\hat{\phi}}^{'} ds$ is coming from the structural stiffness and $\int_0^1 m({k_{m2}}^2 - {k_{m1}}^2) H_{\hat{\phi}}^T H_{\hat{\phi}} ds$ is coming from the rotational effect (Eq. 2.28). According to the requirements of the problem, the term A_4 and GJ involved in torsion actuation and structural stiffening respectively, can be tailored so as to yield maximum torsion actuation and at the same time giving appropriate frequencies. In general, lower torsion stiffness is beneficial for torsion actuation.

In Fig. 8.5, torsion response is plotted with increasing rotation speed. There is no other load on the beam except the smart couple (shear actuation) and the propeller moment caused by centrifugal force. This is compared with the zero loading situation. In the case of torsion actuation, a twist of 2.66° is obtained for the nonrotating case. But, the twist decreases as the rotation speed increases. At rotation speed as high as 100 Hz, a twist of around 0.35° is obtained. However, the decrease in twist with rotation speed is much less when compared to the decrease in bending (observed in previous section). This is because the centrifugal stiffening of the torsion mode is less compared to that of the transverse bending mode. It can be seen from Eq. 2.28 that the term leading to the centrifugal stiffening in the torsion direction is $\int_0^1 m({k_{m2}}^2 - {k_{m1}}^2) H_{\hat{\phi}}^T H_{\hat{\phi}} ds$ whereas in the bending direction is $\int_0^1 F_A H_u^{'T} H_u^{'} ds$. The latter term is much larger in magnitude. In case of a square cross-section, the term $\int_0^1 m({k_{m2}}^2 - {k_{m1}}^2) H_{\hat{\phi}}^T H_{\hat{\phi}} ds$ will reduce to zero. As the width of the cross section increases the term ${k_{m2}}^2$ will become dominant and as the height increases the term ${k_{m1}}^2$ will become dominant. In general, the width is much greater in magnitude than the height. It should be noted that the expressions used here non dimensional quantities.

Thus, it can be said that in case of torsion response, there remains enough twist to enable active control of vibration reduction even at high rotation speeds. Whereas in case of transverse bending, the centrifugal stiffening effect almost nullifies the effect of piezoceramic actuation at high rotation speeds. Shear-based piezoceramic actuation using the d_{15} term appears useful for torsion actuation of rotating beams.

8.5 Shear Actuation with Sinusoidal Loads

The above results were obtained with the only loads acting on the beam being those due to rotation and actuation. Also the deflections were only due to smart actuation and were too small to activate the nonlinear terms, which are motion dependent. It can be seen from Fig. 8.5 that even at 100 Hz a satisfactory amount of torsion deflection is obtainable. Thus, to examine the effects of the nonlinear terms and time varying loading, a sinusoidal tip load is given to a beam rotating at 100 Hz. The linear effects are obtained by suppressing the nonlinear terms in the computer program used for getting the results. Results for the linear and nonlinear cases are obtained for (1) no smart actuation and (2) with smart actuation.

In the framework of torsion actuation, effect of including nonlinear terms in the analysis is shown in Figs. 8.6, 8.7 and 8.8 for torsion, inplane, and transverse bending response respectively (nondimensionalized with respect to the radius R of the blade). As shown in Fig. 8.6, there is considerable change in the torsion response pattern for linear and nonlinear case. The smart loading also follows the same pattern, giving it a shift by some magnitude, showing that with a smart loading a definite amount of change in response can be obtained. But, the magnitude of difference between the nonlinear and the linear cases is very little in the cases of inplane (Fig. 8.7) and transverse (Fig. 8.8) bending response when compared to torsion. The importance of

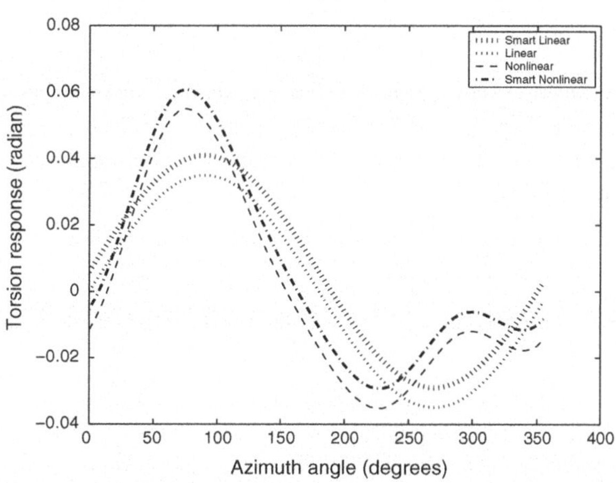

Fig. 8.6 Effect of nonlinear terms on tip torsion response under tip sinusoidal load and rotation speed of 100 Hz with and without smart torsion actuation

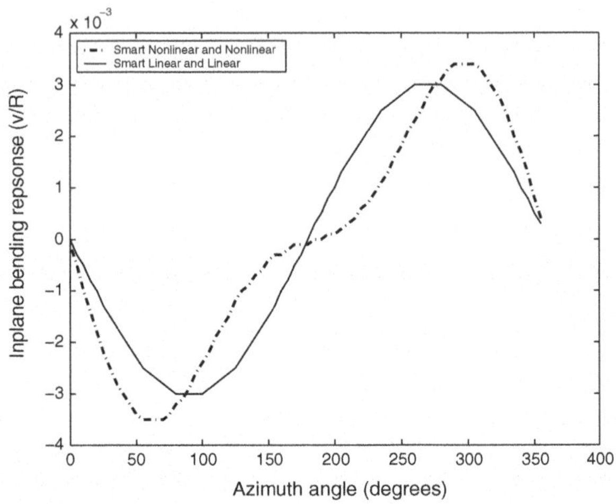

Fig. 8.7 Effect of nonlinear terms on tip inplane bending response under tip sinusoidal load and rotation speed of 100 Hz with and without smart torsion actuation

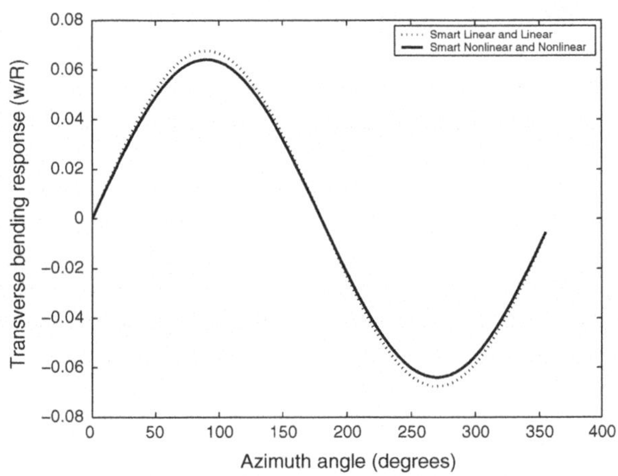

Fig. 8.8 Effect of nonlinear terms on tip transverse bending response under tip sinusoidal load and rotation speed of 100 Hz with and without smart torsion actuation

including nonlinear terms in the analysis is clear, especially for accurate prediction of torsion response. However, the effect of smart loading brings a change in magnitude in the response whereas the geometric nonlinearities also cause a phase change in the response.

8.6 Bending Actuation with Sinusoidal Loads

A similar case as described above is then examined with the smart couple in the form of a bending actuation and the beam rotating at 10 Hz. This rotating speed is selected due to the reason that at this speed some amount of transverse deflection still remains and beyond this speed the deflection falls off steeply (Fig. 8.3). The

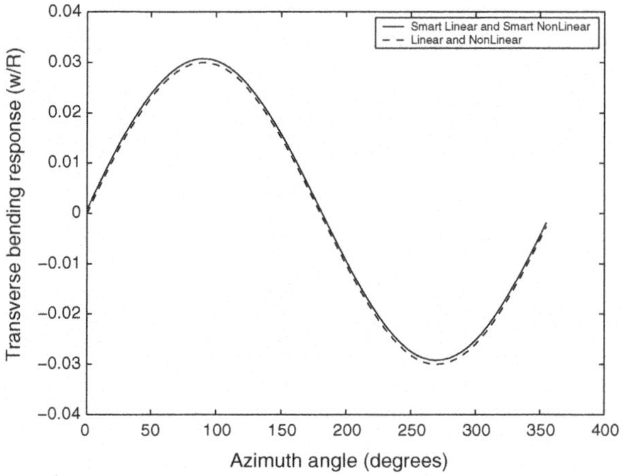

Fig. 8.9 Effect of nonlinear terms on tip transverse bending response under tip sinusoidal load and rotation speed of 10 Hz with and without smart bending actuation

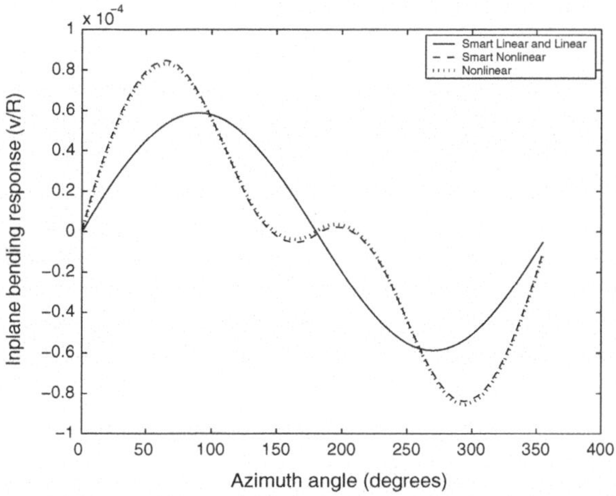

Fig. 8.10 Effect of nonlinear terms on tip inplane bending response under tip sinusoidal load and rotation speed of 10 Hz with and without smart bending actuation

nonlinear and smart effect in the case of transverse bending, inplane, and torsion response are shown in Figs. 8.9, 8.10 and 8.11 respectively.

It could be noted from these figures that, in the presence of bending actuation, inplane bending, transverse bending, and torsion response do not change to a greater extent in magnitude. So, it could be inferred that the need of including nonlinear terms in case of bending actuation is not very significant.

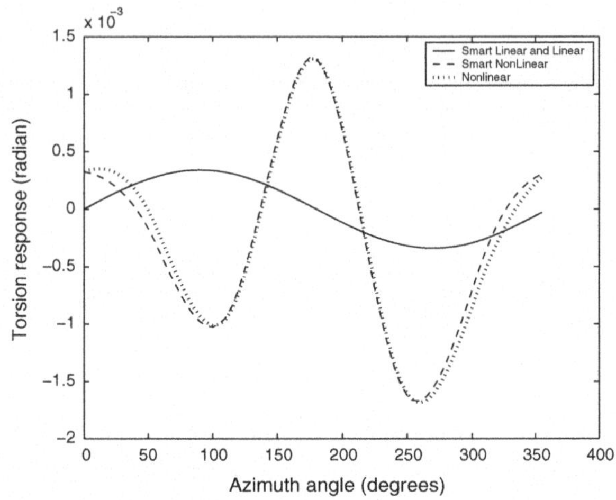

Fig. 8.11 Effect of nonlinear terms on tip torsion response under tip sinusoidal load and rotation speed of 10 Hz with and without smart bending actuation

8.7 Torsion Actuation with Pretwist

Rotating systems create accelerations arising from the elastic structural deformations, Coriolis and centripetal forces. The complexity increases further when initial pretwist is also included in a beam as additional couplings are introduced due to the pretwist which can affect the beam response to static and dynamic loads [16, 17]. Here, an example has been taken of a beam which is pretwisted by (a) 2° (b) −2°. Linear pretwist is considered for these cases. A torsion smart moment is then applied on such a beam through the shear mode as discussed in Sect. 8.4. It can be noted from Fig. 8.12 that at a pretwist of ±2°, the torsion deflection varies approximately from 0 to 2.29° at a rotation speed of 0–100 Hz. But when a smart torsion moment is

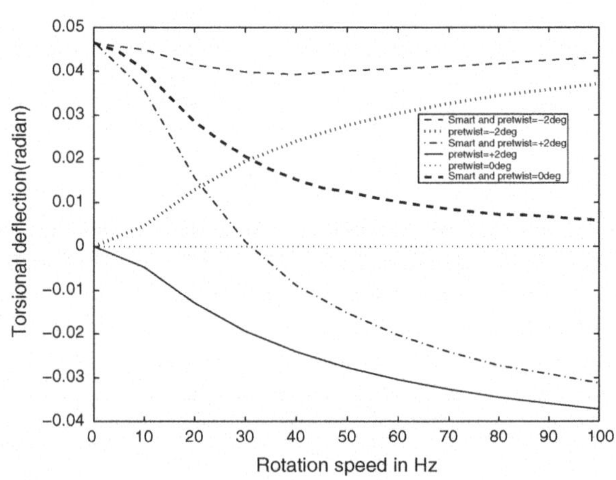

Fig. 8.12 Effect of pretwist and smart torsion actuation tip torsion deflection with increasing rotation speed

applied the trend followed by the torsion deflection changes. There is a difference in the tip torsion response of about 2.69° at $\Omega = 0$, which reduces to about 0.17° at $\Omega = 100\,\text{Hz}$. The effect of the piezoceramic actuation is to increase the torsion deflection of the beam. The increase is maximum in the nonrotating condition and becomes less as rotation speed increases. It can be observed from this example of ±2° of pretwist that combinations of smart actuation and pretwist can be worked out to obtain a desired torsion deflection over a range of rotation speeds. Similar attempts were made on applying bending actuation to a pretwisted beam but the results obtained showed very little change and hence have not been presented here. Thus, it can be said that torsion actuation proves to be a better alternative in this case.

8.8 Summary

The dynamic behavior of rotating beams with piezoceramic actuation is studied using bending (d_{31}) and shear (d_{15}) actuation for application to structures such as helicopter and wind turbine rotor blades. Numerical results show that the centrifugal stiffening effect reduces the tip transverse bending deflection and elastic twist obtained from smart actuation as the rotation speed increases. However, the effect of rotation speed on the tip elastic twist obtained using smart actuation is less pronounced. Results are obtained for the cases with only actuation loads, with actuation loads and added periodic tip loads, and for a pretwisted beam. The importance of nonlinear terms for accurate prediction of torsion, inplane bending and transverse bending response is depicted. Thus, it can be said at this stage that the utility of d_{15} based piezoceramic actuators can be exploited for applications in which torsion mode of response is of importance.

References

1. Wang, C., Ang, K., Ajit, A.: Status of application of smart structures technology to rotorcraft systems. J. Intell. Mater. Syst. Struct. **10**, 164–175 (1999)
2. Librescu, L., Meirovitch, L.: Control of cantilevers vibration via structural tailoring and adaptive materials. Am. Inst. Aeronaut. Astronaut. **35**, 1309–1315 (1997)
3. Bhattacharya, B., Bhatt, M., Krishnamurthy, A.: Dynamics and control of laminated composite structures with surface bonded/embedded sensors and actuators. In: Discussion Meetings on Aerodynamic Testing and Structural Dynamics, pp. 141–156 (1994)
4. Hodges, D., Chung, Y., Shang, X.: Discrete transfer matrix method for nonuniform rotating beams. J. Sound Vib. **240**, 276–283 (1994)
5. Thakkar, D., Ganguli, R.: Dynamic response of rotating beams with piezoceramic actuation. J. Sound Vib. **270**(4–5), 729–753 (2004)
6. Friedmann, P.: Rotary wing aeroelasticity with application to vtol vehicles. In: Proceedings of the 31st AIAA/ASME/ASCE/ASC Structures, Structural Dynamics and Materials Conference, Long Beach, CA, pp. 1624–1667, 2–4 April 1990
7. Jung, S., Chopra, I.: Assessment of composite rotor blade modeling techniques. J. Am. Helicopter Soc. **44**(3), 188–205 (1999)

8. Volovoi, V., Hodges, D., Cesnik, C., Popescu, B.: Assessment of beam modeling methods for rotor blade applications. J. Math. Comput. Model. **33**, 1099–1112 (2001)
9. Park, C., Walz, C., Chopra, I.: Bending and torsion models of beams with induced-strain actuators. Smart Mater. Struct. **5**, 98–113 (1996)
10. Piezo Systems Inc.: 186 Massachussets Avenue Cambridge, MA 02139
11. Sun, C., Zhang, X.: Use of thickness-shear mode in adaptive sandwich structures. Smart Mater. Struc. **4**, 202–206 (1995)
12. Benjeddou, A., Trindade, M., Ohayon, R.: Piezoelectric actuation mechanisms for intelligent sandwich structures. Smart Mater. Struct. **9**, 328–335 (1999)
13. Glazounov, A.E., Zhang, Q.M., Kim, C.: Piezoelectric actuator generating torsional displacement from piezoelectric d15 shear response. Appl. Phys. Lett. **72**(20), 526–2528 (1998)
14. Senthil, V., Batra, R.: Exact solution for the cylindrical bending of laminated plates with embedded piezoelectric shear actuators. Smart Mater. Struct. **10**, 240–251 (2001)
15. Raja, S., Prathap, G., Sinha, P.: Active vibration control of composite sandwich beams with piezoelectric extension-bending and shear actuators. Smart Mater. Struct. **11**, 63–71 (2002)
16. Surace, G., Anghel, V., Mores, C.: Coupled bending-torsion vibration analysis of rotating pretwisted blades: an integral formulation and numerical examples. J. Sound Vib. **206**, 473–486 (1997)
17. Yoo, H., Kwak, J., Chung, J.: Vibration analysis of rotating pre-twisted blades with a concentrated tip mass. J. Sound Vib. **270**, 839–853 (2001)

Chapter 9
Box-Beam Active Rotor Blades

The previous chapter clearly showed that shear-based piezoceramic actuation using d_{15} is feasible for obtaining active twist. The current chapter dwells more on the development and validation of the concept of induced-shear-based piezoceramic actuation in helicopter rotor blades considering a simple two-cell box cross section. It is known that two-cell cross-section model is closer approximation to a realistic rotor blade section and is a good model for preliminary design studies. Section 9.1 discusses the structural model used for the rotor blade analysis. Rotor properties and rotating natural frequencies of the blade are given in this section. Section 9.2 shows the significance of including the piezoceramic nonlinearity in the analysis. A discussion regarding the increase in efficiency due to the consideration of piezoceramic nonlinearity is presented. In Sect. 9.3, an open-loop analysis is performed to obtain the reduction in vibration by applying a smart loading with a suitable phasing. Section 9.4 presents the feedback control algorithm which feeds back the shear strain rate and in turn calculates the amount of voltage required to apply the smart loading. Finally, Sect. 9.5 summarizes the issues addressed in the current chapter.

9.1 Model Details

The basis for the formulation required for this analysis is discussed in Chap. 2. For the aerodynamic modeling blade element theory is used, the reverse flow effect is considered. The unsteady aerodynamics model developed by Leishman and Beddoes is used. The smart blade has a two-cell cross-section with an outer dimension of 165.1×37.0 mm.

The inner details and design of the section are selected such as to provide reasonable frequencies for the rotor blade [1]. The two-cell symmetric box consists of a layer of piezoelectric actuator (1.0937 mm) surface bonded on the top and bottom of the host structure [2]. An adhesive layer (0.437 mm) is used to bond the host

R. Ganguli et al., *Smart Helicopter Rotors*, Advances in Industrial Control,
DOI 10.1007/978-3-319-24768-7_9

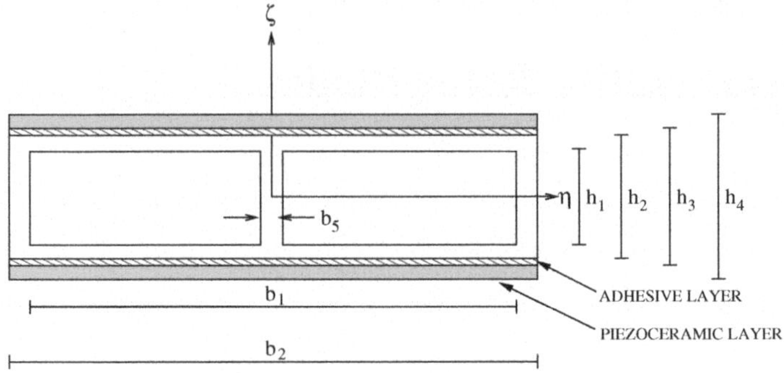

Fig. 9.1 Two-cell box cross section

Table 9.1 Mechanical properties considered for simulation

Dimensions		Properties			
Width (mm)	Height (mm)	PZT-5H	Adhesive	Aluminum	Units
$b_1 = 161.60$	$h_1 = 30.438$	$E = 63$	$E = 3.068$	$E = 68.94$	10^3 MPa
$b_2 = 165.10$	$h_2 = 33.938$	$\nu = 0.3$	$\nu = 0.3485$	$\nu = 0.3$	–
$b_5 = 1.75$	$h_3 = 34.812$	$\rho = 7600$	$\rho = 1000$	$\rho = 2700$	kg/m^3
	$h_4 = 37.00$	$d_{15} = 730$	–	–	10^{-09} mm/V
		$d_{31} = 254$	–	–	10^{-09} mm/V

Table 9.2 Rotor properties considered for simulation

c/R = 0.06	$C_T/\sigma = 0.07$	Mass per unit length, $m_0 =$ 4.912 kg/m
Number of blades, N = 4	Radius, R = 3.81 m	$\beta_p = 0.0$
$\theta_{tw} = 0.0$	$\mu = 0.3$	$\sigma = 0.076$
$\gamma = 5.79$	$\Omega = 40.12$ rad/s	

structure with the piezoceramic layer. Below the smart and adhesive layers underlies the host (aluminum) as depicted in Fig. 9.1. The design and properties of the model used for analysis are listed in Table 9.1. Other blade and rotor characteristics are listed in

Table 9.2. The blade has zero pretwist and precone and can be considered as a rotating active box-beam.

For a 4-bladed soft-inplane hingeless rotor design, the 4/rev (4 times per revolution) loads represent the main source of helicopter vibration. Numerical results are obtained in forward flight at an advance ratio of $\mu = 0.3$ and $C_T/\sigma = 0.07$. The rotating blade natural frequencies calculated for such a rotor are listed in Table 9.3.

Table 9.3 Rotating natural frequencies of the blade

Mode	Frequency (per rev)
First lag	0.54
First flap	1.05
Second flap	2.7
Second lag	3.4
First torsion	5.7

9.2 Piezoceramic Nonlinearity

Initially, shear-based actuators have been proposed and utilized in the form of shear/torque tubes and sandwich structures. These kind of actuators have also been experimentally tested. On testing, it was found that there exists a nonlinear relation between the piezoelectric coefficient d_{15} and the applied electric field, this relation is also compared with the output of the tube actuator and the results seem to match satisfactorily.

Shear response of the piezoceramic material like PZT-5H varies with the amplitude of the driving electric field. Even small amplitudes of applied electric field result in significant changes in the piezoelectric coefficient d_{15} [3]. Hence, understanding the variation is important for analyzing and arriving at a proper design and performance of the actuator. In case of soft piezoceramics like PZT-5H the ac-dependence of the piezoelectric coefficients at fields above the threshold of $E_a = 110\,\text{V/mm}$ could be well approximated as a power law [4]. The onset of nonlinearity occurs at a threshold value of $E_a = 110\,\text{V/mm}$. The limiting value of the amplitude of the applied voltage is 490 V/mm for PZT-5H. Beyond this value, depoling takes place. Strain-polarization dependence which is linear and nonhysteric in lower fields becomes hysteric and nonlinear at higher fields. Here, we utilize the power law and compare the results for linear and nonlinear cases. A polynomial fit for d_{15} in terms of applied voltage is obtained from the data points shown in Fig. 9.2. This graph represents the nonlinear behavior of soft piezoceramic materials on application of an electric field. Number '4' corresponds to PZT-5H and this is considered in the present analysis. The following relation is obtained from the data set for PZT-5H. Here E_a is the amplitude of the applied sinusoidal electric field. Note that the effect of nonlinearity between d_{15} and E_a results in an increase in d_{15} as E_a increases. Therefore, assuming a constant d_{15} results in a conservative estimate in the shear actuation which can be obtained from the material.

$$d_{15} = 737.3411 + 159.699E_a + 102.653{E_a}^2 - 39.006{E_a}^3 + 5.495{E_a}^4. \tag{9.1}$$

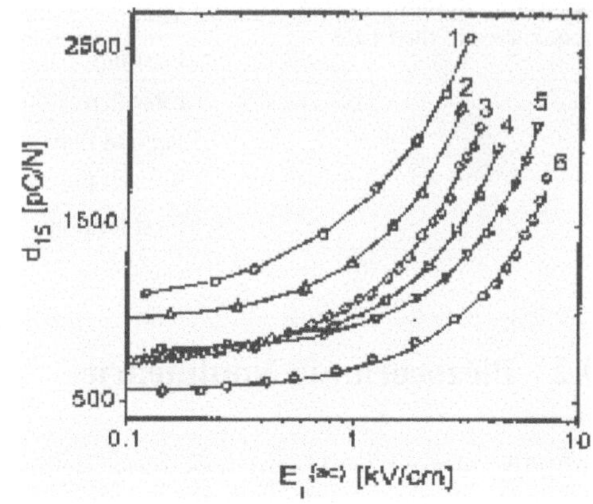

Fig. 9.2 Nonlinear relation between d_{15} and applied electric field

9.3 Open-Loop Control

Vibration reduction being the objective of this book, a parametric study over the variation in the objective function J is done on the application of a time varying 4P smart load. The objective function J used in the analysis can be defined as follows

$$J = \sqrt{(F_{xH}^{4P})^2 + (F_{yH}^{4P})^2 + (F_{zH}^{4P})^2} + \sqrt{(M_{xH}^{4P})^2 + (M_{yH}^{4P})^2 + (M_{zH}^{4P})^2}, \quad (9.2)$$

where the forces ($F_{(..)}$) and moments ($M_{(..)}$) are nondimensionalized with respect to $m_0\Omega^2 R^2$ and $m_0\Omega^2 R^3$, respectively. This form of objective function provides a useful way to convert the three forces and moments into one scalar number and has been used in optimization studies aimed at reducing vibration [5]. Investigation of the behavior of J for both bending and shear actuation is demonstrated in Figs. 9.3, 9.4, and 9.5. These plots indicate that the variation in J for bending actuation is less when compared to that of torsion actuation.

The behavior of J is studied for mainly three cases utilizing torsion actuation: (1) baseline case with no smart loads (2) with linear coefficient (3) with nonlinear coefficients. Same rotor properties and aerodynamic loading is assumed for all three cases. The applied field is of the form $E_a \sin(x\psi + \Phi)$ where x is 3/rev, 4/rev, or 5/rev. So, the applied smart loading is 3/rev, 4/rev, or 5/rev with a certain phasing. The phase Φ is found such that J is minimized, for a given E_a. The percentage reduction in J from the baseline case, both for linear coefficients (case 2) and nonlinear coefficients (case 3) is plotted in Fig. 9.6. It could be deduced from this figure that the reduction in vibration obtained at 350 V in the linear case is achievable at 204 V in the nonlinear case. So, the factor is around 0.58 times of the amplitude of the voltage required in the linear case. Hence, if the nonlinear law is used, applied voltage could be reduced to quite an extent. Using the nonlinear relation, amplitude of the applied voltage required to achieve a certain vibration reduction is much less compared to that in linear case.

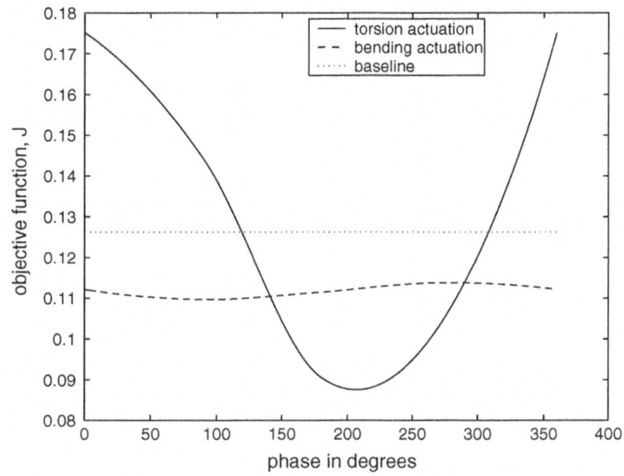

Fig. 9.3 Variation of J with phase at 3/rev smart load for torsion and bending actuation

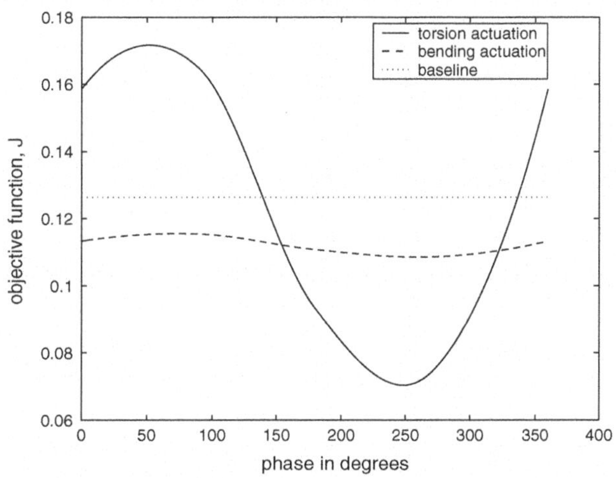

Fig. 9.4 Variation of J with phase at 4/rev smart load for torsion and bending actuation

The peak of the curve of the nonlinear variation (high reductions) corresponds to the regime between 275–325 V. On an increase in amplitude of the applied potential in this regime, very less variation is visible correspondingly in J. Highest reduction being 51.4 % corresponding to an amplitude of 300 V. The comparisons of F_x, F_y, F_z, M_x, M_y, and M_z with smart loading (300 V with nonlinear shear coefficients) and baseline is shown in Fig. 9.7. Here, these forces and moments are normalized by rotor thrust and rotor torque, respectively. Longitudinal, lateral, and vertical hub loads have reduced by about 29.7, 18.2 and 15.8 %, respectively and roll, pitch, and yaw moments have reduced by 40.2, 83.8, and 2.6 %, respectively.

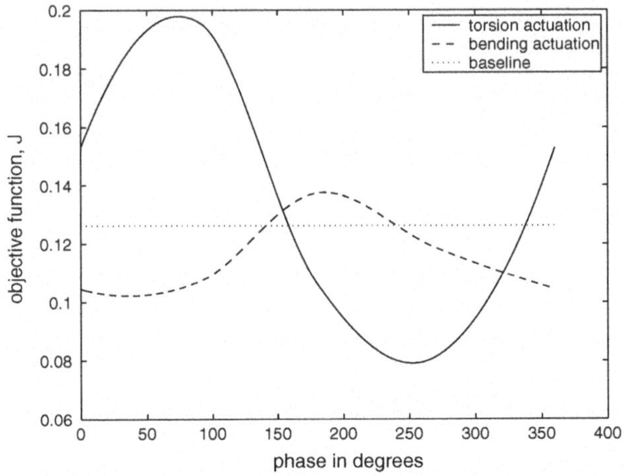

Fig. 9.5 Variation of J with phase at 5/rev smart load for torsion and bending actuation

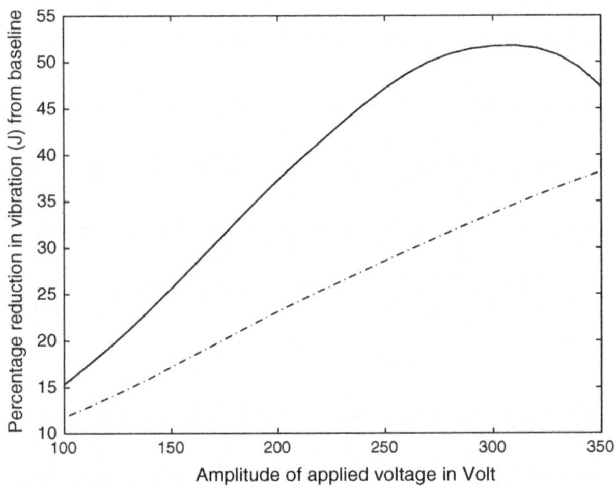

Fig. 9.6 Variation in reduction of vibration with increase in amplitude of voltage-open-loop control

9.4 Closed-Loop Control

The above discussion was based on applying a sinusoidal voltage to the rotor blades. Following is the analysis based on feedback control in which the shear strain rate is sensed and then fed to the actuator, resulting into a smart loading. One of the main advantages of smart structure actuation is the possibility of development of a controller based on measurements in the rotating system. This allows the development of a fixed gain controller. In addition, the self-sensing capability of piezoceramic actuators can be exploited. The upper and lower bounds for the amplitude of the

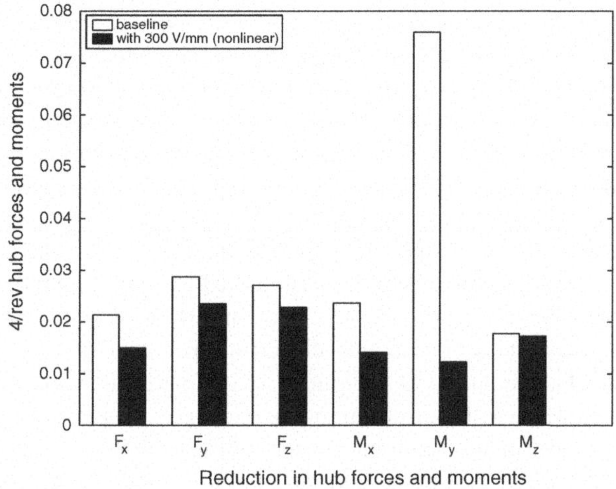

Fig. 9.7 Reduction in hub forces and moments with nonlinear coefficient

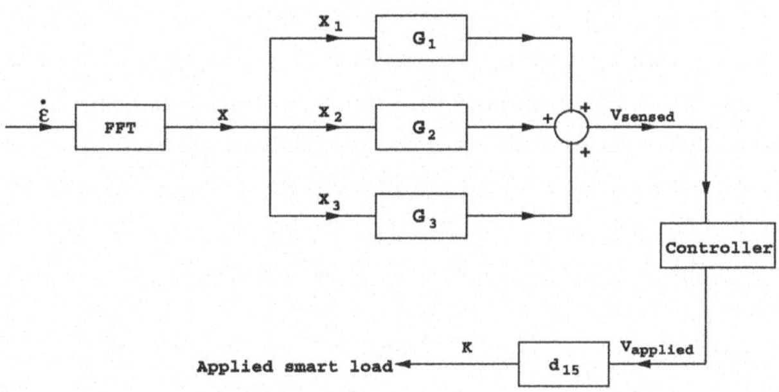

Fig. 9.8 Control—schematic

applied potential are kept as ±400 V, keeping in mind the depoling limit. A simple control algorithm (Fig. 9.8) describes the method by which the closed-loop control works.

The algorithm senses the time rate of change of strain, locations for the sensors could be anywhere on the rotor blade. This being a finite element formulation, Gauss points can be selected as locations for measuring the strain rate. Here, $4l_i/7$ in the second element from the root is selected as the location for the sensor. Here l_i is the elemental length.

It is known that maximum strain exists near the root element. Sensor location studies for active structural response have been performed by [6]. After the time rate of change of strain ($\dot{\varepsilon}$) is obtained, Fast Fourier Transform (FFT) over this is done to obtain the coefficients corresponding to different harmonics. Only 3/rev, 4/rev, 5/rev components are considered as these contain information about the higher harmonic forcing experienced by the rotor. This is the initial step of the control algorithm.

Second is to find out sensed voltage as a function of the gains and the above-mentioned coefficients. Here, three gains are calculated corresponding to 3/rev, 4/rev, 5/rev components to minimize the objective function J while keeping the maximum potential within ± 400 V to prevent depoling. Third is to supply the rms (root mean square) value of the sensed voltage to the nonlinear power law to compute the shear coefficients d_{15}. Finally, after these coefficients are calculated, the smart loading to be applied on the rotor blade is calculated.

Finite element analysis in space and time is done to obtain the corresponding objective function J (measure of vibration) and the hub loads and moments. For the best case comparisons of F_x, F_y, F_z, M_x, M_y, and M_z with smart loading (sensed voltage with nonlinear shear coefficients, closed-loop control) and baseline is shown in Fig. 9.9 for an advance ratio of 0.3. Figure 9.10 compares the variation of torsional deflection at the blade tip with azimuth angle for baseline and closed-loop control. The controller changes the higher harmonic content of the torsion response and by doing so reduces the helicopter vibration. Longitudinal hub load has increased by 3.8 % while lateral and vertical hub loads have reduced by about 1.9 and 18 %, respectively. Roll and pitch moments have reduced by 6 and 89 and yaw moment has increased by 3 %, respectively. An overall reduction of 43 % is obtained. Reduction in hub forces and moments for an advance ratio of 0.2 are shown in Fig. 9.11. Figure 9.12 compares the variation of torsional deflection at the blade tip with azimuth angle for baseline and closed-loop control. In this case, the controlled response shows a considerable increase in higher harmonic content compared to the baseline blade. Note that while the three forces and the yawing moment are less for $\mu = 0.2$ compared to $\mu = 0.3$, the roll and pitch 4/rev moments are larger, requiring a different type of active twist for its suppression. A notable reduction of 30 and 83 % is obtained in roll and pitch moments respectively, with an overall reduction in vibration of 59 %.

In this chapter, the controller was tuned for optimum gains at the different advance ratios. The physics of helicopter flow being complicated, the aeroelastic and nonlinear effects governing helicopter vibration are different at different forward speeds. This requires different controller gains at different forward speeds. However, it should be noted that these gains are fixed and not variable. The implementation of active twist control on a helicopter rotor would therefore require controller gains to be obtained for a variety of flight conditions within the helicopter flight envelope. In this way, it is similar to the implementation of the IBC concept. Chattopadhyay et al. [7] mentions that, in general, among all six vibratory hub loads, the pitching moment and the rolling moment contribute most to the vehicle vibration. As seen from Figs. 9.9 and 9.11 the pitching moment is dominant among the 4/rev hub loads at $\mu = 0.3$ and the roll and pitch moments are dominant at $\mu = 0.2$. The choice of the objective function J is therefore useful for reducing the dominant 4/rev loads as can be seen from the results. The results in Figs. 9.9, 9.10, 9.11 and 9.12 clearly show that the controller gains can be tuned to different flight conditions and give considerable vibration reduction using piezoceramic actuation.

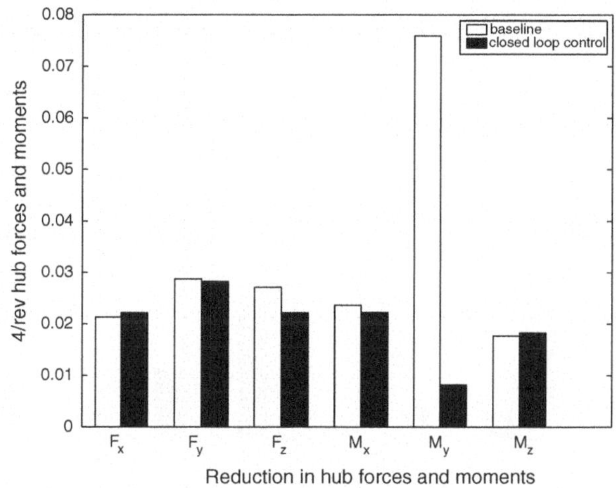

Fig. 9.9 Reduction in hub forces and moments with closed-loop control at advance ratio of 0.3

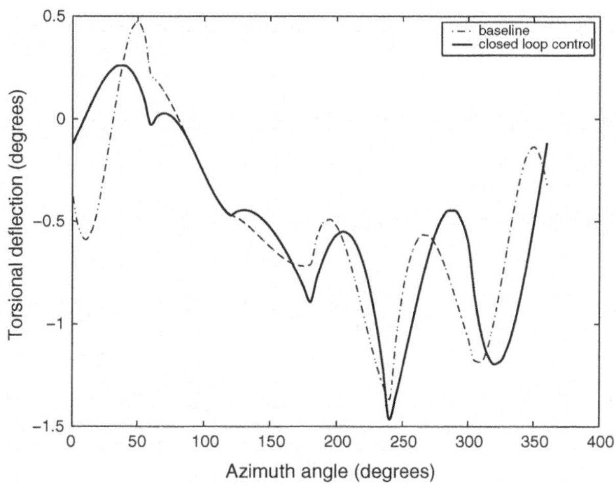

Fig. 9.10 Torsional deflection with closed-loop control at advance ratio of 0.3

9.5 Summary

Based on the governing equations obtained for smart helicopter rotor blades with surface bonded piezoceramic actuators, vibration reduction, piezoceramic nonlinearity, open-loop control, and closed-loop control have been investigated. The nonlinear relationship between the piezoelectric shear coefficient and applied ac-field is represented as a polynomial curve fit. The nonlinear effects are investigated by applying a sinusoidal voltage to the helicopter rotor blade. The rotor blade is modeled as a two-cell box section with piezoelectric layers surface bonded to the top and bottom of the

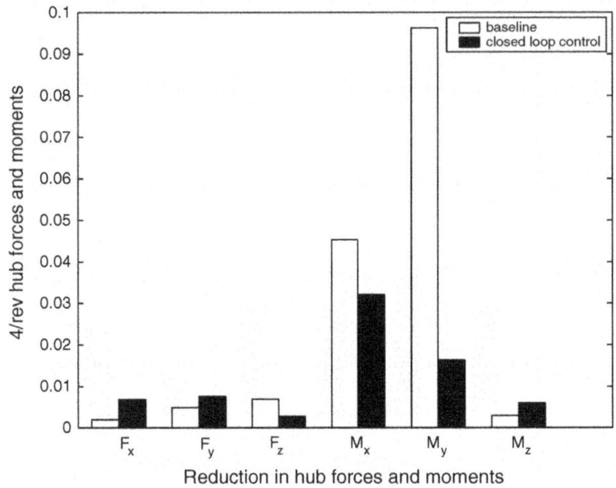

Fig. 9.11 Reduction in hub forces and moments with closed-loop control at advance ratio of 0.2

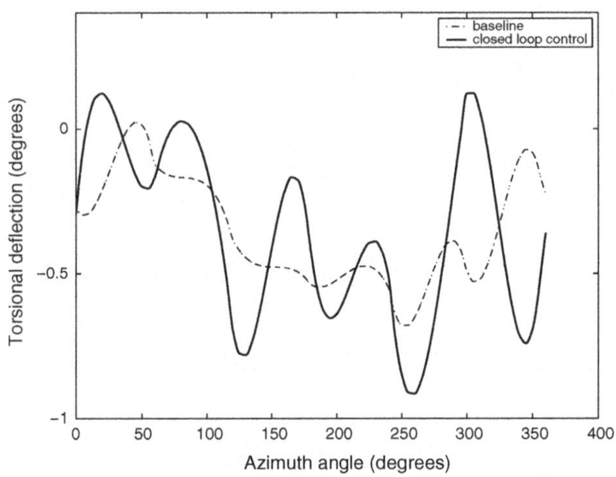

Fig. 9.12 Torsional deflection with closed-loop control at advance ratio of 0.2

blade. Comparison of results with linear and nonlinear shear coefficients is presented. Use of nonlinear relationship (compared to linear) to achieve targeted reductions in strains or displacements result in a reduction in the requirement of applied amplitude of the sinusoidal field. A strain rate feedback control law is implemented which feeds back the higher harmonics of the time rate of change of strain in azimuthal direction. The sensed voltage is then applied to the rotor blade resulting in a vibration reduction for a 4-bladed, soft-inplane hingeless rotor in forward flight. The numerical results obtained in this chapter at different forward speeds clearly show the feasibility of using shear (d_{15}) based actuation for helicopter vibration control.

References

1. Jung, S., Park, I.: Structural behavior of thin- and thick-walled composite blades with multicell sections. J. Am. Inst. Aeronaut. Astronaut. **43**(3), 572–581 (2005)
2. Thakkar, D., Ganguli, R.: Helicopter vibration reduction in forward flight with induced shear based piezoceramic actuation. Smart Mater. Struct. **30**(3), 599–608 (2004)
3. Li, S., Cao, W., Cross, L.: The extrinsic nature of nonlinear behavior observed in lead zirconate titanate ferroelectric ceramic. J. Appl. Phys. **69**(10), 7219–7224 (1991)
4. Mueller, V., Beige, H.: Nonlinearity of soft pzt piezoceramics for shear and torsional actuator applications. In: Proceedings of the IEEE International Symposium on Application of Ferroelectrics, IEEE catalog number: 98CH362455, pp. 459–462 (1998)
5. Ganguli, R.: Optimum design of a helicopter rotor for low vibration using aeroelastic analysis and response surface model. J. Sound Vib. **258**(2), 327–344 (2002)
6. Venkatesan, C., Udayasankar, A.: Selection of sensor locations for active vibration control of helicopter fuselages. J. Aircraft **36**(2), 434–442 (1999)
7. Chattopadhyay, A., Walsh, J.L., Riley, M.F.: Integrated aerodynamic load/dynamic optimization of helicopter rotor blades. J. Aircraft **28**(1), 58–65 (1991)

Chapter 10
Airfoil-Section Rotor Blades

The work done in the previous chapter is further extended to the design of a two-cell airfoil cross-section which is a more realistic model of the helicopter rotor blade. Optimum actuator placement, the effect of the current active twist concept on key characteristics such as blade loads, aeromechanical stability, and trim controls is investigated. Aeroelastic analysis of the rotor blade is very important in the case of any refinement in the structural or aerodynamic model. Since the incorporation of piezoceramic changes the blade structural characteristics, it is important to ensure that the benefits gained due to the refinement should not be detrimental to the other aspects of the rotor such as performance, trim, and stability. Section 10.1 describes the adopted design of the airfoil section its properties. Selected validation studies are performed in Sect. 10.2 for the basic performance of the adopted model. Basic performance includes the investigation for the main rotor power coefficient. Section 10.3 discusses the optimum placement of the actuators along the blade span at advance ratios of 0.2 and 0.3. The yardstick used to determine the optimum placement is the reduction in vibration. In Sect. 10.4, the influence of active twist on flap, lag, and torsion response of the blade is discussed. The amount of twist generated due to the induces shear mechanism is depicted. Section 10.5 discusses the variation in the rotating frame blade root loads and the first six harmonics of the blade due to active twist. The reduction in hub loads due to active twist is discussed in Sect. 10.6. Aeromechanical stability determines whether the system becomes unstable due to the introduction of the new piezoceramic material and is discussed in Sect. 10.7. In Sect. 10.8, the change in the rotor power and trim angles due to active twist is presented. Last, Sect. 10.9 summarizes the work done in this chapter.

10.1 Model Details and Structural Properties

Although many studies consider helicopter blades to be box beams, a more realistic model is to consider it as an airfoil section [1]. Typically, helicopter rotor blades are thin-walled sections except near the blade root where they are thick-walled sections.

R. Ganguli et al., *Smart Helicopter Rotors*, Advances in Industrial Control,
DOI 10.1007/978-3-319-24768-7_10

Table 10.1 Rotor properties considered for simulation

c/R = 0.075	$C_T/\sigma = 0.0708$	Root cutout = 20.45 %
Number of blades, N = 4	Radius, R = 6.6 m	$\beta_p = 2.5°$
$\theta_{tw} = -11.20°$	$\mu = 0.3$	$\sigma = 0.096$
$\gamma = 9.98$	$\Omega = 32.88$ rad/sec	

However, for preliminary design, a thin-walled cross-section model of the blade is approximate. To avoid any unrealistic results due to modeling the blade root as a thin-walled section, we do not consider the root location for actuator placement. A NACA 0012 airfoil section with a two-cell internal structure is developed such that it dynamically represents a real rotor. The first axial, flap, lag, and torsion rotating natural frequencies are matched with that of the Advanced Light Helicopter (ALH) rotor made by Hindustan Aeronautics Limited. The airfoil cross-section (chord = 50 cm) details are shown in Fig. 10.1. It can be observed that the piezoceramic layer is laid between outer skin and the spar in the cross-section, both on top and bottom. This prevents the wear and tear of the piezoceramic layer. The width of the piezoceramic layer is 5.75 cm and height is 0.1 cm. The piezoceramic layer is bonded to the surface of the airfoil by an adhesive layer of thickness 0.025 cm. Rotor blade being nonuniform, mass per unit length, airfoil thickness, and the width of the spar are variables along the span of the rotor blade. For the aerodynamic modeling, a free wake model developed by Bagai–Leishman [2] is used along with the reverse flow and unsteady aerodynamics model developed by Leishman and Beddoes. Blade and rotor characteristics are listed in Table 10.1. Note that the rotor blade has a large radius of 6.6 m and a high lock number of 9.98. In addition, the blade has pretwist and precone.

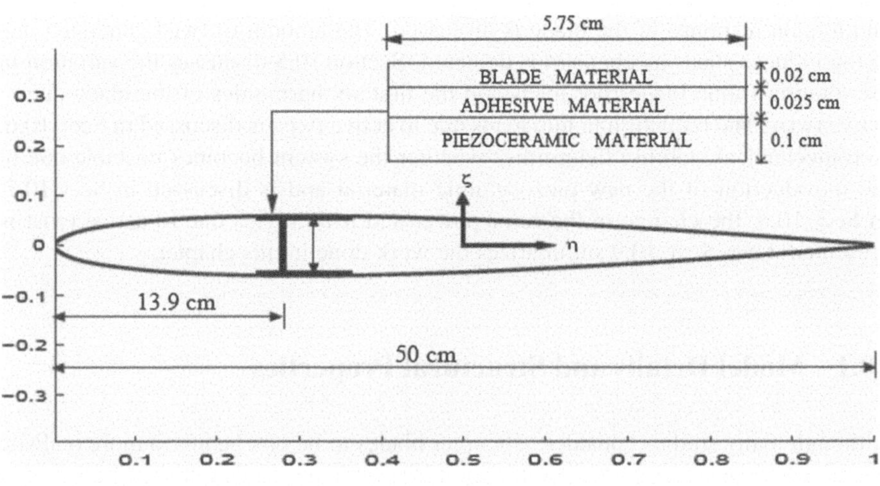

Fig. 10.1 Schematic of cross-section of rotor blade with piezoelectric actuators

Table 10.2 Rotating natural frequencies of the blade

Mode	Frequency (per rev)
First lag	0.68
First flap	1.09
Second flap	2.56
Second lag	3.92
Third flap	4.41
First torsion	4.66
Fourth flap	6.74
Fifth flap	9.76

The rotor blade is divided into five finite elements. The selected finite element is a 15 degree of freedom model, which is explained in detail in Chap. 2. The blade finite element structural properties are a smeared representation of the actual ALH rotor blade. Consequently, the spatial variation of elements is not same due to the dependency on the varying mass and stiffness distribution in the blade. To reduce computational time, the first eight normal modes are used for the modal analysis. Six time finite elements are used for discretization in the azimuthal direction. Stability analysis includes both rotor blade and fuselage degrees of freedom. Numerical results are obtained in forward flight at an advance ratio of $\mu = 0.2, 0.3$ and $C_T/\sigma = 0.0708$. The rotating blade natural frequencies calculated for such a rotor are listed in Table 10.2. We see that the frequencies represent a soft inplane hingeless rotor with a reasonably high-torsion frequency. For most rotors, torsion frequency ranges from 3/rev to 6/rev and has a critical role in active twist concept. The box beam blade considered in the last chapter had a first torsion frequency of 5.7/rev.

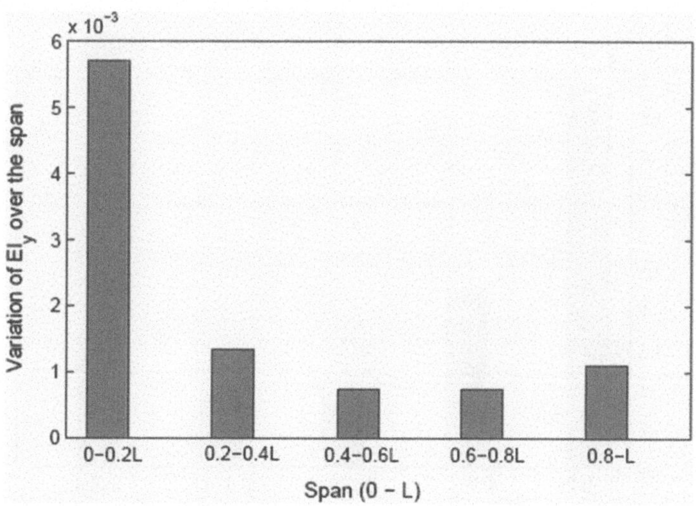

Fig. 10.2 Variation of EI_y for PZT-5H rotor

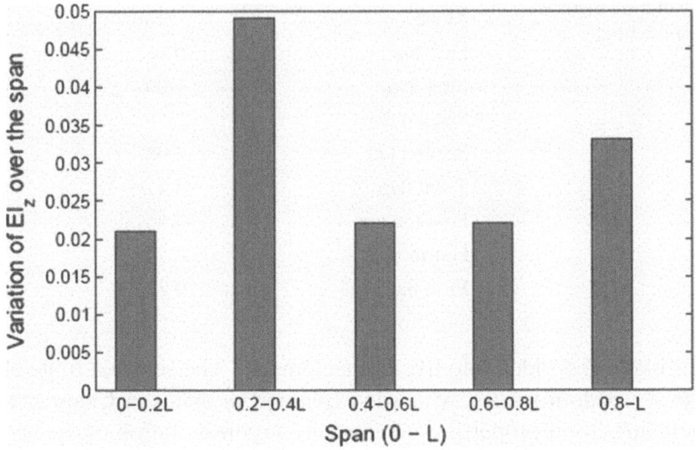

Fig. 10.3 Variation of EI_z for PZT-5H rotor

Span-wise stiffness and mass variation for the PZT-5H-based rotor are depicted in Figs. 10.2, 10.3, 10.4, 10.5, 10.6, 10.7, and 10.8. Chord/radius ratios for all elements except the first element from the root is kept same. The first element is smaller by 17 % than the other four elements. As seen from Fig. 10.2, the flap stiffness is highest at the root and decreases outboard. For the lag stiffness, the lowest stiffness occurs at the root as seen from Fig. 10.3. It could be seen from Fig. 10.4 and that the element spanning from 0.4–0.6L has the least torsion stiffness among all the elements from 0–L, while the torsion stiffness at the root is quite high. The mass distribution in

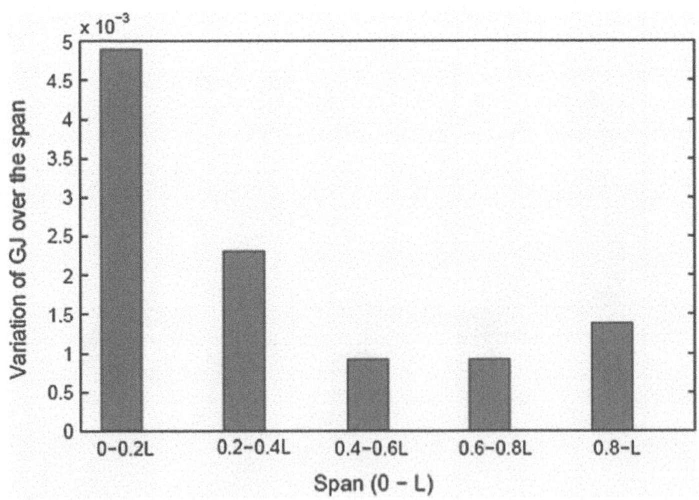

Fig. 10.4 Variation of GJ for PZT-5H rotor

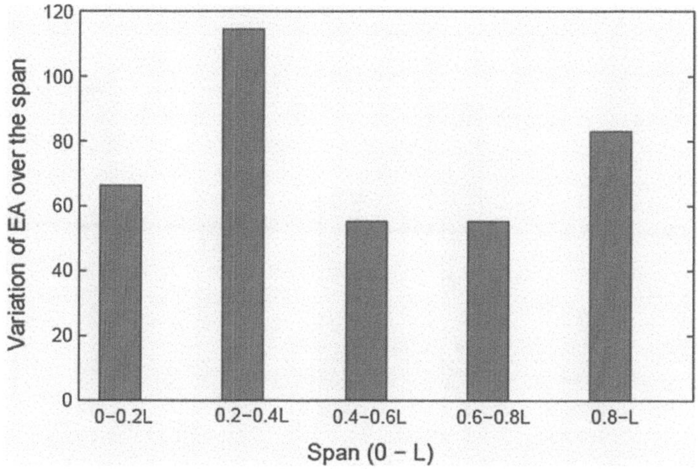

Fig. 10.5 Variation of EA for PZT-5H rotor

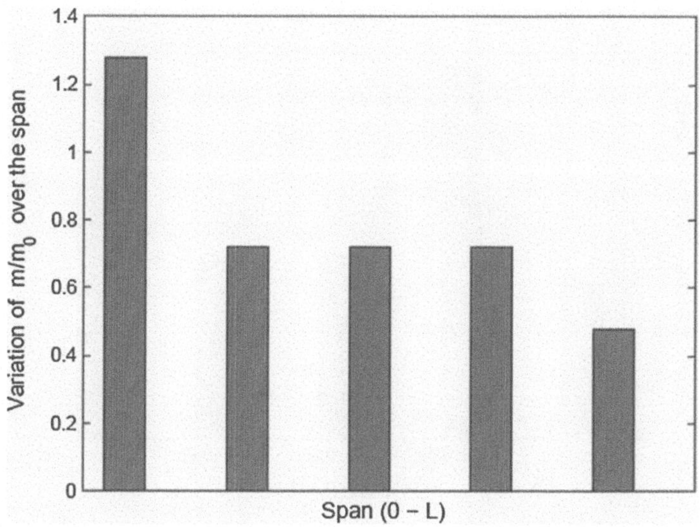

Fig. 10.6 Variation of m/m_0 for PZT-5H rotor

Fig. 10.6 shows high mass at the root and lowest mass at the tip. However, $km_1{}^2$ and $km_2{}^2$ show a low value at the root and at 0.4–0.8L regions and higher values at 0.2–0.4L and at the tip. The important point to note here is that for a real rotor blade, the mass and stiffness properties show a considerably nonuniform distribution along the blade span which can have considerable influence on blade dynamics and aeroelastic behavior.

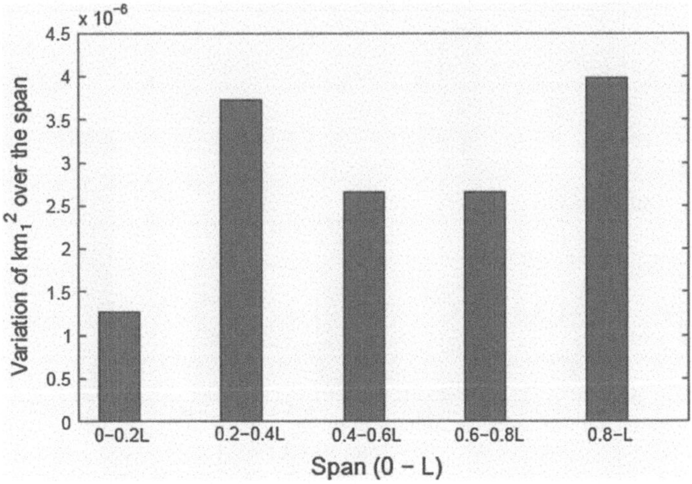

Fig. 10.7 Variation of $km_1{}^2$ for PZT-5H rotor

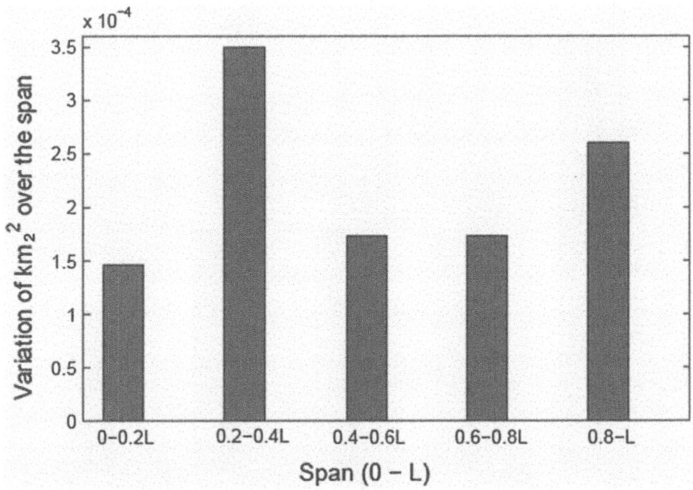

Fig. 10.8 Variation of $km_2{}^2$ for PZT-5H rotor

10.2 Validation Studies

For validating the overall aerodynamics and the rotor blade model, basic performance predictions at steady forward flight are compared with available flight test data for ALH. Main rotor power predictions are shown in Fig. 10.9 for nondimensional thrust coefficient C_T/σ of 0.05848 and 0.0708, respectively. Satisfactory trends in the power coefficients results are obtained at all flight speeds with a decline in the power required at lower speeds followed by an increase at very high speeds. The power required to hover as obtained from closed form relations is also shown. However, there is a slight underprediction at higher speeds and overprediction at low speeds but the trend is correctly captured.

Note, however, that the rotor used here is an academic approximation of a real rotor and has not been developed for a validation study. The purpose of the model is to ensure a rotor blade which is sufficiently similar to a real blade to allow use in a feasibility study.

10.3 Actuator Placement

Actuator placement is investigated with the help of numerical simulation and using the closed-loop controller discussed in the previous chapter. For the closed-loop controller shear strain rate is sensed and then fed to the actuator, resulting into a smart loading. The upper and lower bounds for the amplitude of the applied potential are kept as ±400 V. The control algorithm shown in Fig. 9.8 of Chap. 9 describes the method by which the controller works.

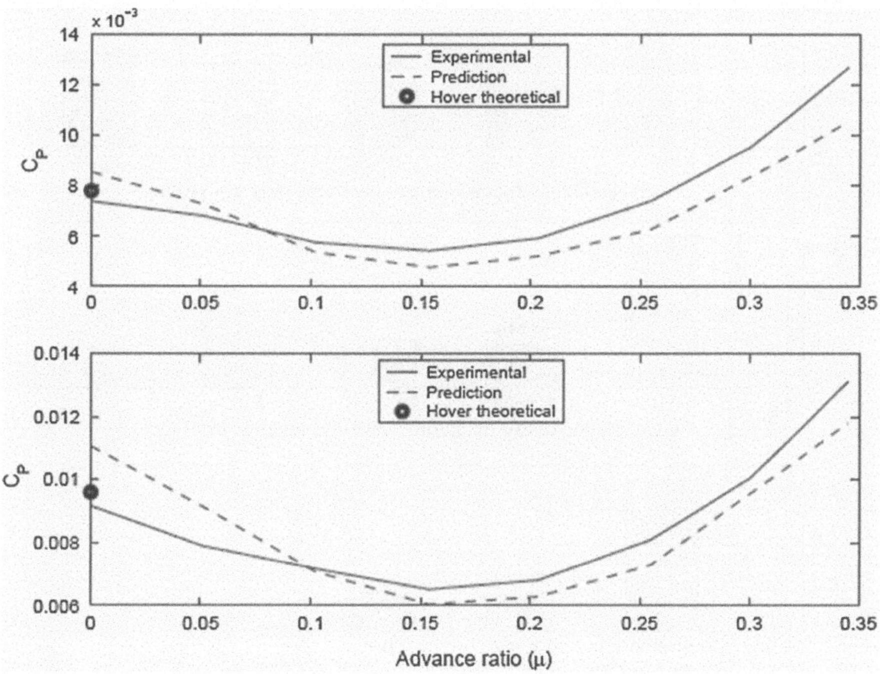

Fig. 10.9 Nondimensional main rotor power coefficient with respect to advance ratio

Optimal placement of actuators is investigated by comparing the amount of reduction obtained on varying the placement of actuators. Here placement means that a particular actuator is active, i.e., a voltage is applied to it and the nonactive actuator is dormant., i.e., no voltage is applied to it. Note, that removing an actuator physically would change the baseline blade dynamics due to mass and stiffness changes. As shown in Figs. 10.10 and 10.11, three sets of analysis were performed for achieving the optimum location of the actuators for advance ratios of 0.2 and 0.3.

The first set involves the placement of a single actuator in the rotor blade such that it attains maximum reduction in vibration. The first element from the root is not included in the analysis because of the difficulty of placement of actuator near the root area. The root area combines a thick-walled section leading to flexure-like structure. Housing of an actuator section in this area is difficult and is avoided to obtain a conservative design. Here, the blade is divided into five spatial elements (including the root-cutout area). Among the five spatial elements (0–L), in which the actuators could be placed, the third element from the root (0.4–0.6L) is found to be effective. This results in a vibration reduction of 45 and 36 percent for advance ratios of 0.2 and 0.3, respectively. The third element spans 40–60 % of the blade length from the root end. The second set in the analysis contributes to finding the efficient locations

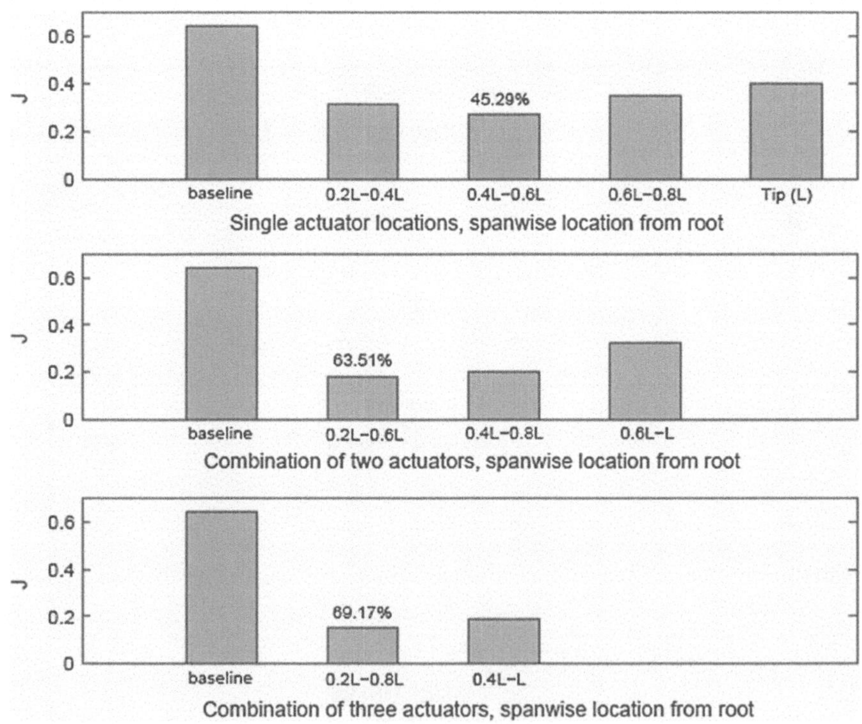

Fig. 10.10 Optimal location of actuators at advance ratio of 0.2

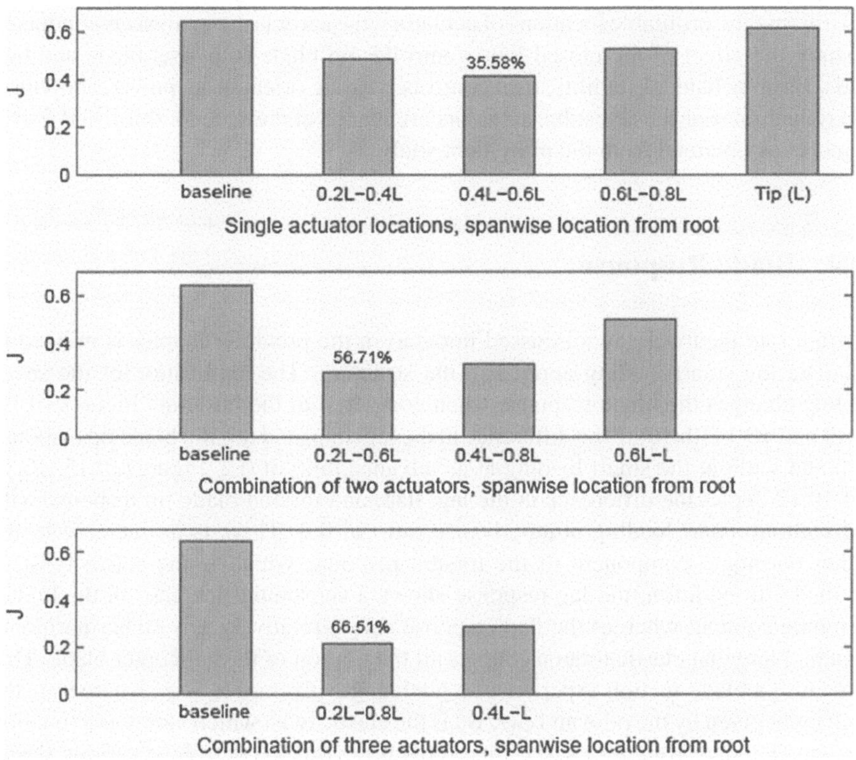

Fig. 10.11 Optimal location of actuators at advance ratio of 0.3

for two actuators. The placement of actuators in the second and third elements is found to be better than other locations, leading to a reduction in vibration of about 63.5 and 56.7 % for advance ratio of 0.2 and 0.3, respectively. The second and third elements span are 20–60 % of the blade length. The analysis for the last set involves the process of locating the best locations for three actuators. Placements in second, third, and fourth elements were found to be the most profitable in terms of vibration reduction. A vibration reduction of 69 and 66 % for advance ratio of 0.2 and 0.3, respectively, was achieved for this case. Here, the elements span are 20–80 % of the blade length. Apart from the amount of reduction, it is important to notice that with the increase in the number of actuators from one to two has a considerable difference in reduction when compared to increment from two to three. Hence this analysis helps in concluding that the positioning of actuators in the second, third, and fourth number of elements from the root is useful. In general, for the rotor studied here it appears that the center locations are better than the locations near the root or the tip. This could be because the torsional stiffness at these elements is lower than at the root, and these locations are structurally more significant than the tip elements.

Utilizing the profitable locations of actuators, the aeroelastic analysis is conducted to study the effect of the closed-loop controller on blade response, blade and hub loads, aeromechanical stability, trim controls, vehicle orientation, power, and vibration reduction. For these results, actuators are placed at the second, third, and fourth elements as obtained from the placement study.

10.4 Blade Response

A strain rate feedback law discussed in detail in the previous chapter is utilized to calculate the smart loading applied to the structure. The application of the smart loading changes the blade response when compared to the baseline. Figures 10.12, 10.13 and 10.14 illustrate the difference in the lag, flap, and torsion blade tip response with and without the smart loading at an advance ratio of 0.2. Figures 10.15, 10.16 and 10.17 depict the difference in the lag, flap, and torsion blade tip response with and without smart loading at an advance ratio of 0.3. There is an increase in the higher harmonic component of the torsion response which is the active twist of the blade. In addition, the lag response shows a substantial increase in the higher harmonic content, whereas the flap response shows relatively less higher harmonic content. Note that elastic torsion couples all the motion of the helicopter blade. This is because a blade section experiences a pitch angle given as $\theta_0 + \hat{\phi}$, where θ_0 is the pitch input given by the pilot and $\hat{\phi}(r, \psi)$ is the elastic twist which varies as a function of r and ψ. Controlling the twist $\hat{\phi}$ therefore allows one to control the various elastic displacements of the blade because the lift, drag, and pitching moment at a blade

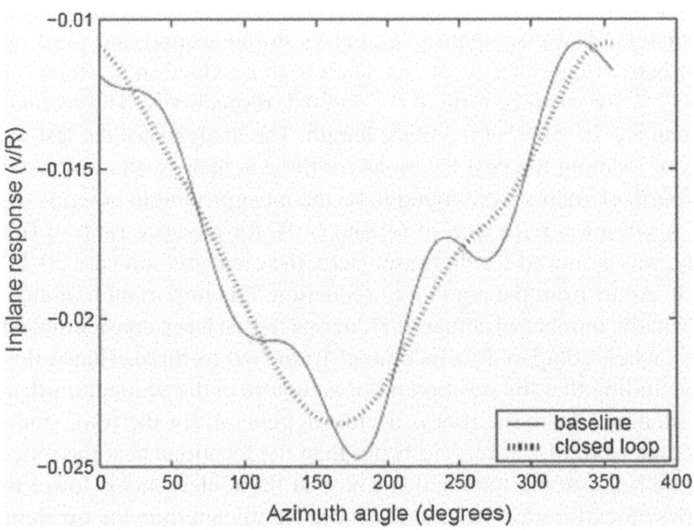

Fig. 10.12 Lag response at an advance ratio of 0.2 for baseline and closed-loop control cases

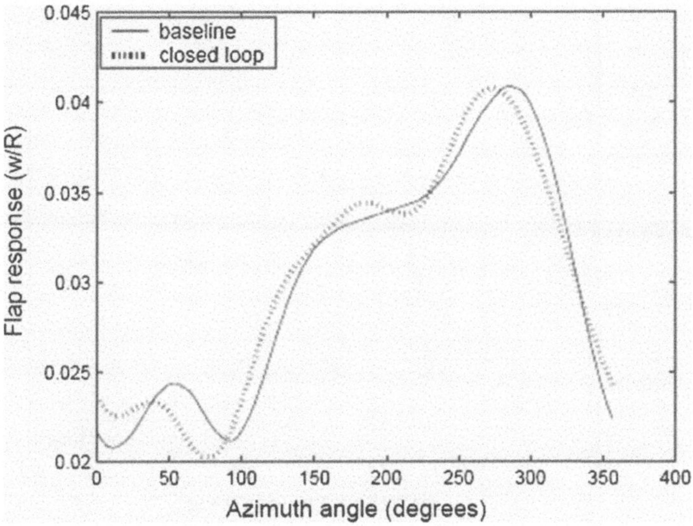

Fig. 10.13 Flap response at an advance ratio of 0.2 for baseline and closed-loop control cases

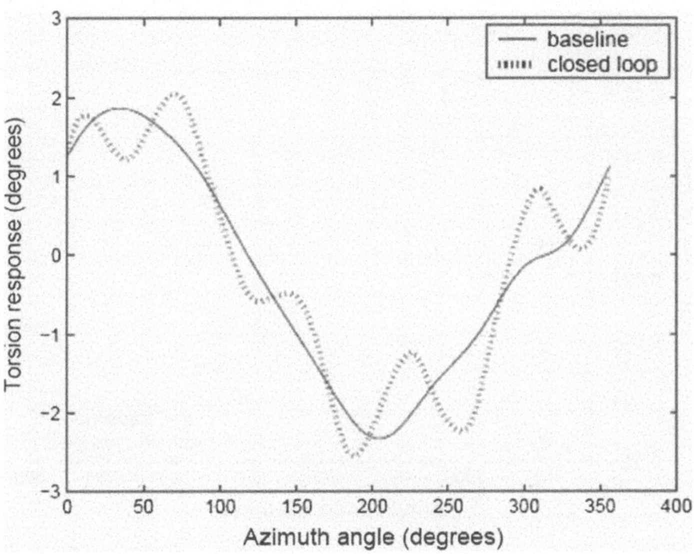

Fig. 10.14 Torsion response at an advance ratio of 0.2 for baseline and closed-loop control cases

section directly depend on the pitch. We also note that according to Nitzsche and Breitbach, the lag and torsion mode are critical to helicopter vibration. We also see that active twist is able to change the lag and torsion response and therefore effect the 4/rev loads. In general, the lag and torsion response are less sensitive to centrifugal

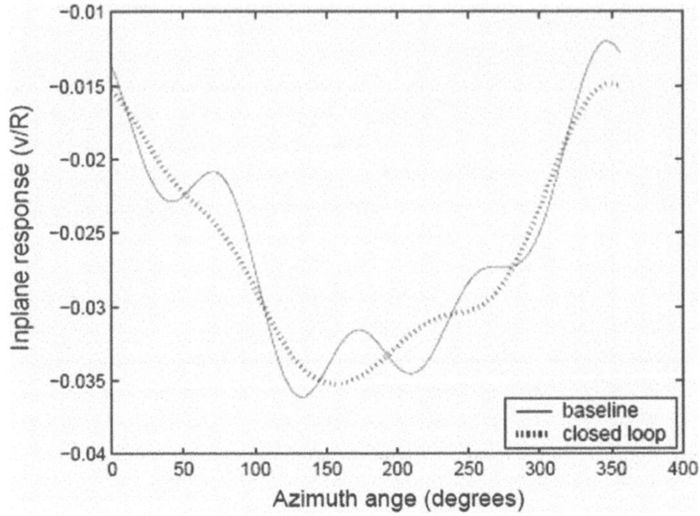

Fig. 10.15 Lag response at an advance ratio of 0.3 for baseline and closed-loop control cases

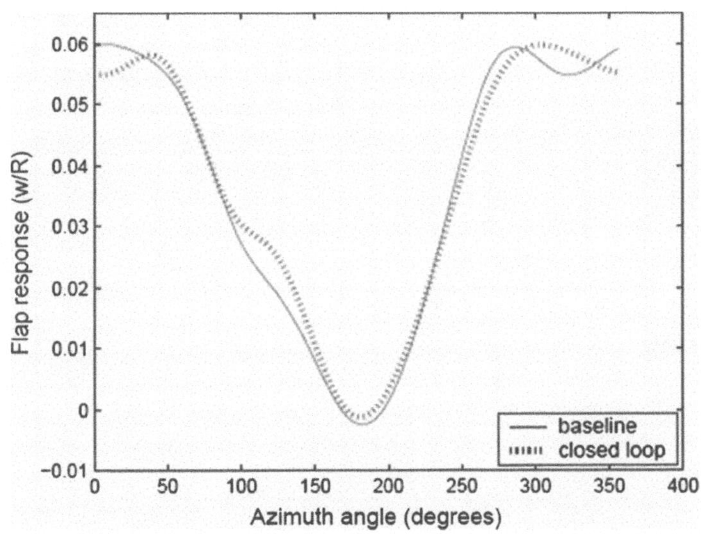

Fig. 10.16 Flap response at an advance ratio of 0.3 for baseline and closed-loop control cases

stiffening compared to flap response so changes in rotor aeroelastic behavior is easier to obtain by changing them.

Figures 10.18 and 10.19 show the active twist obtained (difference in the torsion response between baseline case and closed-loop control case) at advance ratio of 0.2 and 0.3, respectively. An active twist ranging from $-1°$ to $0.8°$ is obtained at $\mu =$

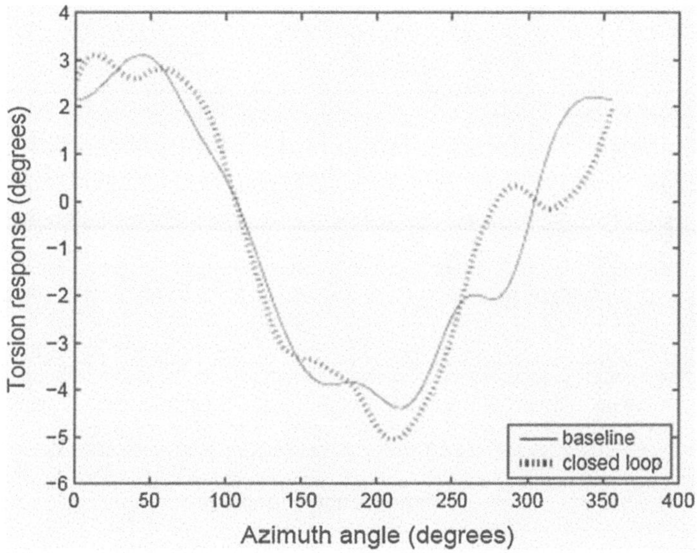

Fig. 10.17 Torsion response at an advance ratio of 0.3 for baseline and closed-loop control cases

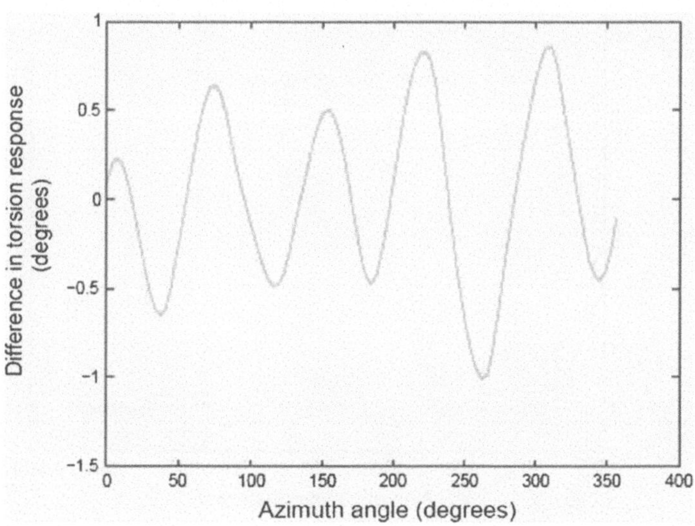

Fig. 10.18 Difference in torsion response at an advance ratio of 0.2 for baseline and closed-loop control cases

0.2 and ranging from $-2°$ to $2°$ is obtained at $\mu = 0.3$. In addition, the active twist consists of higher harmonics of the rotor speed which results in cancelation of the higher harmonic airloads acting on the rotor. The closed-loop controller inherently considers the nonlinear relation between the amplitude of the driving electric field and

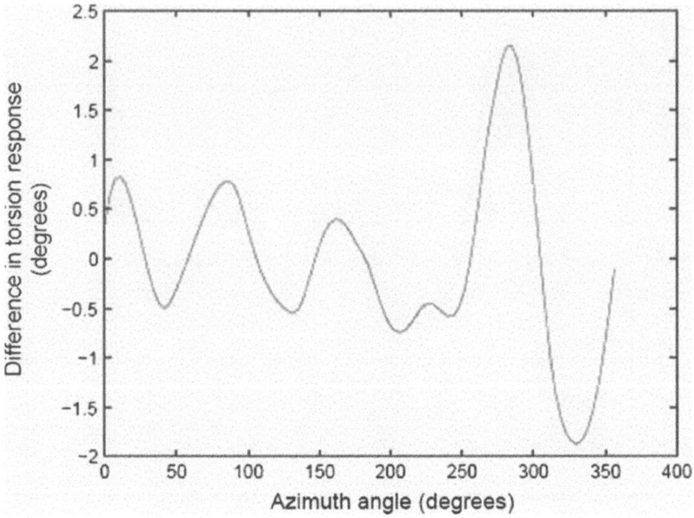

Fig. 10.19 Difference in torsion response at an advance ratio of 0.3 for baseline and closed-loop control cases

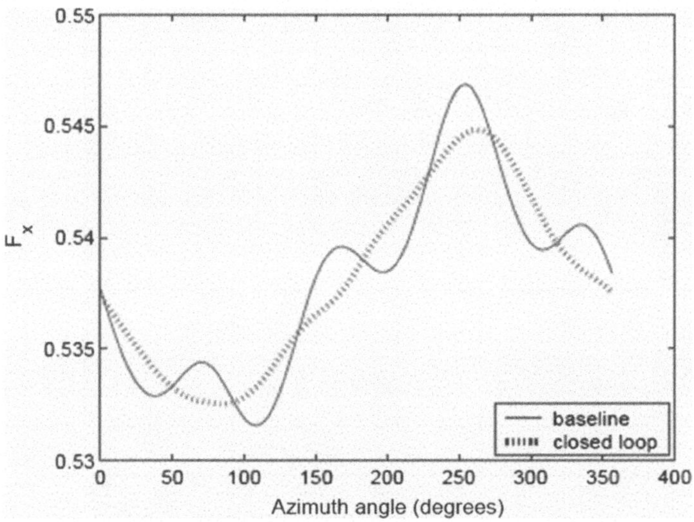

Fig. 10.20 Rotating frame blade root load F_x at an advance ratio of 0.2 for baseline and closed-loop control cases

the piezoelectric coefficient d_{15}. A strong nonlinear increase in d_{15} with the applied electric field results in considerable strain actuation resulting in a moderately high-active twist.

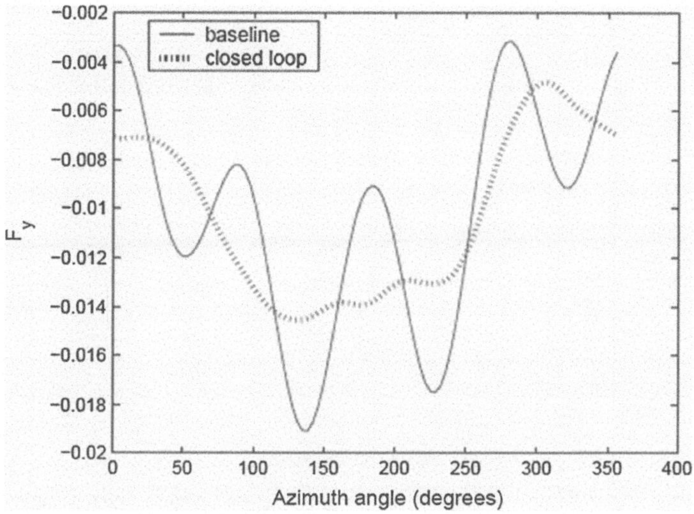

Fig. 10.21 Rotating frame blade root load F_y at an advance ratio of 0.2 for baseline and closed-loop control cases

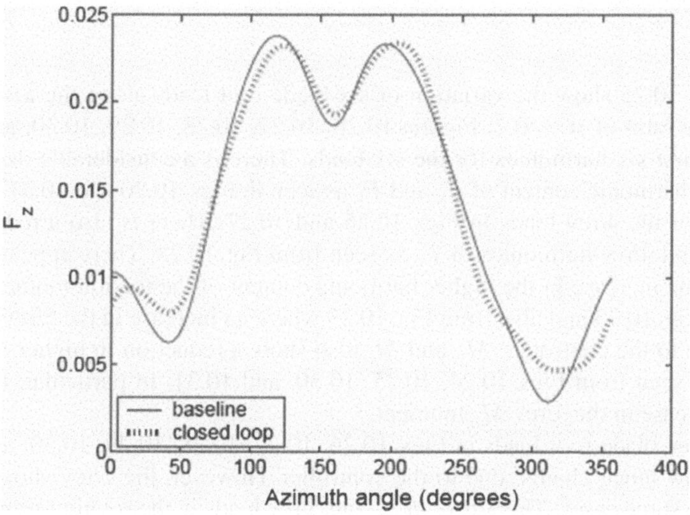

Fig. 10.22 Rotating frame blade root load F_z at an advance ratio of 0.2 for baseline and closed-loop control cases

10.5 Blade Loads

The rotating frame blade loads are first compared with and without feedback control for advance ratios of 0.2. This gives an idea regarding the variation in rotating frame hub loads when the controller is switched on. Figures 10.20, 10.21, 10.22, 10.23,

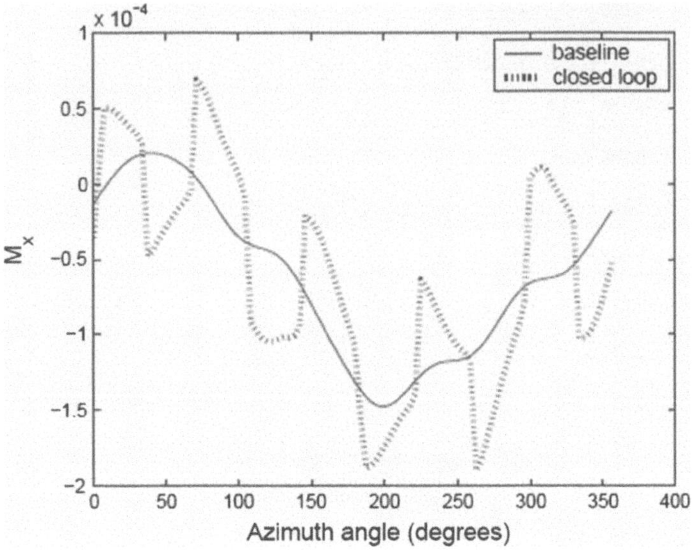

Fig. 10.23 Rotating frame blade root load M_x at an advance ratio of 0.2 for baseline and closed-loop control cases

10.24, and 10.25 show the variation of six blade root loads along the azimuth for an advance ratio of $\mu = 0.2$. Figures 10.26, 10.27, 10.28, 10.29, 10.30, and 10.31 show the first six harmonics for the six loads. There is a considerable decrease in the higher harmonic content of F_x and F_y as seen in Figs. 10.20 and 10.21 from the reduction of the 4/rev loads in Figs. 10.26 and 10.27. There is also a reduction in the 3/rev and 4/rev harmonics of F_z as seen from Fig. 10.28. There appears to be a considerable increase in the higher harmonic content of the torsion moment M_x as shown in Fig. 10.23 and also from Fig. 10.29 where an increase in the 5/rev moment occurs due to the controller. M_y and M_z also show a reduction in higher harmonic content as seen from Figs. 10.24, 10.25, 10.30, and 10.31. In particular, there is a larger decrease in the 4/rev M_z moment.

The 1/rev blade root loads in Figs. 10.26, 10.27, 10.28, 10.29, 10.30, and 10.31 do not show much change due to the controller. However, the 2/rev show a small increase in some cases. The 3/rev, 4/rev, and 5/rev loads in the rotating frame which are the source of 4/rev loads in the fixed frame are almost universally reduced with considerable reduction in the 4/rev F_x and F_y loads and the 4/rev M_z moment.

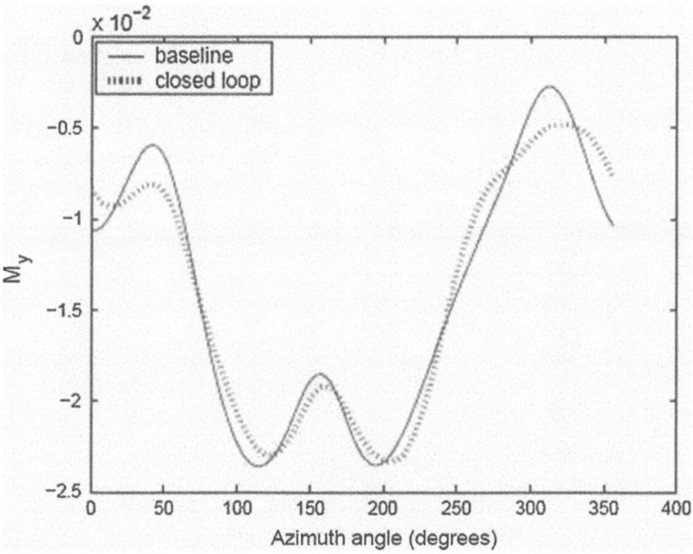

Fig. 10.24 Rotating frame blade root load M_y at an advance ratio of 0.2 for baseline and closed-loop control cases

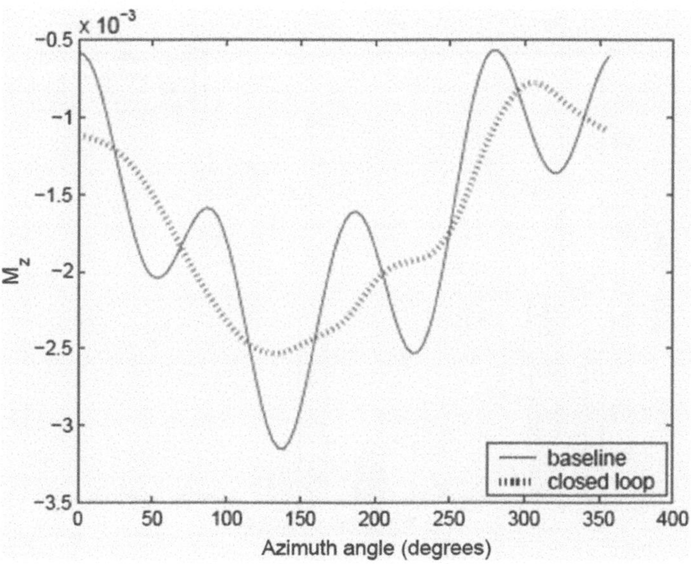

Fig. 10.25 Rotating frame blade root load M_z at an advance ratio of 0.2 for baseline and closed-loop control cases

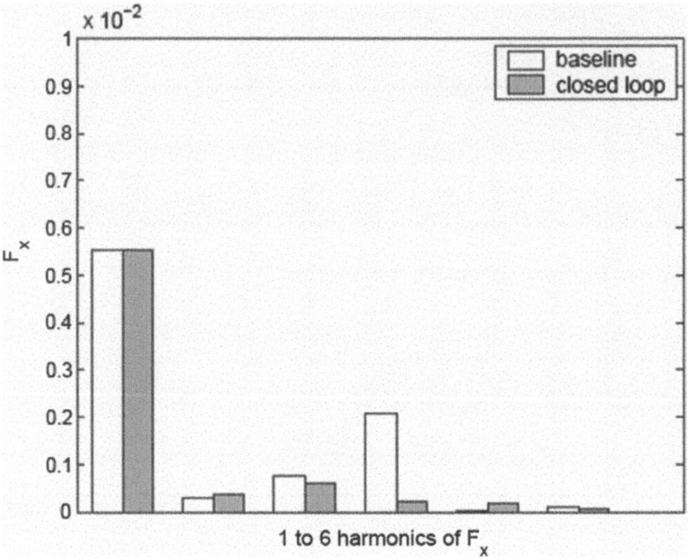

Fig. 10.26 Harmonics of blade root rotating frame load F_x at an advance ratio of 0.2 for baseline and closed-loop control cases

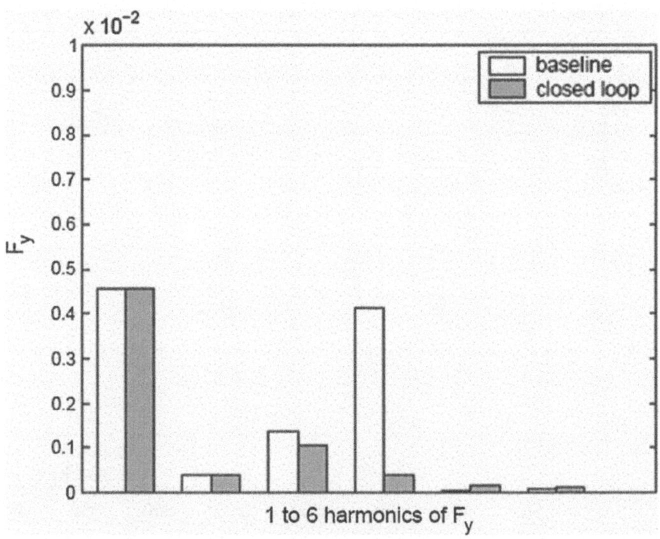

Fig. 10.27 Harmonics of blade root rotating frame load F_y at an advance ratio of 0.2 for baseline and closed-loop control cases

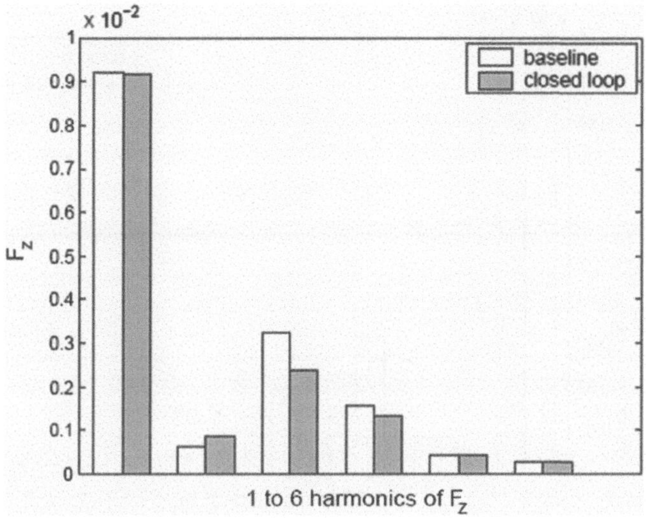

Fig. 10.28 Harmonics of blade root rotating frame load F_z at an advance ratio of 0.2 for baseline and closed-loop control cases

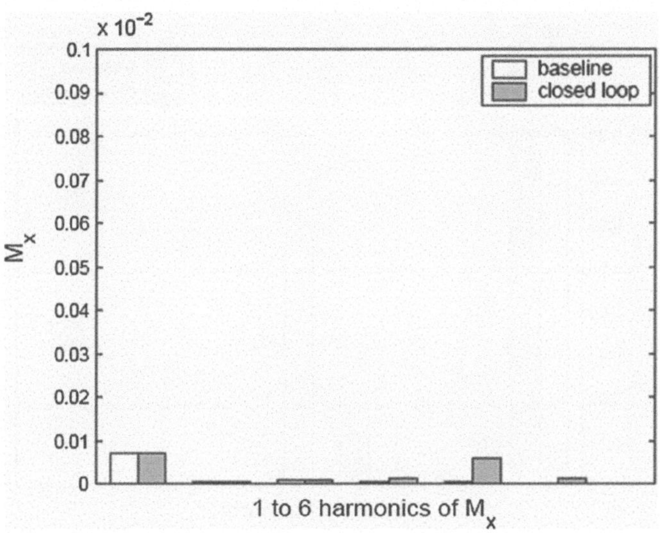

Fig. 10.29 Harmonics of blade root rotating frame load M_x at an advance ratio of 0.2 for baseline and closed-loop control cases

Next, results are obtained at $\mu = 0.3$. The six blade root loads are shown in Figs. 10.32, 10.33, 10.34, 10.35, 10.36, and 10.37 along the azimuth. Figures 10.38, 10.39, 10.40, 10.41, 10.42, and 10.43 show the first six harmonics of the six blade root loads. It can be seen that at $\mu = 0.3$, the 1/rev and 2/rev F_z and M_y loads

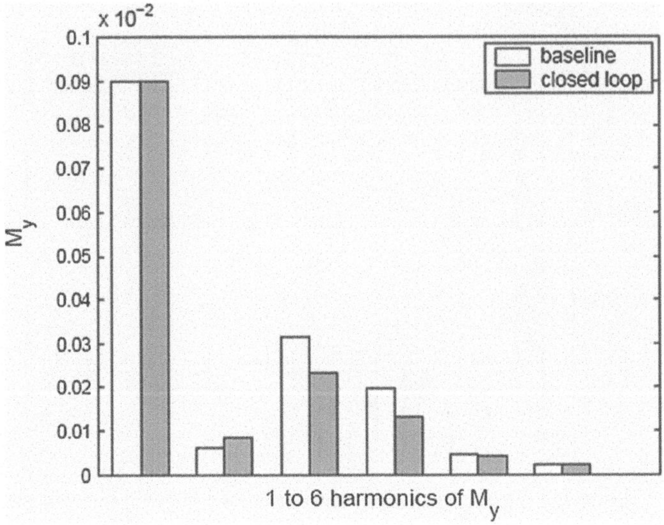

Fig. 10.30 Harmonics of blade root rotating frame load M_y at an advance ratio of 0.2 for baseline and closed-loop control cases

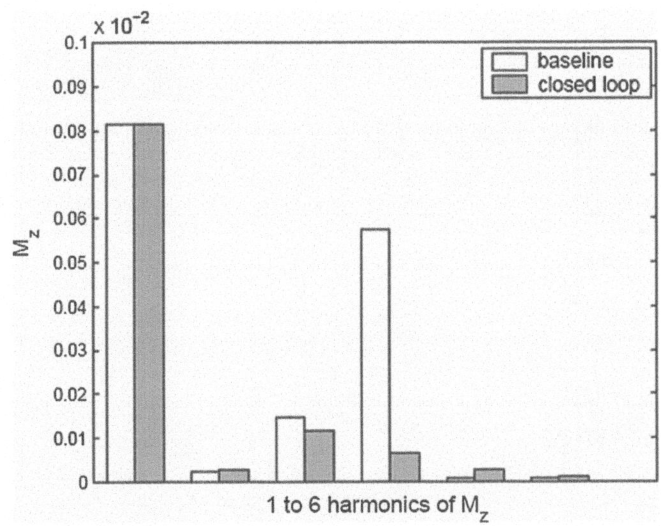

Fig. 10.31 Harmonics of blade root rotating frame load M_z at an advance ratio of 0.2 for baseline and closed-loop control cases

show an increase, which can result in high dynamic stresses at the blade root which is where the highest dynamic stresses occur for hingeless rotors. The 4/rev F_x and F_y loads and the 4/rev M_z moment are substantially reduced at $\mu = 0.3$ as shown in Figs. 10.38, 10.39, and 10.43, in a manner similar to the results at $\mu = 0.2$ in

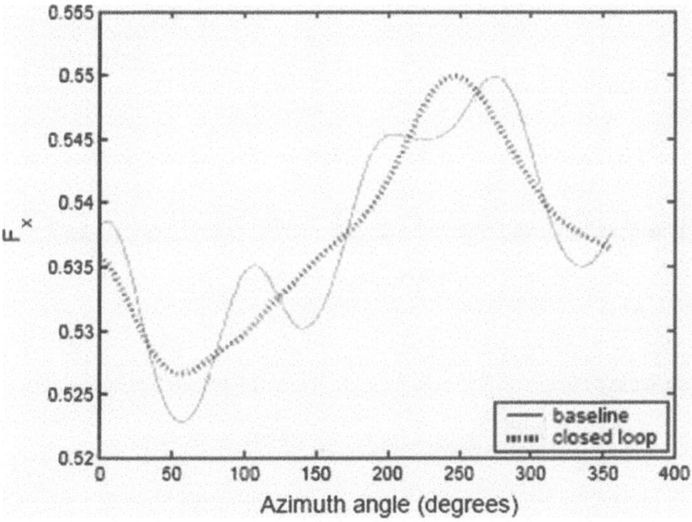

Fig. 10.32 Rotating frame blade root load F_x at an advance ratio of 0.3 for baseline and closed-loop control cases

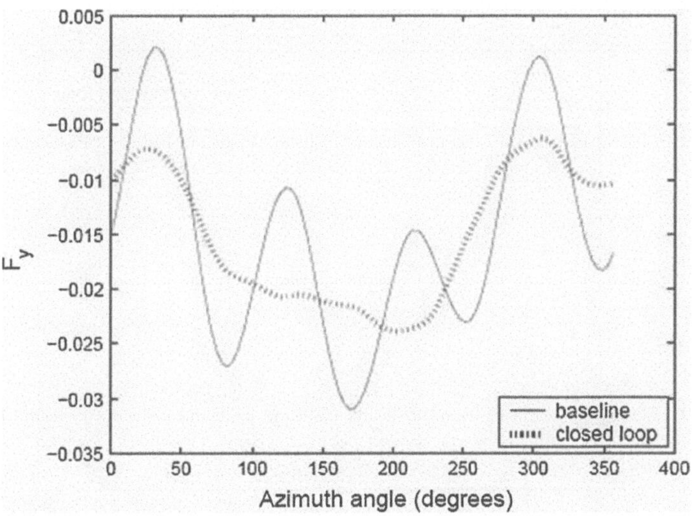

Fig. 10.33 Rotating frame blade root load F_y at an advance ratio of 0.3 for baseline and closed-loop control cases

Figs. 10.26, 10.27, and 10.31. The higher harmonic content of M_x, however, goes up as seen in Figs. 10.35 and 10.41. It should be noted that since the magnitude of M_x is much lesser than M_y and M_z for this rotor, there is an increase in the 4/rev moment. The active twist rotor therefore leads to higher harmonic torsional moments on the blade which can result in increased pitch-link loads. These issues of higher dynamic

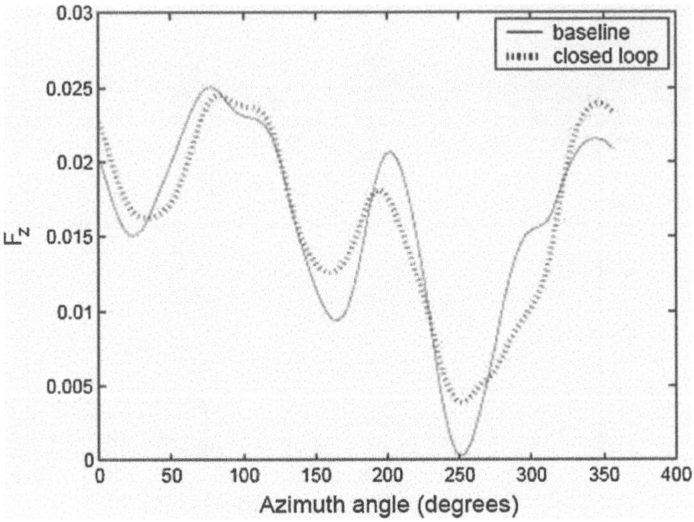

Fig. 10.34 Rotating frame blade root load F_z at an advance ratio of 0.3 for baseline and closed-loop control cases

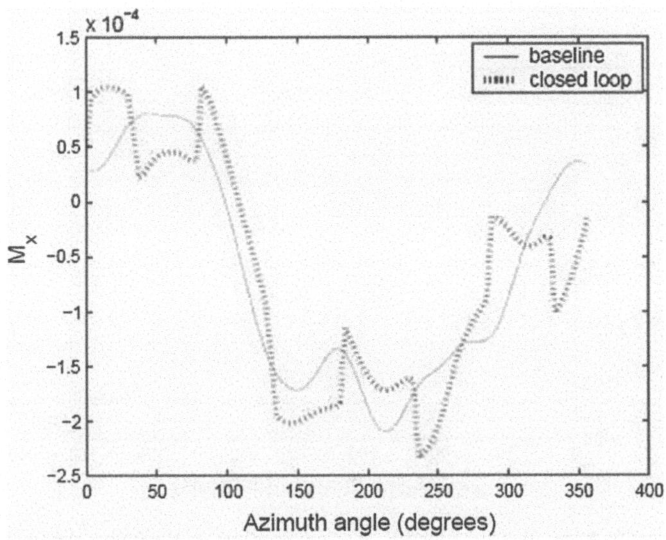

Fig. 10.35 Rotating frame blade root load M_x at an advance ratio of 0.3 for baseline and closed-loop control cases

stresses and increased pitch-link loads need to be addressed during the structural design of an active twist rotor.

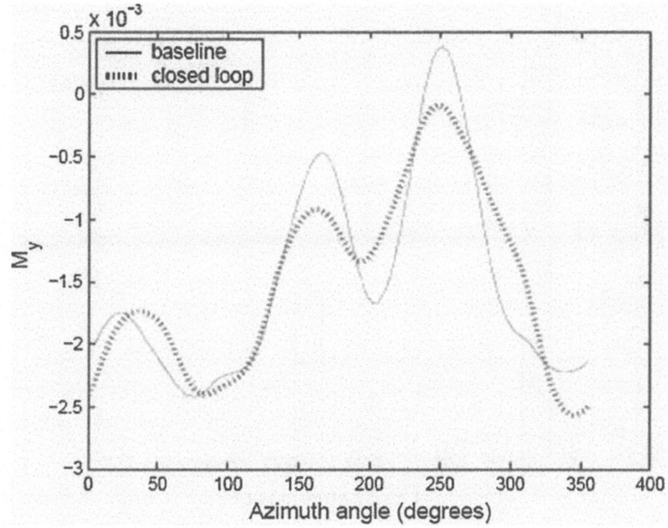

Fig. 10.36 Rotating frame blade root load M_y at an advance ratio of 0.3 for baseline and closed-loop control cases

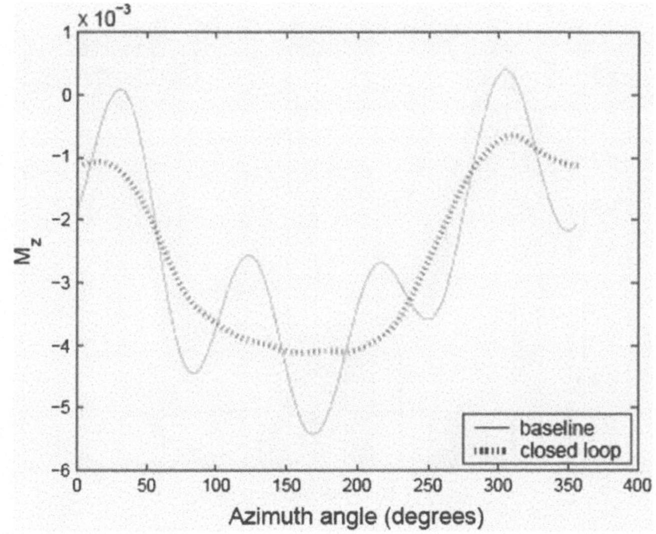

Fig. 10.37 Rotating frame blade root load M_z at an advance ratio of 0.3 for baseline and closed-loop control cases

10.6 Hub Loads

Minimization of the objective function J is done to calculate controller gains. The objective J was discussed in detail in Sects. 9.3 and 9.4 of the previous chapter and

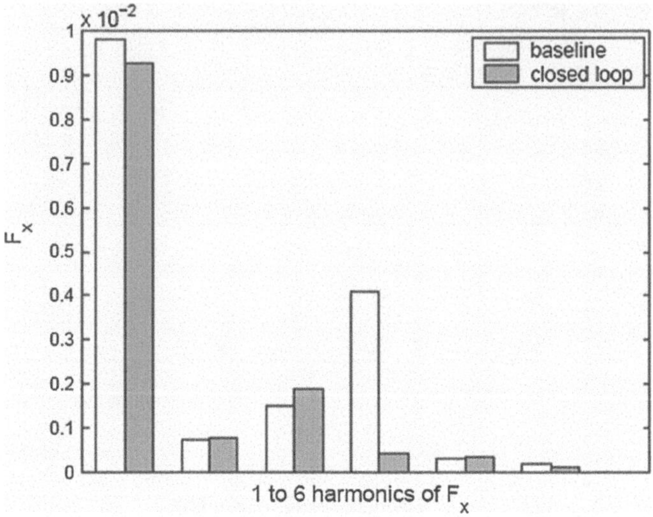

Fig. 10.38 Harmonics of blade root rotating frame load F_x at an advance ratio of 0.3 for baseline and closed-loop control cases

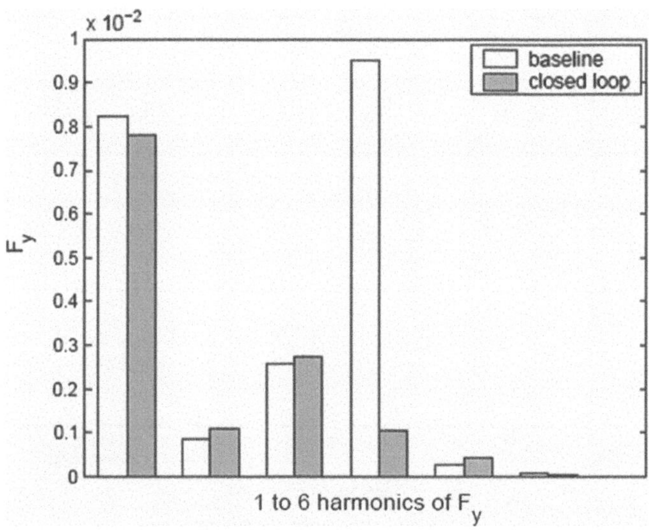

Fig. 10.39 Harmonics of blade root rotating frame load F_y at an advance ratio of 0.3 for baseline and closed-loop control cases

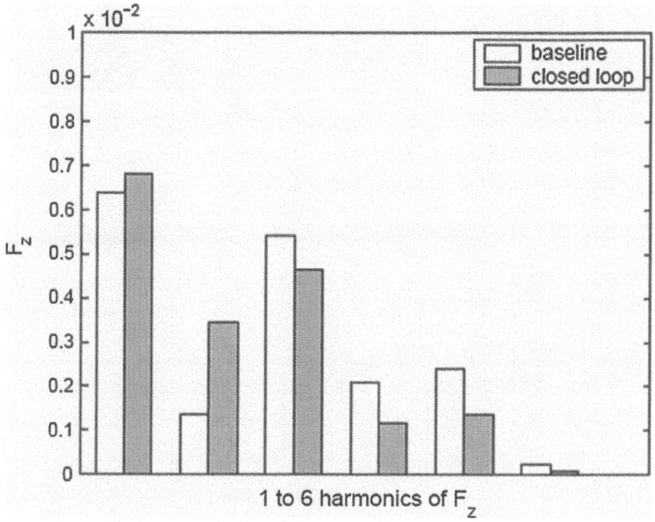

Fig. 10.40 Harmonics of blade root rotating frame load F_z at an advance ratio of 0.3 for baseline and closed-loop control cases

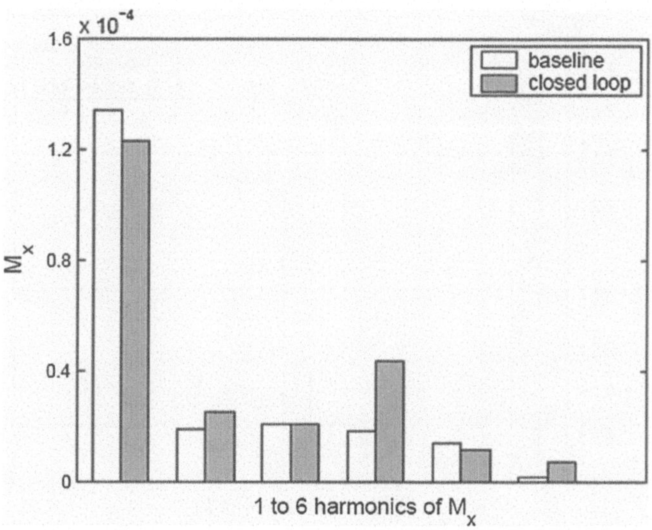

Fig. 10.41 Harmonics of blade root rotating frame load M_x at an advance ratio of 0.3 for baseline and closed-loop control cases

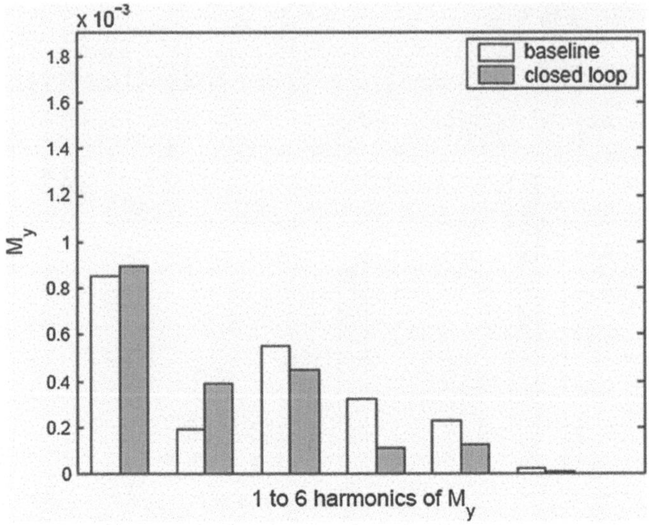

Fig. 10.42 Harmonics of blade root rotating frame load M_y at an advance ratio of 0.3 for baseline and closed-loop control cases

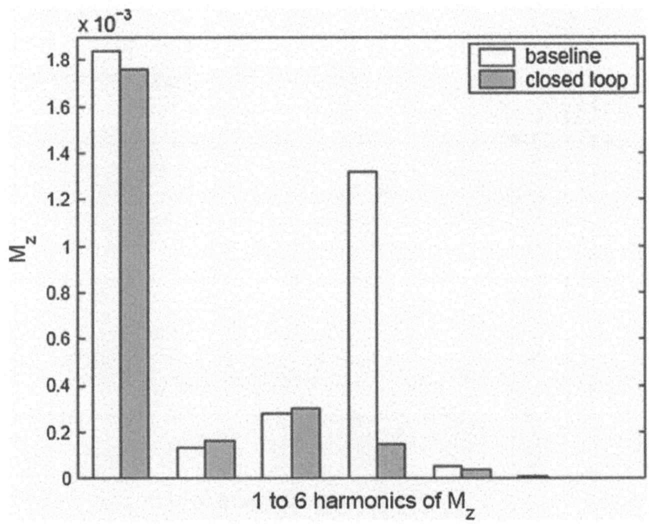

Fig. 10.43 Harmonics of blade root rotating frame load M_z at an advance ratio of 0.3 for baseline and closed-loop control cases

represents a measure of vibratory hub loads. Figure 10.44 depicts the reduction in the hub forces for advance ratio of 0.2 and 0.3, respectively. Note that for the helicopter rotor used in this chapter, the three 4/rev forces and yawing moment is more dominant. The 4/rev hub vertical shear is the dominant source of vibration in many helicopters. Individual reductions of 16.33, 23, 14.26, 12.10, 44.3, and 88.3 % take place in F_{xH}, F_{yH}, F_{zH}, M_{xH}, M_{yH}, M_{zH}, respectively, for an advance ratio of 0.2. Reductions of

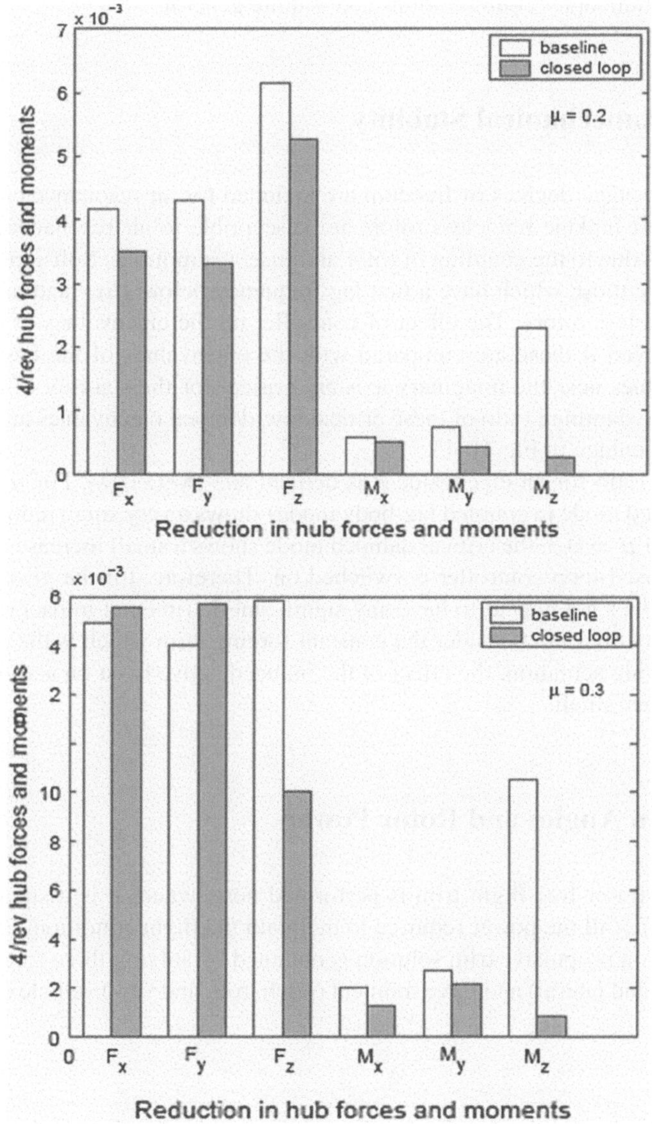

Fig. 10.44 Comparison of 4/rev hub loads between baseline and closed-loop control cases at $\mu = 0.2$ and $\mu = 0.3$

44, 36.6, 20.3, and 91.6 take place in F_{zH}, M_{xH}, M_{yH}, and M_{zH}, respectively, with a slight increment of 3.9 and 10.6 %, respectively in F_{xH} and F_{yH} for an advance ratio of 0.3. Therefore, considerable reduction in hub loads can be obtained using active twist. Note that in this chapter, we have considered a rotor which is quite different compared to the box beam blade used in the previous chapter. However, for both these rotors, the active twist concept has shown a substantial reduction in the 4/rev hub loads. This shows that the control approach developed in this book can be used for different helicopter configurations and is quite general.

10.7 Aeromechanical Stability

Rotor and fuselage degrees of freedom are included for air resonance calculations. Typically, soft-inplane hingeless rotors are susceptible to air resonance instability which occurs due to the coupling of rotor and fuselage motions. Soft-inplane hingeless rotors are those which have a first lag frequency below 1/rev and are the main class of hingeless rotors. The effect of controller on the eigenvalues of the system can be observed if these are compared with the eigenvalues of the baseline case. The eigenvalues near the imaginary axis are critical for the stability of the system and hence the damping ratio of these critical low-damped eigenvalues are plotted in terms of percentage in Fig. 10.45.

Damping ratio for an eigenvalue λ is defined as $-Re(\lambda)/|\lambda|$. For $\mu = 0.2$ the lowest damped mode (a coupled lag-body mode) shows a very small reduction in the damping. For $\mu = 0.3$, the critical damped mode shows a small increase in damping when the closed-loop controller is switched on. Therefore, for the given rotor, the active twist does not appear to have any significant detrimental impact on stability. Since stability does not consider the constant forcing term which is the main effect of piezoceramic actuation, the effect of the induced active twist on aeromechanical stability is very small.

10.8 Trim Angles and Rotor Power

Propulsive trim or free flight trim is performed here, where it is assumed that the engine supplies all the power required to maintain the flight condition. For a steady flight condition propulsive trim solution is obtained by solving three force (vertical, longitudinal and lateral) and three moment (pitch, roll, and yaw) vehicle equilibrium

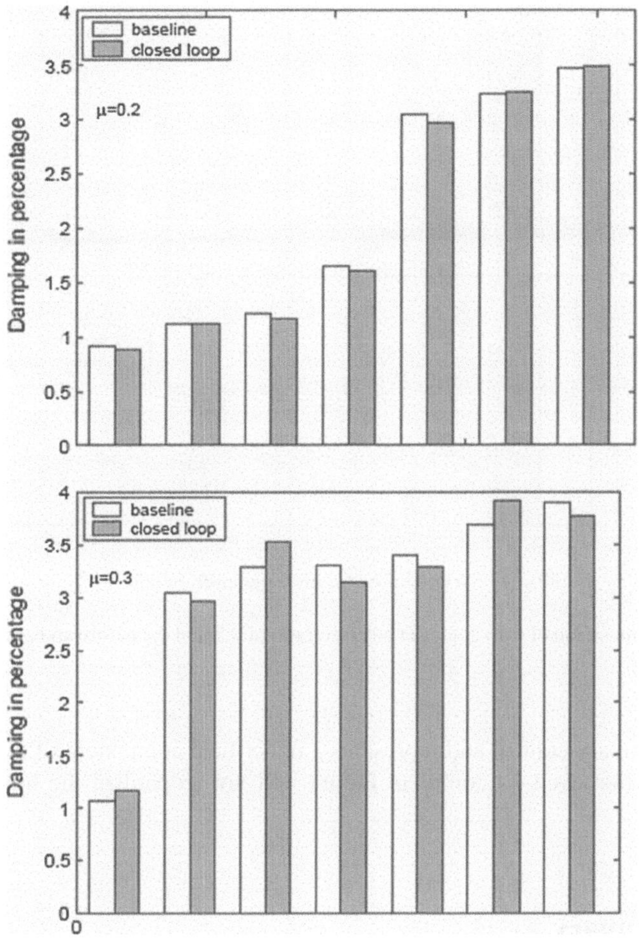

Fig. 10.45 Change in damping between baseline and closed-loop control cases for first seven lowest damped modes at $\mu = 0.2$ and $\mu = 0.3$

equations. Further, for a specified gross weight and level flight speed the trim solution gives the rotor pitch controls: collective θ_0, cyclic θ_{1c}, cyclic θ_{1s}, and tail rotor pitch; collective θ_t (tail). Required main rotor power is also calculated. No significant difference is found between the rotor power required by the baseline case when compared to closed-loop control case. Comparison of the difference in values in control angles for the baseline case with that of the closed-loop control case shows no appreciable change as can be seen from the Fig. 10.46. The magnitude of control angles is plotted in terms of degrees. Therefore, the pilot has to make only very small adjustments to the trim control angles to adjust for the active twist controller. The reason for this behavior is that controller only gives higher harmonic inputs which do not affect the steady loads. In addition, the active twist controller is separate from the

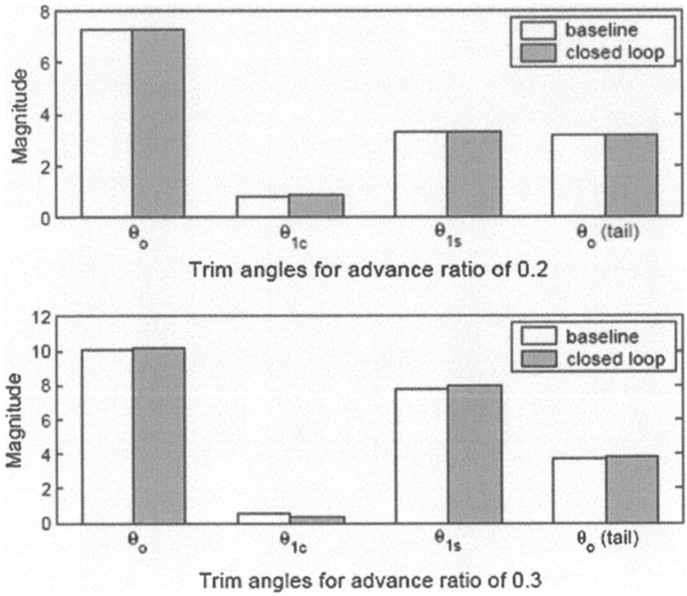

Fig. 10.46 Comparison of trim angles at advance ratio of 0.2 and 0.3 before and after closed-loop control

helicopter primary control and represents a noncritical component of the helicopter in the sense that even its complete failure will not jeopardize the mission of the helicopter.

10.9 Summary

Induced shear-based mechanism is used for attaining active twist in a soft-inplane hingeless rotor with a two-cell NACA 0012 airfoil cross-section. The rotor properties dynamically represent a real soft-inplane hingeless rotor. Finite element analysis of a rotor blade is done in space and time domain in the framework of a time domain unsteady aerodynamic environment and a free wake model. A closed-loop controller is developed to obtain the optimum voltage required to be given to the rotor blade for obtaining twist using strain rate feedback. The significant nonlinear relationship which exists between piezoelectric shear coefficient and applied electric field is included in the controller design. The piezoelectric actuators are placed on the top and bottom of the web, inside the skin of the two-cell airfoil section. Actuator placement locations are explored at an advance ratio of 0.2 and 0.3 using closed-loop control with only one, two, and three actuators, respectively. Here, each actuator spans are 20 % of the blade length. The aeroelastic properties of the active twist controlled rotor are studied and the effects of the controller on blade response, blade and hub loads, aeromechanical stability, and trim control angles are evaluated. It is found that active twist designed for vibration reduction does not have much influence on aeromechanical stability and trim control angles which increase the feasibility for applications. This chapter therefore clearly shows the possibility of shear-induced active twist as a strong contender for vibration reduction in helicopter rotors in forward flight.

References

1. Thakkar, D., Ganguli, R.: Induced shear actuation of helicopter rotor blade for active twist control. Thin Walled Struct. **45**(1), 111–121 (2007)
2. Bagai, A., Leishman, J.G.: Rotor free-wake modeling using a pseudo-implicit technique—including comparisons with experimental data. J. Am. Helicopter Soc. **40**(3), 29–41 (1995)

Chapter 11
Dynamic Stall Alleviation Using Active Twist

In the previous chapters, moderate speed regimes of the helicopter have been explored, where flow separation and stall is not so important. However, it is possible that substantial flow separation on the rotor may cause very high levels of vibrations which active twist may not be able to reduce. The current chapter presents a methodology to directly suppress the flow separation-induced vibration in a helicopter rotor. As we shall see in this chapter, large stroke is needed to suppress vibration induced by stalled flow which then needs application of new and novel piezoceramics. Emerging materials such as single-crystal piezoceramics have a potential to meet the increasing demand of stroke in smart material applications. This chapter addresses the problem of vibration suppression using both single-crystal piezoceramics and soft piezoceramics. Section 11.1 introduces the single-crystal piezoceramics and highlights the importance of these materials over conventional soft piezoceramics. Section 11.2 presents the two airfoil models used for aeroelastic analysis. Section 11.3 discusses the structural properties of the rotor blade. In Sect. 11.4, importance of including dynamic stall and flow separation in the aeroelastic analysis is highlighted. Sections 11.5 and 11.6 discusses the optimum placement of the actuators in the presence of dynamic stall, unsteady flow separation, and free wake at high-speed forward flight. Section 11.7 discusses about the robustness of the feedback loop control algorithm. Robustness of the controller is verified by introducing random noise to the sensed variable. Reduction in vibration obtained using the shear actuation mechanism with both PZT-5H and PZN-8%PT ⟨111⟩ is discussed in Sect. 11.8. In Sect. 11.9, angle of attack distributions over the rotor disk is investigated and twice per revolution control is implemented to achieve the reduction in angle of attack over the rotor disk and to reduce the boundaries of the dynamic stall region. Actuator failure/degradation is one of the practical challenges which could occur in flight. Loss in reduction due to this shortcoming is addressed in Sect. 11.10. Finally, Sect. 11.11 closes by summarizing the important contributions made in this chapter.

R. Ganguli et al., *Smart Helicopter Rotors*, Advances in Industrial Control,
DOI 10.1007/978-3-319-24768-7_11

11.1 Single-Crystal Piezoceramics

Typical piezoelectric materials such as PZT-5H have stroke limitations. There are limits to the use of high electric fields to increase actuator authority due to increasing complexity and bulky nature of the required electronic equipment and the risk of depoling. Relaxor-based single-crystal piezoelectrics have been developed in recent years and promise substantial increase in bandwidth, sensitivity, and a high energy density. Piezoelectric properties of PZN-PT type single crystals with $\langle 001\rangle$-orientation have been found to possess large direct piezoelectric coefficients (d_{33}, d_{31}) which are almost of the order of 10 times that of a soft piezoceramic-like PZT-5H. However, the shear coefficient, d_{15} of these materials is relatively poor as compared to soft piezoceramics-like PZT-5H, and their applications have been limited to those requiring direct strain. Of recent, Liu et al. [1] reported that the piezoelectric shear coefficient d_{15} of PZN-4.5 %PT and PZN-8%PT crystals oriented in $\langle 111\rangle$ is very high as compared to the conventional soft piezoceramic materials. At the same applied electric field, PZN-8%PT was found to have a much higher d_{15} than PZT-5H. Thus, PZN-8%PT was proposed to be a promising material for applications involving shear transducers. Interestingly, this new material fits in very well with our concept of active twist blade. The shear mode properties of these lead-based single crystals have also been investigated by Peng et al. [2]. PZN-8%PT is linear within the range of 210 V/mm. For this material, d_{15} is calculated to be 18750e-9 mm/V (at 200 Volt). This value is almost 10 times the value of d_{15} for PZT-5H at 400 V and 13 times the value at 200 Volt. As a result, the amount of strain generated with the help of this kind of single crystal is appreciable and proves to be very useful in generating a higher amount of twist leading to a higher vibration reduction, especially at higher advance ratios. Large size single crystals are now commercially available at reduced costs [3]. The Young's modulus of PZN-8%PT is almost double as that of aluminum and PZT-5H. Density is marginally higher as compared to that of PZT-5H.

11.2 Rotor Blade Model

In the previous chapter, a NACA 0012 airfoil section with a two-cell internal structure was developed such that it dynamically represents a real rotor. The first flap, lag, and torsion rotating natural frequencies were matched with that of the Advanced Light Helicopter (ALH) rotor (4 bladed hingeless rotor) for the PZT-5H material. The airfoil cross-section (chord=50 cm) details are shown in Fig. 10.1. Rotor properties are the same as mentioned in Table 10.1 in the previous chapter. Note that the rotor blade has a large radius of 6.6 m and a high lock number of 9.98. In addition, the blade has pre-twist and pre-cone. It can observed that the piezoceramic layer is laid between outer skin and the spar in the cross-section, both on top and bottom. This prevents the wear and tear of the piezoceramic layer. The width of the piezoceramic layer is 5.75 cm and height is 0.1 cm. The piezoceramic layer is bonded to the surface of the

Table 11.1 Properties considered for simulation

	PZN-8%PT	PZT-5H	Adhesive	Aluminum
Young's modulus (10^3 MPa)	114.7	63	3.068	68.94
Poisson's ratio	0.3	0.3	0.3485	0.3
Density (kg/m^3)	8000	7600	1000	2700
d_{15} (10^{-09} mm/V)	18750 (200 Volt)	1928.9 (400 Volt)	–	–

Table 11.2 Rotating blade natural frequencies and vibration reduction

Mode	Frequency (per rev)		
	2-cell box	Airfoil with PZT-5H	Airfoil with PZN-8T
First Lag	0.54	0.677	0.6708
First Flap	1.05	1.09	1.08
Second flap	2.7	2.56	2.59
Second lag	3.4	3.92	3.74
Third flap	4.8	4.41	4.65
First Torsion	5.7	4.65	4.54
Vibration reduction at $\mu = 0.3$	43 %	67 %	74 %

airfoil by an adhesive layer of thickness 0.025 cm. Rotor blade being nonuniform, mass per unit length, airfoil thickness, and the width of the spar are variables along the span of the rotor blade. For the aerodynamic modeling, a free wake model developed by Bagai–Leishman is used along with the reverse flow and unsteady aerodynamics model developed by Leishman and Beddoes. Therefore, two different rotor models are developed due to the use of two different materials for the same NACA 0012 airfoil section model. Single-crystal piezoceramics-like PZN-8%PT ⟨111⟩ and soft piezoceramic-like PZT-5H have different mechanical properties (Table 11.1), resulting into two different nonuniform rotors having dissimilar natural frequencies as shown in Table 11.2. Here the blade with the PZT-5H which is discussed in detail in the previous chapter matches the ALH frequencies more closely. However, instead of changing the structural characteristics of the PZN-8%PT blade, the host structure of blade is kept the same and only the piezoceramic material properties have been changed. Since both the density and the stiffness of PZN-8%PT are higher than PZT-5H, the effects of the dynamic properties is mixed with an increase in the higher flap modes and a decrease in the second lag and first torsion mode. The effect of the difference between the two rotors can be seen from the difference in their baseline properties. The effect of the piezoceramic actuation can then be evaluated by the reduction in vibration from the individual baseline blades [4, 5].

11.3 Influence of Structural Properties

The aeroelastic analysis performed here considered two types of rotor designs, one with PZT-5H and other with PZN-8%PT. Both these designs consider the airfoil section to be made of aluminum. This is reasonable in the current book whose focus is more on deciding optimum actuator locations, controller development, stall alleviation, and the use of nonlinear d_{15} for PZT-5H and novel single-crystal piezo-ceramics. The main effect of including composites in the analysis will be to change the stiffness and mass properties of the blade, in particular the torsion stiffness. In this book, three different rotor blades were considered including a uniform two-cell cross-section blade and two nonuniform cross-section blades. Chord/radius ratios for all elements except the first element form the root are kept same. The first element is smaller by 17 % than other four elements. Span-wise stiffness and mass variation for the PZT-5H-based rotor are depicted in Figs. 10.1, 10.2, 10.3, 10.4 and 10.5. Similarly, for the PZN-8%PT based rotor, they are shown in Figs. 11.1, 11.2, 11.3, 11.4, 11.5, 11.6, and 11.7. Figure 11.1 shows the flap stiffness is highest at the root and decreases outboard. The lag stiffness is least between 0.4–0.8L and highest at 0.2–0.4L (Fig. 11.2). The torsion stiffness is highest at the root and decreases outboard as seen in Fig. 11.3. It can also be seen from Figs. 11.1 to 11.3 that the element spanning from 0.4–0.6L has the least stiffness among all the elements from 0–L. This could be suggested as a probable reason for 0.4–0.6L being a potential actuator location. From Fig. 11.5, it can be seen that the mass is highest at the root and decreases outboard. Figure 11.6 shows that ${km_1}^2$ is lowest at the root and highest at the tip. Figure 11.7 shows that ${km_2}^2$ is highest at 0.2–0.4L and lower at the root and 0.4–0.8L. Also note that ${km_2}^2$ values are much greater than ${km_1}^2$ values for rotor blade cross sections. A

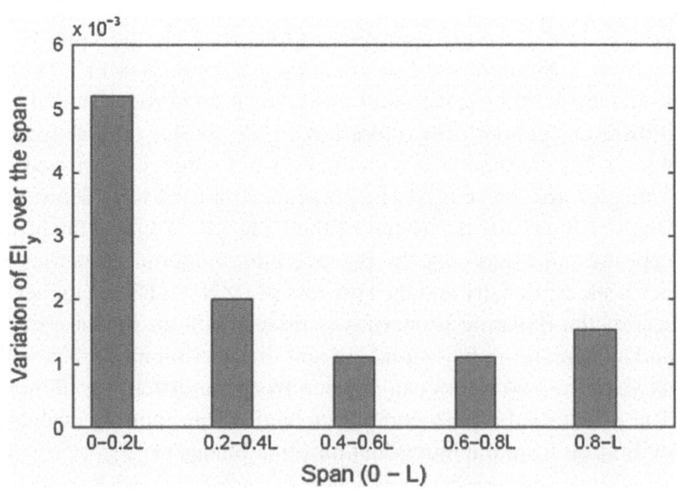

Fig. 11.1 Variation of EI_y for PZN-8%PT rotor

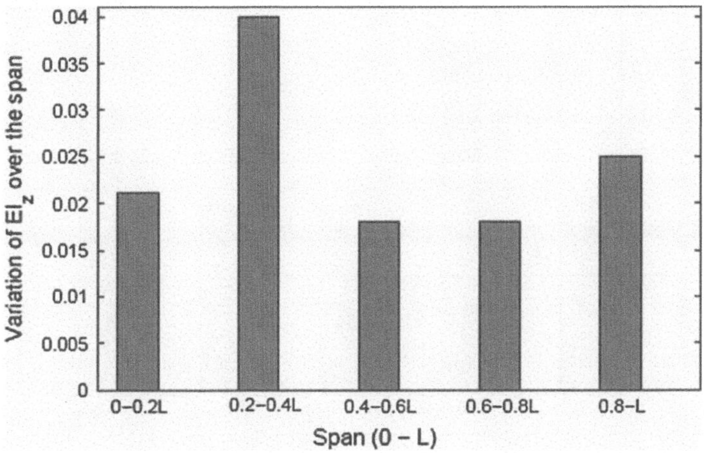

Fig. 11.2 Variation of EI_z for PZN-8%PT rotor

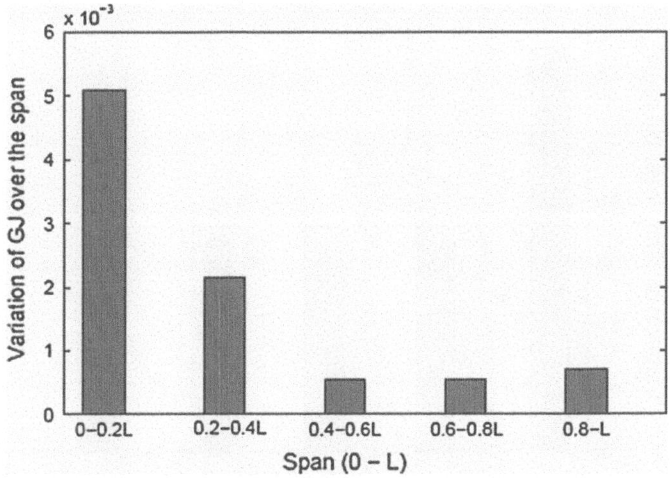

Fig. 11.3 Variation of GJ for PZN-8%PT rotor

close comparison between the spanwise properties of the PZT-5H and PZN-8%PT rotor shows the trends to be similar with a slight difference in the magnitudes due to the different actuator material.

Table 11.2 shows the frequencies of these three different blade configurations and the vibration reduction at $\mu = 0.3$ obtained using the closed-loop controller. These blades span a range of frequencies and all show substantial vibration reductions.

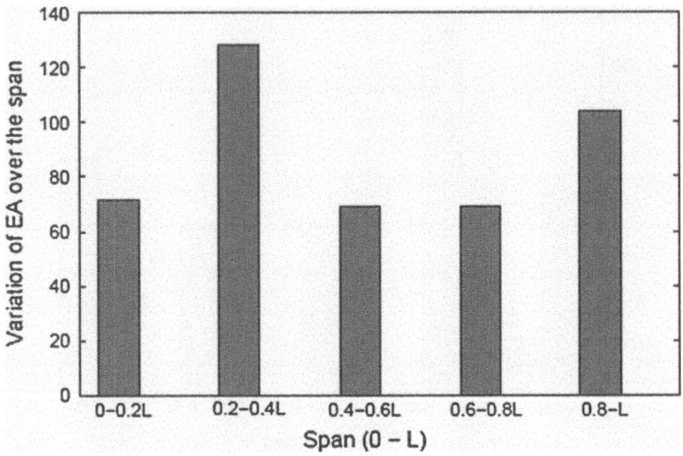

Fig. 11.4 Variation of EA for PZN-8%PT rotor

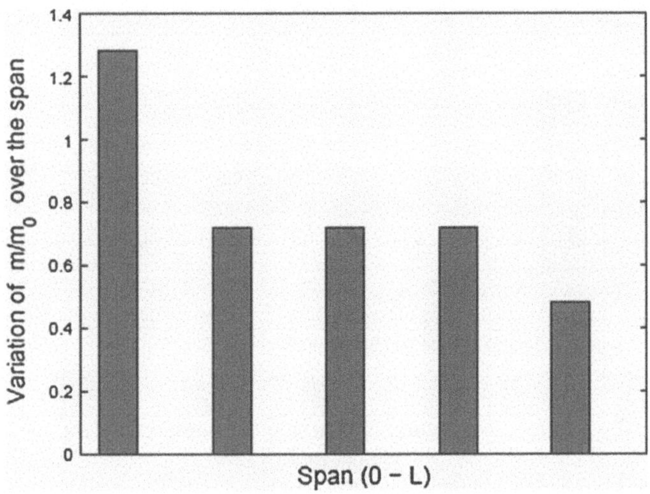

Fig. 11.5 Variation of m/mo for PZN-8%PT rotor

Therefore, it can be concluded that the active twist concept will also work well for a composite blade whose baseline frequencies may be different from the isotropic blade. However, composites lead to other nonclassical effects such as transverse shear, warping, and couplings which need to be thoroughly investigated.

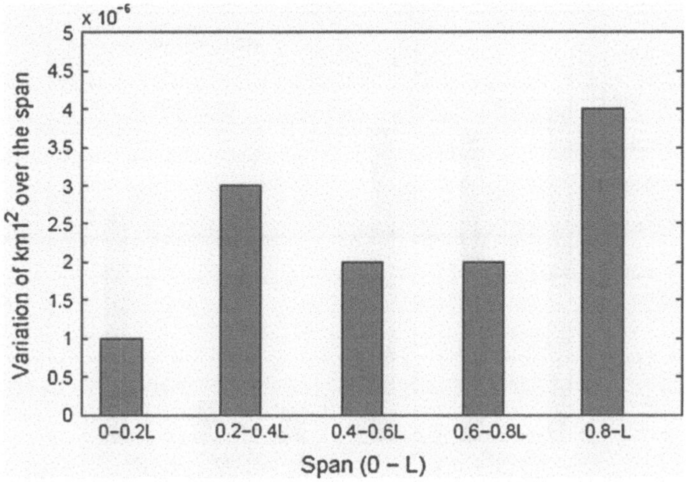

Fig. 11.6 Variation of km_1^2 for PZN-8%PT rotor

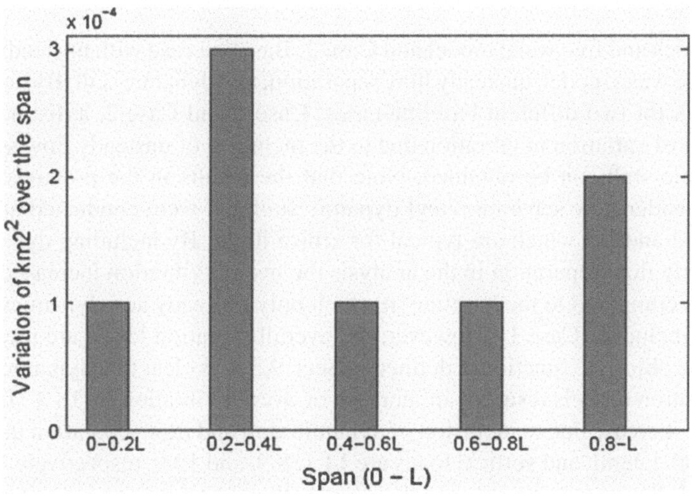

Fig. 11.7 Variation of km_2^2 for PZN-8%PT rotor

11.4 Importance of Dynamic Stall and Unsteady Flow Separation

The rotor blade with PZT-5H is used to evaluate the importance of dynamic stall and flow separation at high-speed forward flight. A comparison of two cases is done at an advance ratio of 0.35 to understand the effect of including dynamic stall and flow separation in the analysis. Here Case 1: Baseline case with only unsteady

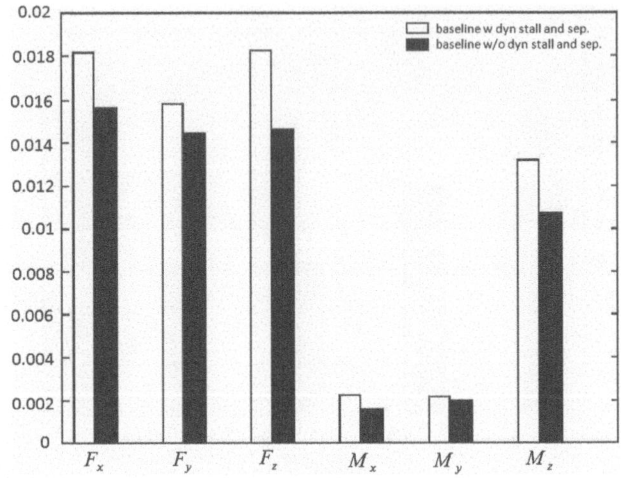

Fig. 11.8 Effect of dynamic stall and flow separation on 4/rev hub loads at $\mu = 0.35$

aerodynamics and free wake model and Case 2: Baseline case with unsteady aerodynamics, free wake model, unsteady flow separation, and dynamic stall. By performing the analysis for two different baseline cases, Case 1 and Case 2, a distinct idea of the amount of variation in vibration due to the inclusion of unsteady flow separation and dynamic stall can be obtained. Note that the results in the previous chapters did not consider flow separation and dynamic stall and were conducted at advance ratios of 0.2 and 0.3 which are typical for cruise flight. By including dynamic stall and unsteady flow separation in the analysis the overall vibration increased by 18 % (Case 2) as compared to the baseline in which only unsteady aerodynamics and free wake was included (Case 1). However, the overall vibration levels are measured in terms of the objective function J defined in Sect. 9.3. It is clear that dynamic stall and flow separation models result in an increase of overall vibration by 18 %. Individual percentage increase due to inclusion of dynamic stall and flow separation in the 4/rev longitudinal, lateral, and vertical loads are 14.1, 8.9, and 19.9, respectively, as shown in Fig. 11.8. Percentage increase in the 4/rev rolling, pitching, and yawing moment are 26.5, 5.2, and 18.7, respectively. Hence, including stall in the analysis prevents the underprediction of helicopter vibration at high-speed flights where significant flow separation exists on the rotor disk.

11.5 Actuator Placement for PZT-5H

Placement of actuators is investigated using the strain rate feedback controller in the presence of unsteady aerodynamics, free wake, flow separation, and dynamic stall environment. This is done by comparing the amount of reduction obtained on varying the placement of actuators by setting the electric field to zero for nonactive actuators. There are five possible locations of the actuator corresponding to five finite elements (from root(0) to tip(L))located over the blade span. Each of these element has a span of 20 % of the blade length. Three sets of analysis were performed for achieving the optimum location of actuators. The PZT-5H actuators are given an electric field of 400 Volt. The nonlinearity of the d_{15} value of PZT-5H is accounted for using the relationship demonstrated in [3].

$$d_{15} = 737.3411 + 159.699E_a + 102.653{E_a}^2 - 39.006{E_a}^3 + 5.495{E_a}^4. \quad (11.1)$$

The variation in vibration (J) from their baselines corresponding to different configurations of actuators is given in Table 11.3 for PZT-5H. There are two occurrences of an increase in vibration from the baseline in Table 11.3 and are indicated by ↑. All other cases in Table 11.3 mark reductions from the baseline in percentage. It is important to notice that the increase in number of actuators from one to two has a considerable effect on vibration reduction when compared to increment from two to three. In general, for the rotor studied here, it appears that the center locations are better than the locations near the root or the tip. This is most likely due to the low-torsion stiffness of the finite elements in the center region. We see that at $\mu = 0.3$, substantial vibration reduction of 39 % can be obtained by even one actuator placed between 60 to 80 % blade span. Again, two actuators placed between 20 and 60 % along the blade result in a vibration reduction of about 45 % at $\mu = 0.3$. Finally, if the above two best cases are considered and actuators were placed between 20 and 80 % of the

Table 11.3 Variation in vibration for different actuator configurations with PZT-5H in the presence of dynamic stall and flow separation

One actuator		
Location on blade	Percentage variation in vibration (J)	
	$\mu = 0.3$	$\mu = 0.35$
0.2L–0.4L	23.17	7.18 ↑
0.4L–0.6L	34.29	12.01
0.6L–0.8L	38.91	1.80
0.8L–L (Tip)	9.81	3.5 ↑
Two actuator		
Location on blade	Percentage variation in vibration (J)	
0.2L–0.6L	44.67	12.2
0.4L–0.8L	39.56	10.40
0.6L–L (Tip)	33.01	5.56
Three actuator		
Location on blade	Percentage variation in vibration (J)	
0.2L–0.8L	48.70	21.02
0.4L–L	44.51	17.98

blade span, a vibration reduction of about 49 % is obtained at $\mu = 0.3$. This means that soft PZT-5H can be used to reduce vibration levels by 50 % at typical moderate forward speed of $\mu = 0.3$. Next, we look at the results for $\mu = 0.35$ in Table 11.3. For a single actuator, the best vibration reduction of about 12 % is obtained for the actuator placed between 40 and 60 % of the blade span. Two actuators placed at 20 to 60 % of the blade span also result in a maximum reduction of about 12 %. Even with three actuators placed between 20 to 80 % of the blade span, the vibration reduction is only 21 % at $\mu = 0.35$. Thus the maximum vibration reduction obtained using PZT-5H has deteriorated from about 50 % at $\mu = 0.3$ to only about 20 % at $\mu = 0.35$.

As seen in Table 11.3, vibration reduction at advance ratios of 0.3 and 0.35 is maximum with three actuators spanning 0.2L–0.8L. With the optimum locations of the actuators (0.2L–0.8L), a feedback control is done to study the effect of dynamic stall and unsteady flow separation on vibration levels. A reduction of 48.7 and 21.02 % is obtained at advance ratios of 0.3 and 0.35, respectively. Vibration reduction obtained was less at $\mu = 0.35$ compared to $\mu = 0.3$ due to the fact that very high-unsteady loads are generated due to dynamic stall and unsteady separation and a higher amount of twist is required to overcome this at $\mu = 0.35$. At $\mu = 0.3$, if dynamic stall and flow separation model is not used, the vibration reduction is about 67 %. These results indicate a limitation of the PZT-5H material which is not able to generate the amount of unsteady airloads required through active twist to reduce a large amount of vibration even on applying high electric fields and using the strong nonlinear increase of d_{15} with increase in electric field.

11.6 Actuator Placement for PZN-8 % PT

The results in the previous section suggest the need of using materials other than PZT-5H which can generate higher amplitudes of induced shear strain. In general, low actuation stroke is a limitation of piezoceramic materials which inhibit their use in many actuator applications. Piezoelectric shear strain coefficient d_{15} is the key measure in case of shear strain-induced active control mechanism. It was proposed very recently by Liu et al. that single-crystal ceramics like PZN-8%PT ⟨111⟩ have a very high-shear strain coefficient d_{15}. It is important to mention here that d_{15} of single-crystal ceramics like PZN-8%PT ⟨111⟩ at 200 V is almost 13 times that of PZT-5H at 400 Volt. In order to obtain higher reduction of vibration, there is a need to investigate the capability of such new materials. Table 11.4 describes a parametric study of the variation in vibration (J) with different actuator locations for the newly evolved material PZN-8PT. For all the results in this chapter, PZT-5H is actuated at 400 V and PZN-8%PT at 200 Volt. First, considering the results at $\mu = 0.3$ with a single actuator, the vibration reduction is about 40 % with actuators placed between 40 and 60 % of the blade span which is different than for PZT-5H (see Table 11.3). For two actuators, the best location is between 40 and 80 % giving a vibration reduction

Table 11.4 Variation in vibration for different actuator configurations with PZN-8%PT in the presence of dynamic stall and flow separation

One actuator		
Location on blade	Percentage variation in vibration (J)	
	$\mu = 0.3$	$\mu = 0.35$
0.2L–0.4L	5.77	10.31
0.4L–0.6L	40.134	28.35
0.6L–0.8L	24.79	22.31
0.8L–L (Tip)	3.2	1.09
Two actuator		
Location on blade	Percentage variation in vibration (J)	
0.2L–0.6L	52.73	34.3
0.4L–0.8L	63.51	47.72
0.6L–L (Tip)	33.52	21.0
Three actuator		
Location on blade	Percentage variation in vibration (J)	
0.2L–0.8L	70.50	59.34
0.4L–L	65.66	46.93

of about 64 %. Again, this is different compared to PZT-5H where 20 to 60 % was the best location. But, with three actuators, the location between 20 to 80 % of the blade span is found to be best, which is the same as was obtained for PZT-5H. We note that at $\mu = 0.3$, the maximum vibration reduction with one, two, and three actuators increases from 39, 45, and 49 % for PZT-5H to about 40, 64, and 71 % of PZN-8%PT, a percent improvement of 1, 19, and 22, respectively. However, the really large improvement using the single-crystal actuator is seen at $\mu = 0.35$ in Table 11.5. The actuator locations for the best performance are the same as at $\mu = 0.3$. The maximum reduction in vibration for one, two, and three actuators increases from 12, 12, and 21 % for PZT-5H to about 28, 48, and 59 % for PZN-8%PT, a percent increase of 16, 36, and 38, respectively. Optimum location is being 0.2L–0.8L, considering 3 actuators of span 20 % of the total blade length.

On comparing the results obtained for PZT-5H (Table 11.3) and PZN-8%PT (Table 11.4), it can be observed that for most cases, vibration reduction obtained by PZN-8%PT is higher than PZT-5H. With the optimum locations of 0.2L–0.8L, vibration reduction obtained at an advance ratio of 0.3 for PZT-5H and PZN-8%PT was 48.7 and 70.5 %, respectively. At an advance ratio of 0.35, a reduction of about 21 and 59 % was obtained for PZT-5H and PZN-8%PT, respectively. These results lead toward marking PZN-8%PT as an efficient material as compared to PZT-5H. It is worthy to mention here that experimental investigations should be carried out before any conclusive point can be made about the use of single-crystal piezoceramics in helicopter rotor blades. The theoretical analysis and findings presented here

Table 11.5 Failure of a single actuator in PZN-8%PT rotor

$\mu = \mathbf{0.3}$				
0.2L–0.4L	0.4L–0.6L	0.6L–0.8L	$J_{optimum}$ %	J %
$\neq$			70.48	63.5
	$\neq$		70.48	40.8
		$\neq$	70.48	52.7
$\mu = \mathbf{0.35}$				
0.2L–0.4L	0.4L–0.6L	0.6L–0.8L	$J_{optimum}$ %	J %
$\neq$			59.3	47.7
	$\neq$		59.3	32.8
		$\neq$	59.3	41

propose single-crystal piezoceramics-like PZN-8%PT $\langle 111 \rangle$ to be used for experimental studies.

In the absence of dynamic stall and flow separation models, vibration reduction obtained by PZN-8%PT at an advance ratio of 0.3 was 74.36 %. This was marginally higher than that obtained by PZT-5H (67 %). Also, it should be noted that vibration reduction falls by about 4 % (74.36 to 70.50 %) in case of PZN-8%PT by including dynamic stall and unsteady flow separation in the analysis as compared to 18 (67 to 49) % in PZT-5H at $\mu = 0.3$. This indicates that the presence of dynamic stall and flow separation demands higher stroke/generation of unsteady loads to alleviate the vibration. So in such cases, the need of PZN-8%PT-like materials arises.

11.7 Robustness of Controller

Here the attempt is to investigate the robustness of the closed-loop controller. Robustness is assessed by evaluating its performance on addition of random noise to the sensed values. This simulates a condition where the sensor output is contaminated with noise. Suppose, $\dot{\varepsilon}$ is the strain rate predicted by the aeroelastic simulation then noise is added in the following way to find the new noisy value $\dot{\varepsilon}^{(n)} = \dot{\varepsilon}\,(1 + \alpha_n R_n)$, where α_n is the measure of the noise level in the simulated data and R_n is the random number generated between -1 and 1. Here, α_n characterizes the deviation in percentage from the optimum-sensed variables. Figure 11.9 shows the comparison of three cases. Case 1: J with accurately sensed variables, Case 2: Mean value of J for 20 cases, and Case 3: worst case of J out of 20 cases. The results show that the vibration reduction is only slightly affected by the presence of noise in the sensor measurements thereby showing that the controller is robust to the presence of noise in the data. Even for a high noise level of 20 %, the worst case result shows a degradation of only 3.4 % in the vibration reduction showing that the feedback controller

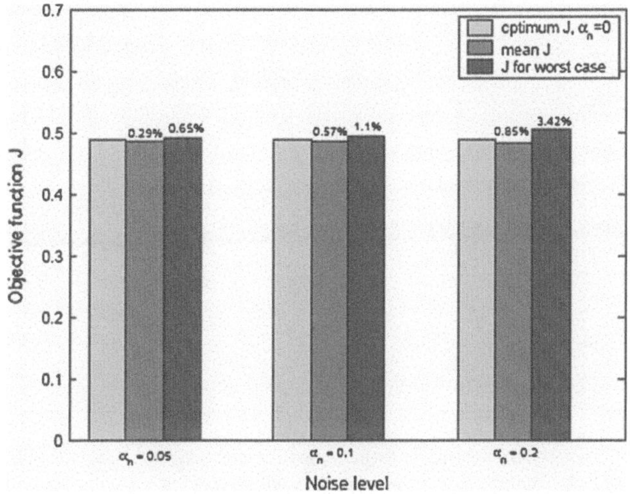

Fig. 11.9 Comparison of variation in J with different values of noise level α_n added to simulated data

is robust for use in the noisy helicopter rotor environment. Here the rotor blade with PZT-5H was used for evaluating the performance of the controller and the advance ratio was $\mu = 0.3$.

11.8 Vibration Reduction Due to PZT-5H and PZN-8%PT

With the optimum locations of the actuators (0.2L-0.8L), an investigation to study the effect of piezoceramic actuation on vibration reduction is carried out. Cases corresponding to advance ratios of 0.3 and 0.35 for both PZT-5H and PZN-8PT are investigated in greater detail by studying their torsion response and hub load variation from the baseline. These allow comparison of performance between the two materials selected for piezoceramic actuation.

At $\mu = 0.3$, relative vibration reduction from baseline for PZT-5H and PZN-8%PT $\langle 111 \rangle$ is 48.7 and 70.5 %, respectively. Figure 11.10 demonstrates that a maximum twist of 2.5° and 4.1° is generated at $\mu = 0.3$ for PZT-5H and PZN-8PT, respectively. Here $\Delta\phi$ is the difference between the torsion response of the controlled blade and the baseline blade. Also, as expected there is substantial higher harmonic content in the active twist. It is clear that PZN-8%PT can generate the high levels of twist required for feasibility of the active twist rotor concept.

At $\mu = 0.35$, relative vibration reduction from baseline for PZT-5H and PZN-8%PT $\langle 111 \rangle$ is 21.02 and 59.34 %, respectively. The active twist generated at $\mu = 0.35$ is shown in Fig. 11.11. This figure depicts that a maximum twist of 1.7° and 4.4° is obtained for PZT-5H and PZN-8%PT, respectively. From Figs. 11.10 and 11.11,

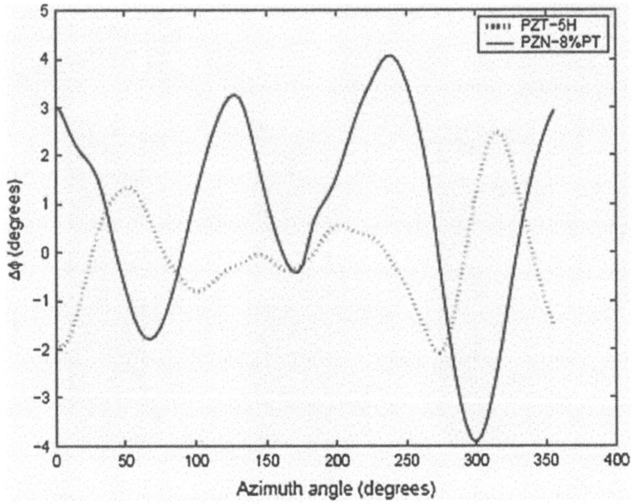

Fig. 11.10 Active twist at $\mu = 0.3$ in the presence of dynamic stall and flow separation

we see a significant negative twist of the blade on the retreating side near an azimuth angle of 300°. Since the retreating blade experiences the highest levels of dynamic stall and flow separation, the controller alleviates these problems by providing the negative twist in this region. It is clear that PZN-8%PT is able to provide a negative twist of upto $-4°$ in contrast to about $-2°$ by PZT-5H.

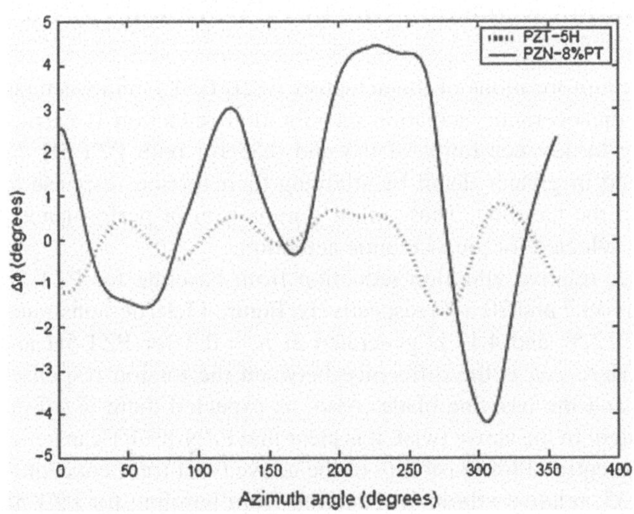

Fig. 11.11 Active twist at $\mu = 0.35$ in the presence of dynamic stall and flow separation

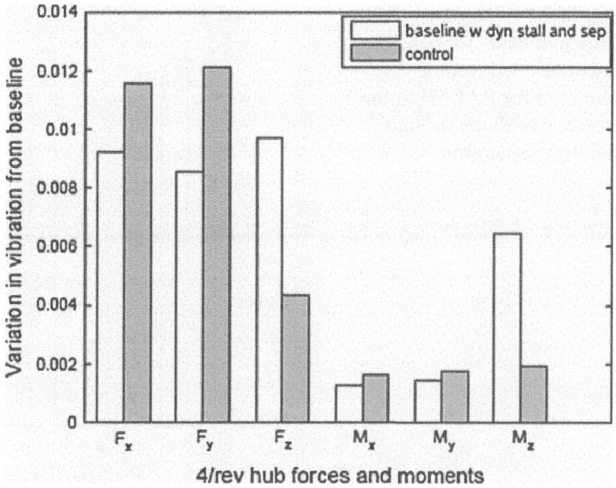

Fig. 11.12 Comparison of 4/rev hub loads for baseline and controlled rotor at $\mu = 0.3$ for PZT-5H in the presence of dynamic stall and flow separation

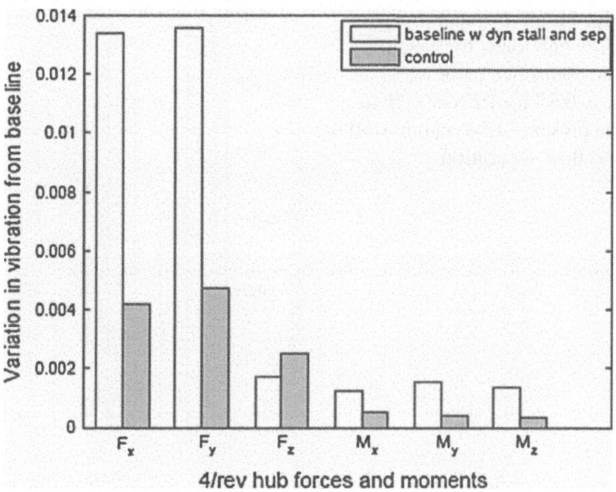

Fig. 11.13 Comparison of 4/rev hub loads for baseline and controlled rotor at $\mu = 0.3$ for PZN-8%PT in the presence of dynamic stall and flow separation

Individual comparison of all 4/rev hub forces with that of their respective baseline cases for PZT-5H and PZN-8%PT $\langle 111 \rangle$ at an advance ratio of 0.3 is depicted in Figs. 11.12 and 11.13, respectively. As can be seen in Fig. 11.12, a reduction of 55.6 and 69.7 % occurs for the PZT-5H rotor in F_z and M_z, respectively. But, increments of 20.4, 28.9, 20.1, and 18.8 % occur in F_x, F_y, M_x and M_y, respectively.

Figure 11.13 shows a reduction of 68.9, 65.4, 67.5, 76.9 and 75.4 % for the PZN-8%PT rotor in F_x, F_y, M_x, M_y and M_z, respectively. An increment in F_z of 29.6 % is observed.

Comparison of the 4/rev hub loads with baseline at an advance ratio of $\mu = 0.35$ is shown in Fig. 11.14 for PZT-5H. Reductions of 1.1, 40.4, 19.2, 3.2, and 23.5 % occur for the PZT-5H rotor in F_x, F_z, M_x, M_y and M_z, respectively. An increment of 0.31 % occurs in F_y. Figure 11.15 depicts the variation in hub forces for PZN-8%PT

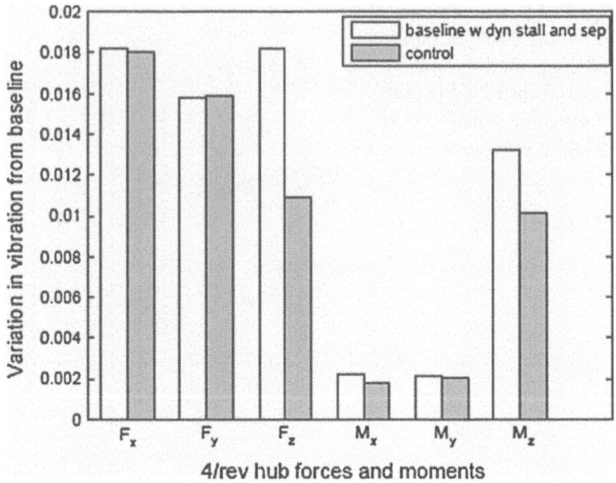

Fig. 11.14 Comparison of 4/rev hub loads for baseline and controlled rotor at $\mu = 0.35$ for PZT-5H in the presence of dynamic stall and flow separation

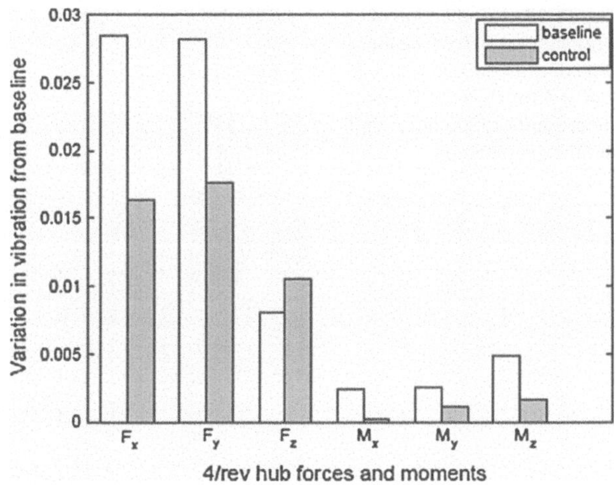

Fig. 11.15 Comparison of 4/rev hub loads for baseline and controlled rotor at $\mu = 0.35$ for PZN-8%PT in the presence of dynamic stall and flow separation

$\langle 111 \rangle$ at $\mu = 0.35$. Reductions of 42.64, 37.3, 90.9, 74.2, 57.8, and 66.9 % take place for the PZN-8%PT rotor in F_x, F_y, M_x, M_y and M_z, respectively. An increment by 54.3 % occurs in F_z.

Note that because of difference in density and especially Young's modulus between PZT-5H and PZN-8%PT (Table 11.2), the dynamics of the two blades are different as seen from the rotating blade frequencies in Table 11.3. The baseline vibration levels in Figs. 11.12 and 11.14 for the PZT-5H blade is dominated by F_x, F_y, F_z and M_z, which leads to the controller reducing these loads in particular. In the case of the PZN-8%PT blade, F_x, F_y are the dominant loads and the controller seeks to reduce them as shown in Figs. 11.13 and 11.15. In all these cases, the controller gains are obtained such that J is minimized which results in the reduction of the dominant loads.

11.9 Variations in Angle of Attack for PZN-8%PT

It was noticed and observed in the above sections that the single-crystal PZN-8%PT performs better than PZT-5H especially at high-speed flight. Therefore, we use PZN-8%PT to demonstrate further results. The blade advancing into the relative wind in forward flight faces increasing velocities as the forward speed increases. Due to the opposite nature of forward speed and rotation speed, as the forward speed increases the retreating blade faces decreasing velocities. As a result, the angle of attack has to be increased on the retreating side to maintain the roll moment leading to the retreating blade stall phenomenon.

As mentioned earlier, the blade is discretized into five finite elements. For the angle of attack (AOA) distributions the first two elements from the root are omitted. AOA distributions are plotted only for the last three elements around the azimuth. The aim is to investigate if any reductions in the angles of attack occur after the open-loop control is applied.

It is known in helicopter dynamics that 2/rev blade actuation is best for reducing stall area, whereas 3/rev, 4/rev and 5/rev actuation is best for reducing vibration in a 4 bladed rotor. To reduce the stall area, we therefore use 2/rev actuation. First, feedback control analysis with 2 gains associated to 2P sin and 2P cosine component (2P is twice per revolution) of the applied control load was attempted using the objective function J for vibration minimization (Eq. 9.2). But, it was observed that this objective function as expected minimizes only vibration and does not necessarily reduce the angle of attack distributions over the rotor disk for the configuration which minimizes vibration. This is because vibration depends on the higher harmonic forcing on the blade which is due to a complicated combination of time varying angle of attack, unsteady aerodynamics, stall, and nonuniform inflow. In fact, as shown recently in a validation study by Datta and Chopra [6], nonuniform inflow due to free wake model plays an important role not only in low-speed flight but also at high-speed flight. Hence, it was found necessary to develop another objective function for AOA distributions and then investigate the factors like reductions in angle of attack, vibration reduction, and main rotor power reduction.

To simplify the analysis open-loop control is selected. The sum of the angles of attack over the azimuth is individually minimized for each of the three finite elements of the blade using 2P open-loop control and an optimum phase is derived. Optimum phase is the phase at which maximum vibration reduction occur. The function to be minimized is given in Eq. 11.2. Here the third Gauss point is utilized to calculate the summations over the azimuth.

$$F_\alpha = \sum_{\psi=0^\circ}^{360^\circ} \alpha_{l\,i}^2 \qquad l_i = 3, 4, 5. \tag{11.2}$$

Figure 11.16 depicts the angle of attack distribution plot over the rotor disk for a baseline case at an advance ratio of 0.3. Maximum angle of attack in the baseline case is 19.46°. Boundaries of the dynamic stall region for baseline case extended from

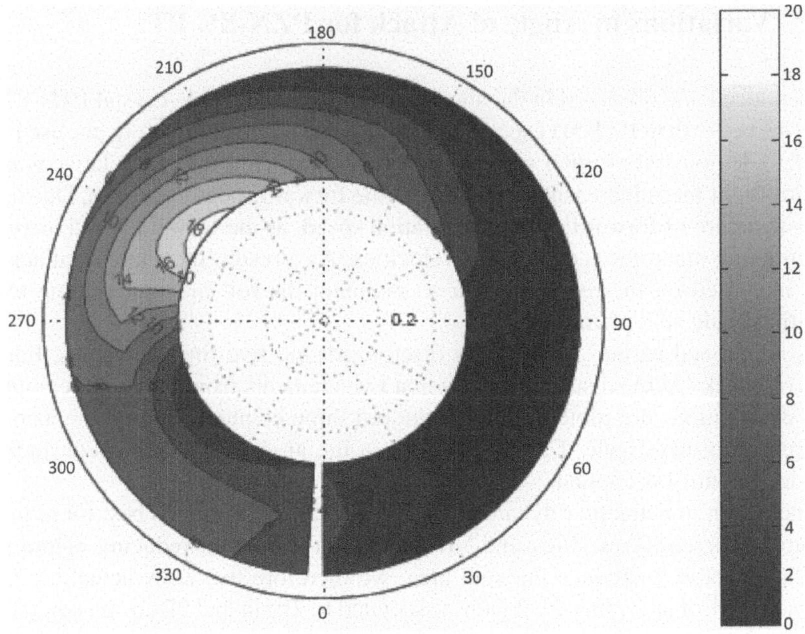

Fig. 11.16 Baseline angle of attack variation at $\mu = 0.3$ for PZN-8%PT rotor

$198° \leq \psi \leq 255°$. On performing the control analysis, it was found that there were reductions in angle of attack over the rotor disk for a baseline case with PZN-8%PT at an advance ratio of 0.3 and a optimum-phased 2P smart load. The maximum AOA was reduced to 13.15° on applying the 2P control load.

Figure 11.17 depicts the angle of attack distribution plot over the rotor disk "with control" at an advance ratio of 0.3. Along with the minimization of the angle of attack distribution, it was also found that the vibration reduction for this case was 19 %. Main rotor power reductions of 2.2 % occurred. It is important to mention that the maximum AOA after the application of 2P control reduced to 13.15° from 19.46°. The controller brings down the AOA below 14° over the entire rotor disk. Hence, the dynamic stall boundaries reduce to zero at cruise speed. Here the static stall angle is considered to be 14° and the region having angle of attack greater than 14° is considered as stalled region. The active twist in the presence of 2P control can be seen in Fig. 11.18 and 4/rev hub loads and moments are depicted in Fig. 11.19. The high F_x and F_y forces in the baseline rotor are reduced but F_z increases.

Similarly the cases with and without control were investigated at an advance ratio of 0.35. Figure 11.20 depicts the angle of attack distribution plot over the rotor disk for a baseline case at an advance ratio of 0.35. Maximum angle of attack in this baseline case is 28.3° and large parts of the rotor disk are in stall. Applying a properly phased 2P control which has minimized the sum of AOA over the rotor disk resulted in reducing the maximum angle of attack to 19.6°.

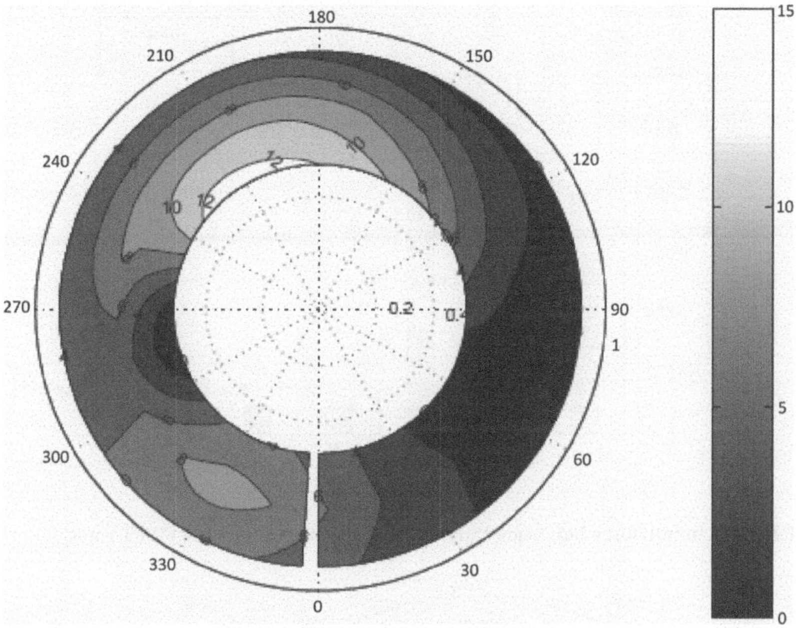

Fig. 11.17 Open-loop control–angle of attack variation at $\mu = 0.3$ for PZN-8%PT rotor

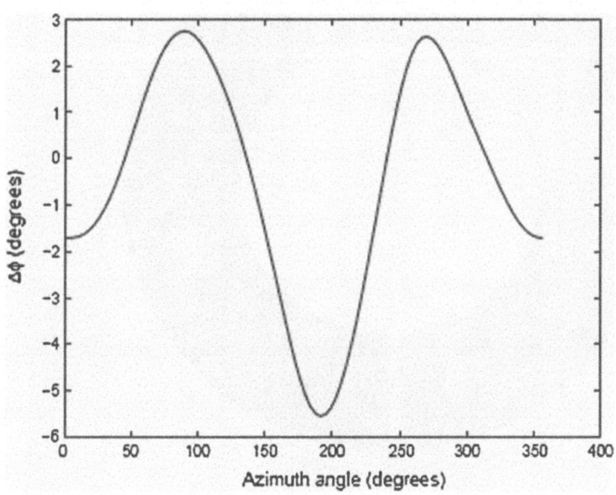

Fig. 11.18 2P control–torsion variation at $\mu = 0.3$ for PZN-8%PT rotor

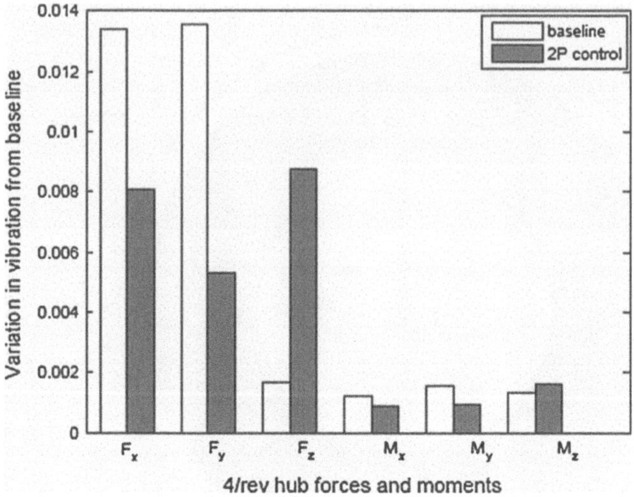

Fig. 11.19 2P control–4/rev hub loads and moments at $\mu = 0.3$ for PZN-8%PT rotor

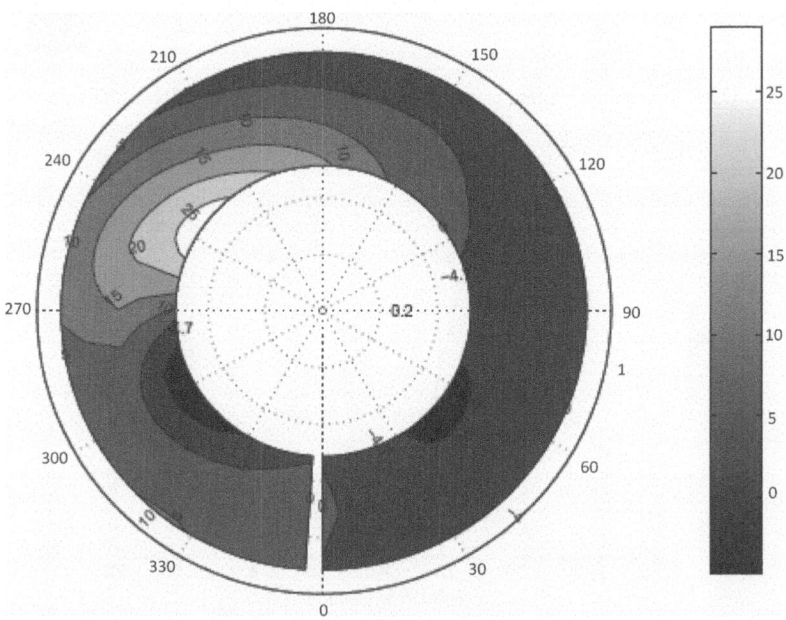

Fig. 11.20 Baseline angle of attack variation at $\mu = 0.35$ for PZN-8%PT rotor

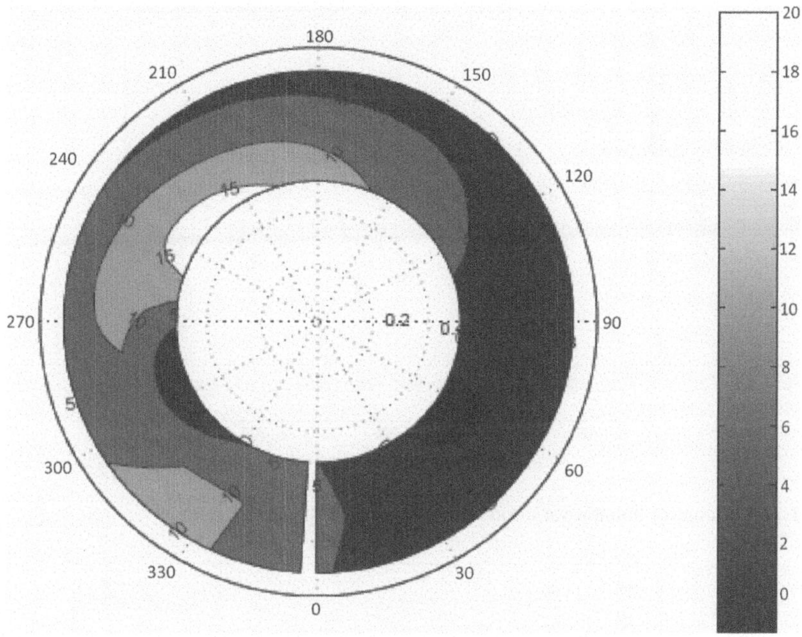

Fig. 11.21 Open-loop control–angle of attack variation at $\mu = 0.35$ for PZN-8%PT rotor

Figure 11.21 depicts the angle of attack distribution plot over the rotor disk “with control" at an advance ratio of 0.35. Main rotor power reduction for this case amounted to 4.8 %. A vibration reduction of 56 % was obtained. In the case of advance ratio of 0.35, boundaries of the dynamic stall region for baseline case extended from $172° \leq \psi \leq 262°$. This was reduced to $183° \leq \psi \leq 256°$ on applying the 2P control load thus leading to reducing the dynamic stall region by 29°. Thus, the results clearly show that 2P control can substantially reduce dynamic stall over the rotor disk using single-crystal piezoceramics. In addition, the reduction in stall area appears to have a collateral effect of reduction in helicopter vibration and main rotor power. At $\mu = 0.35$, the active twist in the presence of 2P control can be seen in Fig. 11.22 and 4/rev hub loads and moments are depicted in Fig. 11.23. Again, it can be observed that there is an increase in the 2/rev active twist and substantial reductions in the dominant F_x and F_y forces for the PZN-8%PT rotor.

11.10 Actuator Failure and Degradation

The results obtained until now have shown the possibility of vibration reduction using the piezoceramic shear actuation. Single-crystal type piezoceramic materials can prove to be excellent actuators for the active twist concept due to their capacity to

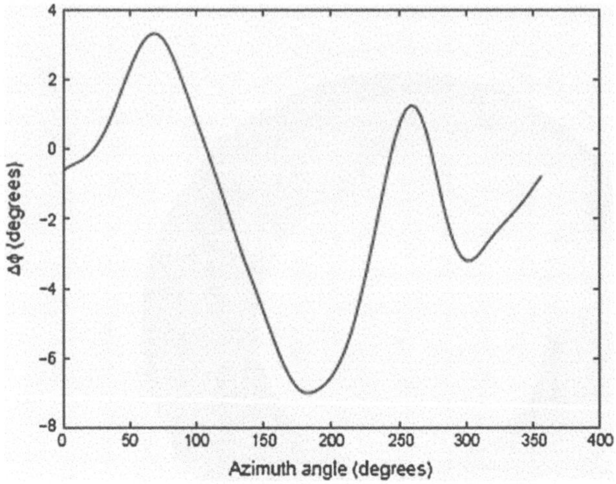

Fig. 11.22 2P control–torsion variation at $\mu = 0.35$ for PZN-8%PT rotor

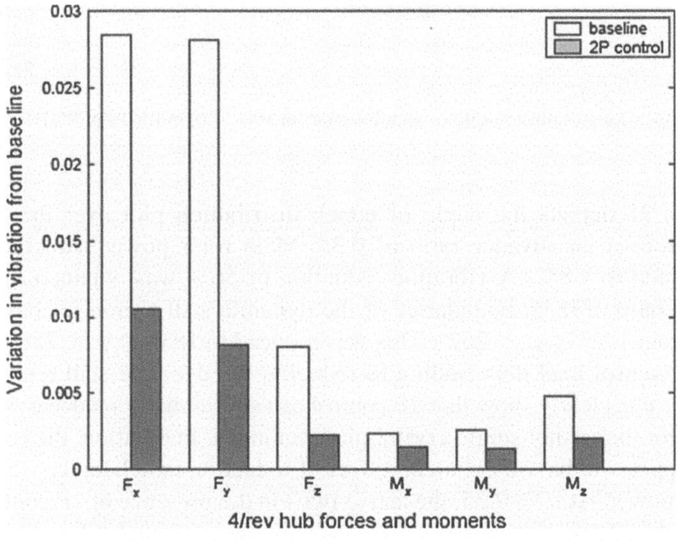

Fig. 11.23 2P control–4/rev hub loads and moments at $\mu = 0.35$ for PZN-8%PT rotor

generate a higher stroke. Nevertheless, there are operational hazards like degradation and failure of actuators in which the stroke generation capacity would taper off with the degradation/failure of actuators and as a result the attainable reduction in vibration would also come down. In order to assess the efficiency of the active controller, reduction in vibration is measured in situations like failure of one/two of the three actuators located over the span of 0.2–0.8L and in case of degradation of actuators upto 30 % using PZN-8%PT $\langle 111 \rangle$ as the actuating material. The 30 % reduction in

d_{15} for a given actuator simulates losses which occur in piezoceramics because of which their induced strain actuation capability is reduced. Such losses can be due to the degradation of the smart material, bonding layer imperfections resulting in less effective strain transfer than previously assumed (100 %) and environmental effects such as temperature which can effect the piezoelectric strain coefficients. Complete failure of the actuator is simulated by setting d_{15} to zero in which case the actuator retains its mass and stiffness characteristics but is inert in terms of actuation. This could simulate a case of electrical malfunction in one actuator.

Maintaining the same tuning for the active controller obtained previously, effect of actuator degradation and failure is analyzed. Table 11.5 shows the fall in vibration reduction in J due to failure of one of the three actuators lying between 0.2 L and 0.8L. When the actuator spanning from 0.2L–0.4L fails a marginal loss of 7–12 (range for $\mu = 0.3$ and $\mu = 0.35$) % in vibration reduction is observed. In case if the 0.4L–0.6L spanning actuator fails a major loss of 27–30 % in vibration reduction takes place. Whereas, in case of 0.6L–0.8L spanning actuator a loss of about 18 % takes place. All these losses are compared with the optimum case where maximum reduction in vibration is obtained, i.e., "No actuator has failed situation" which is represented by $J_{optimum}$ in Table 11.5. It is good to note that in case of any single actuator failure, at least a vibration reduction ranging from 33 to 41 % is obtained at advance ratios of 0.3 and 0.35, respectively.

Looking into the worst case, investigation for the case of two simultaneous actuator failures is addressed. Table 11.6 shows the amount of vibration reduction obtained for the above-mentioned case. It can be seen from this table that the actuator spanning from 0.4–0.6L is the most important and leads to maximum reduction in vibration individually out of actuators located between 0.2–0.8L. Investigating the amount of twist generated in the case with 0.2–0.8L being active in comparison to only 0.4–0.6L being active allows us to figure out the levels of twist generated with respect to baseline. It can be seen from Figs. 11.24 and 11.25 that in case 0.2–0.8L being active, higher levels of active twist are generated as compared to 0.4–0.6L. It should

Table 11.6 Failure of two actuators in PZN-8%PT rotor

$\mu = \mathbf{0.3}$				
0.2L–0.4L	0.4L–0.6L	0.6L–0.8L	$J_{optimum}$ %	J %
$\neq$	$\neq$		70.48	24.8
	$\neq$	$\neq$	70.48	5.8
$\neq$		$\neq$	70.48	40.1
$\mu = \mathbf{0.35}$				
0.2L–0.4L	0.4L–0.6L	0.6L–0.8L	$J_{optimum}$ %	J %
$\neq$	$\neq$		59.3	22.3
	$\neq$	$\neq$	59.3	10.4
$\neq$		$\neq$	59.3	28.5

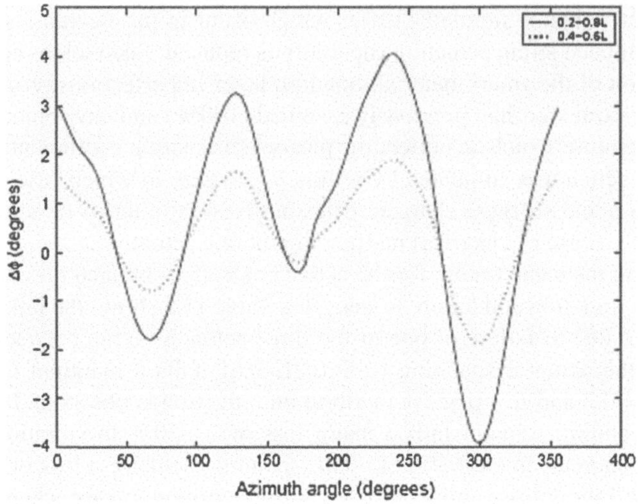

Fig. 11.24 Comparison of active twist at $\mu = 0.3$ for optimum and 2 actuator failure cases

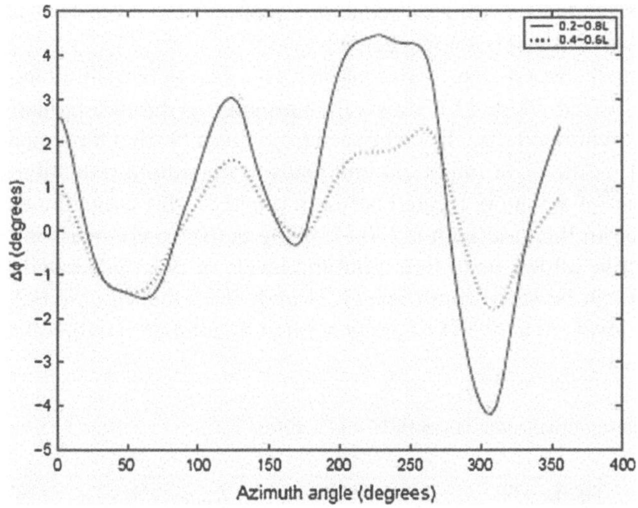

Fig. 11.25 Comparison of twist at $\mu = 0.35$ for optimum and 2 actuator failure cases

be noticed that, even if only one actuator (0.4–0.6L) is active and others have failed, sufficient levels of active twist are still generated to lead to reduction in vibration.

On closely looking at Figs. 11.26 and 11.27, one can note the variation in 4/rev hub loads for the optimum case with respect to the situation when only 0.4–0.6L is active. It can be observed that F_z and M_x are the loads which are affected most by the 0.4–0.6L actuator, leading to a reduction in vibration of 40 and 28 % respectively, at advance ratios of 0.3 and 0.35. The increment in F_z which takes place due to

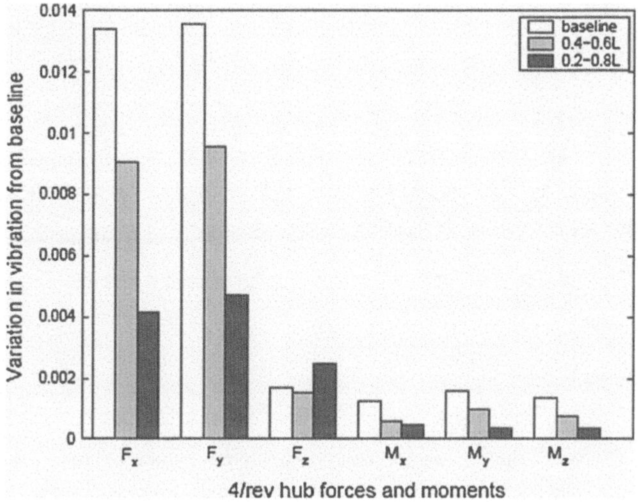

Fig. 11.26 Comparison of 4/rev hub loads at $\mu = 0.3$ for different cases

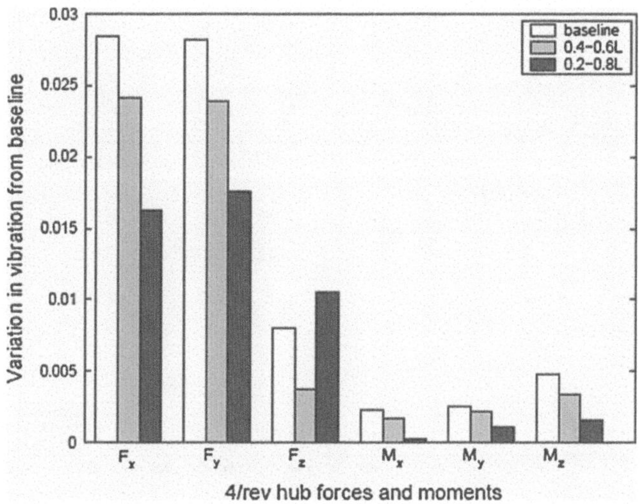

Fig. 11.27 Comparison of 4/rev hub loads at $\mu = 0.35$ for different cases

0.2–0.8L being active is eliminated when only 0.4–0.6L is active. Also, reduction in M_x is only marginally more when 0.2–0.8L are active as compared to 0.4–0.6L being active. This contributes to 0.4–0.6L being a good potential location. The variation of F_z and M_x with time over rotor disk is shown in Figs. 11.28, 11.29, 11.30, and 11.31 for advance ratios of 0.3 and 0.35, respectively. In Fig. 11.28, baseline variation of F_z with time shows a dominant 2P harmonic content. The other cases also show almost a 2P dominant harmonic content with a change in magnitudes, which results

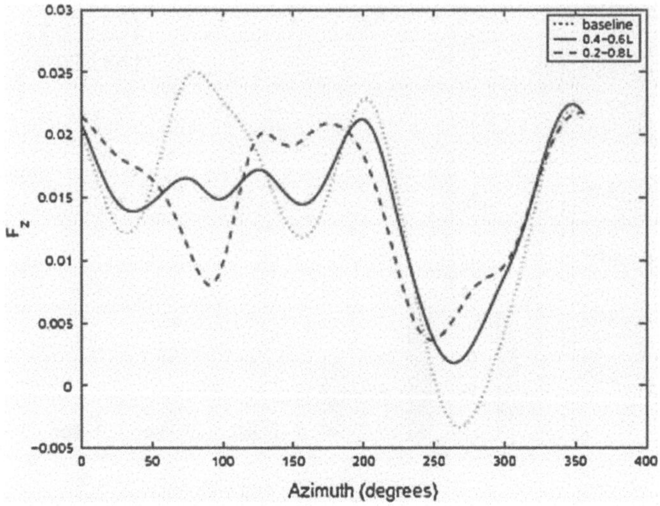

Fig. 11.28 Comparison of F_z at $\mu = 0.3$ for different cases

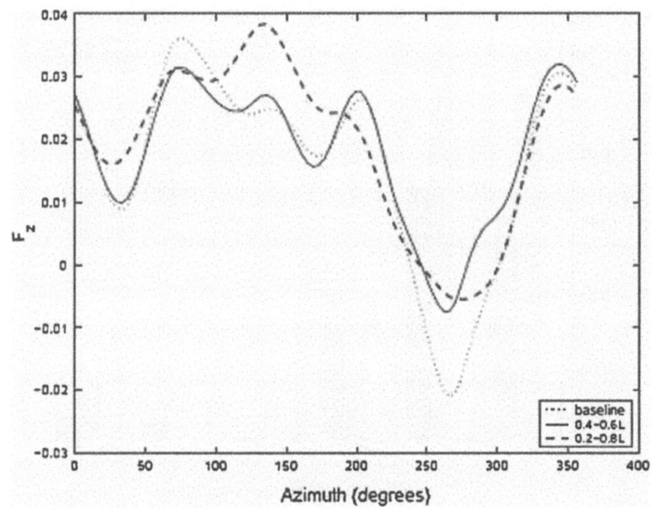

Fig. 11.29 Comparison of F_z at $\mu = 0.35$ for different cases

in reduction in vibration. Similar is the case for Fig. 11.29 with an advance ratio of 0.35. In Fig. 11.30, the baseline variation of M_x with time shows a dominant 1P component. The control cases reduce the amplitude at certain places which leads to reduction in vibration but are, however, not able to produce a smooth behavior.

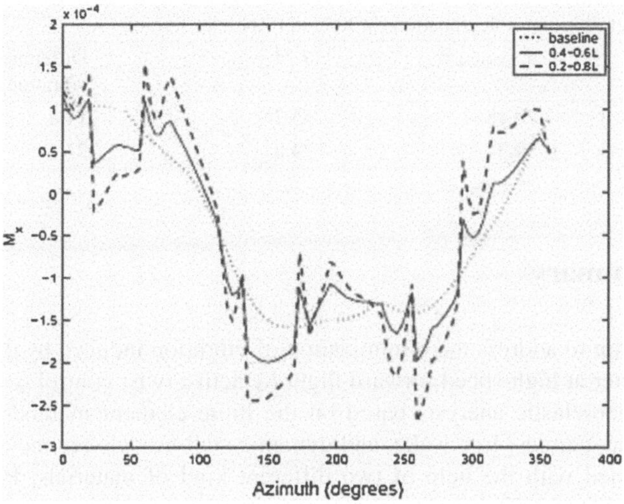

Fig. 11.30 Comparison of M_x at $\mu = 0.3$ for different cases

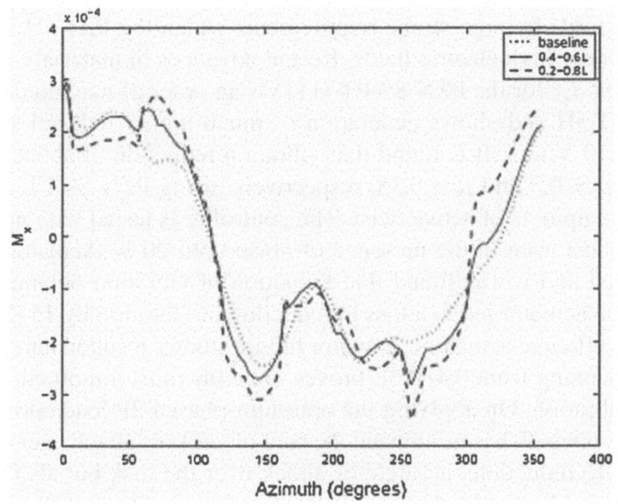

Fig. 11.31 Comparison of M_x at $\mu = 0.35$ for different cases

Table 11.7 shows the effect of an overall degradation in actuator authority by 30 % on vibration reduction. A loss in vibration reduction of 14.8 and 24.5 % is observed due to the degradation of the actuators at advance ratios of 0.3 and 0.35, respectively. Again, note that a vibration reduction of 56 and 35 % is possible at $\mu = 0.3$ and $\mu = 0.35$ even after considerable actuator degradation.

Table 11.7 Overall actuator degradation by 30 % in PZN-8%PT rotor

	$J_{optimum}$ %	J %	Difference (Loss in vibration reduction) %
$\mu = \mathbf{0.3}$	70.48	55.7	14.8
$\mu = \mathbf{0.35}$	59.3	34.8	24.5

11.11 Summary

A direct attempt to address the minimization of vibration induced by dynamic stall in the helicopter at high-speed forward flight by active twist control is made in this chapter. An aeroelastic analysis based on the finite element method is used and unsteady aerodynamics, free wake, and dynamic stall models are included. Active twist is obtained with the help of two different kind of materials; PZT-5H (soft ceramic) and PZN-8%PT (single-crystal ceramic). A strain rate feedback controller is developed. Optimum locations of the piezoceramic actuators are obtained to attain active twist in the rotor blade. It is found that the presence of dynamic stall at high-forward speed leads to large stroke requirements which the PZT-5H is not able to provide even under high electric fields. Recent advances in materials show that the shear coefficient d_{15} for the PZN-8%PT $\langle 111 \rangle$ is an order of magnitude higher than that of the PZT-5H and allows generation of much higher induced strain even at low fields of 210 V/mm. It is found that vibration reduction of about 70 and 59 % is obtained at μ = 0.3 and μ = 0.35, respectively, using PZN-8%PT which is also able to generate upto 4^0 of active twist. The controller is tested with noisy data and found to be robust even in the presence of noise upto 20 %. Actuator degradation was investigated and it was found that reduction in vibration obtained with 70 % efficiency of the actuator led to a loss in reduction in vibration by 15–25 % as compared to 100 % efficient actuators. Actuator failure studies resulted into a finding that the actuator spanning from 0.4–0.6L proves to be the most important for obtaining reduction in vibration. On applying the optimum-phased 2P load corresponding to the configuration which has minimized the sum of angle of attack over the rotor disk not only leads to reductions in angle of attack over the disk but also proves to be beneficial by providing reduction in vibration and contraction of the dynamic stall boundaries.

References

1. Li, S., Cao, W., Cross, L.: The extrinsic nature of nonlinear behavior observed in lead zirconate titanate ferroelectric ceramic. J. Appl. Phys. **69**(10), 7219–7224 (1991)
2. Peng, J., Chen, J., et al.: Shear-mode piezoelectric properties of 0:69$Pb(Mg_{1/3}Nb_{2/3})O_3 -$ 0.31$PbTiO_3$ single crystals. Solid State Commun. **130**, 53–57 (2004)

3. Park, S., Hackenberger, W.: High performance single crystal piezoelectrics: applications and issues. Curr. Opin. Solid State Mater. Sci. **6**, 11–18 (2002)
4. Thakkar, D., Ganguli, R.: Single crystal piezoceramics for dynamic stall suppression. Sens. Actuators: A Phys. **122**(1), 151–157 (2006)
5. Thakkar, D., Ganguli, R.: Use of single crystal and soft piezoceramics for alleviation of flow separation induced vibration in smart helicopter rotor. Smart Mater. Struct. **15**, 331–341 (2006)
6. Datta, A., Chopra, I.: Validation and understanding of UH-60A vibratory loads in steady level flight. In: 58th Annual Forum of the American Helicopter Society, Montreal, Canada (2002)

Appendix A
Kinetic and Strain Energy Terms

$$\frac{\delta U_{iso}}{m_0\Omega^2 R^3} = \int_0^1 (U_{u'_e}\delta u'_e + U_{v'}\delta v' + U_{w'}\delta w' + U_{v''}\delta v'' + U_{w''}\delta w'' + U_{\hat{\phi}}\delta\hat{\phi}$$
$$+ U_{\hat{\phi}'}\delta\hat{\phi}' + U_{\hat{\phi}''}\delta\hat{\phi}'')dx$$

$$\begin{aligned}
where;\quad U_{u'_e} &= EA\left[u'_e + k_A^2\theta'_0(\hat{\phi}' + w'v'') + k_A^2\frac{\hat{\phi}'^2}{2}\right] \\
&\quad - EAe_A[v''(\cos\theta_0 - \hat{\phi}\sin\theta_0) + w''(\sin\theta_0 + \hat{\phi}\cos\theta_0) \\
U_{v''} &= v''(EI_z\cos^2\theta_0 + EI_y\sin^2\theta_0) + w''(EI_z - EI_y)\cos\theta_0\sin\theta_0 \\
&\quad - EAe_Au'_e(\cos\theta_0 - \hat{\phi}\sin\theta_0) - \hat{\phi}'EB_2\theta'_0\cos\theta_0 \\
&\quad + w''\hat{\phi}(EI_z - EI_y)\cos 2\theta_0 - v''\hat{\phi}(EI_z - EI_y)\sin 2\theta_0 \\
&\quad + (GJ + EB_1\theta_0'^2)\hat{\phi}'w' + EAk_A^2\theta'_0w'u'_e \\
U_{w'} &= (GJ + EB_1\theta_0'^2)\hat{\phi}'v'' + EAk_A^2\theta'_0v''u'_e \\
U_{w''} &= w''(EI_y\cos^2\theta_0 + EI_z\sin^2\theta_0) + v''(EI_z - EI_y)\cos\theta_0\sin\theta_0 \\
&\quad - EAe_Au'_e(\sin\theta_0 + \hat{\phi}\cos\theta_0) - \hat{\phi}'EB_2\theta'_0\sin\theta_0 \\
&\quad + w''\hat{\phi}(EI_z - EI_y)\sin 2\theta_0 + v''\hat{\phi}(EI_z - EI_y)\cos 2\theta_0 \\
U_{\hat{\phi}} &= w''^2(EI_z - EI_y)\sin\theta_0\cos\theta_0 + v''w''(EI_z - EI_y)\cos 2\theta_0 \\
&\quad - v''^2(EI_z - EI_y)\sin\theta_0\cos\theta_0 \\
U_{\hat{\phi}'} &= GJ(\hat{\phi}' + w'v'') + EB_1\theta_0'^2\hat{\phi}' + EAk_A^2(\theta'_0 + \hat{\phi}')u'_e \\
&\quad - EB_2\theta'(v''\cos\theta_0 + w''\sin\theta_0) \\
U_{\hat{\phi}''} &= EC_1\hat{\phi}'' + EC_2(w''\cos\theta_0 - v''\sin\theta_0)
\end{aligned}$$

R. Ganguli et al., *Smart Helicopter Rotors*, Advances in Industrial Control,
DOI 10.1007/978-3-319-24768-7

$$\frac{\delta T}{m_0\Omega^2 R^3} = \int_0^1 m(T_{u_e}\delta u_e + T_v\delta v + T_w\delta w + T_{v'}\delta v' + T_{w'}\delta w' + T_\phi\delta\phi + T_F)dx$$

$$where; \quad T_{u_e} = x + u_e + 2\dot{v} - \ddot{u}_e$$

$$T_v = e_g(\cos\theta_0 + \ddot{\theta}_0\sin\theta_0) + v - \hat{\phi}e_g\sin\theta_0 + 2\dot{w}\beta_p + 2\dot{v}'e_g\cos\theta_0$$

$$+ 2\dot{w}'e_g\sin\theta_0 - \ddot{v} + \hat{\ddot{\phi}}e_g\sin\theta_0 - 2\dot{u}_e + 2\int_0^x (v'\dot{v}' + w'\dot{w}')d\xi$$

$$T_{v'} = -e_g(x\cos\theta_0 - \hat{\phi}x\sin\theta_0 + 2\dot{v}\cos\theta_0)$$

$$T_w = -x\beta_p - \ddot{\theta}_0 e_g\cos\theta_0 - 2\dot{v}\beta_p - \ddot{w} - \hat{\ddot{\phi}}e_g\cos\theta_0$$

$$T_{w'} = -e_g(x\sin\theta_0 + \hat{\phi}x\cos\theta_0 + 2\dot{v}\sin\theta_0)$$

$$T_{\hat{\phi}} = -k_m^2\hat{\ddot{\phi}} - (k_{m2}^2 - k_{m1}^2)\cos\theta_0\sin\theta_0 - x\beta_p e_g\cos\theta_0 - ve_g\sin\theta_0$$

$$+ v'xe_g\sin\theta_0 - w'xe_g\cos\theta_0 + \ddot{v}e_g\sin\theta_0 - \hat{\phi}(k_{m2}^2 - k_{m1}^2)\cos 2\theta_0$$

$$- \ddot{w}e_g\cos\theta_0 - k_m^2\ddot{\theta}_0$$

$$T_F = -(x + 2\dot{v})\int_0^x (v'\delta v' + w'\delta w')d\xi$$

The radial shortening term T_F can be written as

$$-\int_0^1 mT_F dx = \int_0^1 m(x + 2\dot{v})\left[\int_0^x (v'\delta v' + w'\delta w')d\xi\right]dx$$

Integrating by parts leads to a more convenient form

$$-\int_0^1 mT_F dx = \int_0^1 F_A(v'\delta v' + w'\delta w')dx + \int_0^1 (v'\delta v' + w'\delta w')\left[\int_x^1 2m\dot{v}\, d\xi\right]dx$$

where the centrifugal force F_A is defined as

$$F_A(x) = \int_1^x mx\, d\xi$$

which gives the "centrifugal stiffening" effect on the transverse bending and inplane bending equations.

Appendix B
Section Properties

$$A_0 = \int_\eta \int_\zeta E\ P\ d\eta d\zeta$$

$$A_1 = \int_\eta \int_\zeta E\ P(\eta^2 + \zeta^2)\ d\eta d\zeta$$

$$A_2 = \int_\eta \int_\zeta (E\ P\ \eta)\ d\eta d\zeta$$

$$A_3 = \int_\eta \int_\zeta (E\ P\ \zeta)\ d\eta d\zeta$$

$$A_4 = \int_\eta \int_\zeta (2\ G_{piezo}\ P_1\ \zeta)\ d\eta d\zeta$$

$$A_5 = \int_\eta \int_\zeta (E\ P\ \eta\ \zeta)\ d\eta d\zeta$$

$$EI_y = \int_\eta \int_\zeta E\zeta^2\ d\eta d\zeta$$

$$EI_z = \int_\eta \int_\zeta E\eta^2\ d\eta d\zeta$$

$$EA = \int_\eta \int_\zeta E\ d\eta d\zeta$$

$$GJ = \int_\eta \int_\zeta G(\hat{\eta}^2 + \hat{\zeta}^2)\ d\eta d\zeta$$

$$EB_1 = \int_\eta \int_\zeta E(\eta^2 + \zeta^2)\ d\eta d\zeta$$

$$EC_1 = \int_\eta \int_\zeta E{\lambda_T}^2\ d\eta d\zeta$$

$$EAe_A = \int_\eta \int_\zeta E\eta\ d\eta d\zeta$$

R. Ganguli et al., *Smart Helicopter Rotors*, Advances in Industrial Control,
DOI 10.1007/978-3-319-24768-7

$$EAk_A{}^2 = \int_\eta \int_\zeta E(\eta^2 + \zeta^2)\, d\eta d\zeta$$

$$EB_2 = \int_\eta \int_\zeta E\eta(\eta^2 + \zeta^2)\, d\eta d\zeta$$

$$EC_2 = \int_\eta \int_\zeta E\zeta\lambda_T\, d\eta d\zeta$$

where P and P_1 depend on piezoelectric properties and are given as $P = d_{31}E_3$ and $P_1 = d_{15}E_2$. However, A_0, A_1, A_2, A_3, A_4, A_5 depend on piezoelectric and section properties.

Blade sectional integrals are defined as

$$m = \int \int_A \rho_s d\eta d\zeta$$

$$me_g = \int \int_A \rho_s \eta d\eta d\zeta$$

$$mk_{m1}{}^2 = \int \int_A \rho_s \zeta^2 d\eta d\zeta$$

$$mk_{m2}{}^2 = \int \int_A \rho_s \eta^2 d\eta d\zeta$$

$$mk_m{}^2 = mk_{m1}{}^2 + mk_{m2}{}^2$$

The cross-sectional terms are nondimensionalized as follows:

$$\frac{EI_y}{m_0\Omega^2 R^4}$$

$$\frac{EI_z}{m_0\Omega^2 R^4}$$

$$\frac{GJ}{m_0\Omega^2 R^4}$$

$$\frac{EB_2}{m_0\Omega^2 R^5}$$

$$\frac{EC_2}{m_0\Omega^2 R^5}$$

$$\frac{EA}{m_0\Omega^2 R^2}$$

$$\frac{k_{m1}}{R}$$

$$\frac{k_{m2}}{R}$$

Appendix C
Shape Functions for Space

$$u(s) = \begin{Bmatrix} u(s) \\ v(s) \\ w(s) \\ \hat{\phi}(s) \end{Bmatrix} = \begin{bmatrix} H_u & 0 & 0 & 0 \\ 0 & H & 0 & 0 \\ 0 & 0 & H & 0 \\ 0 & 0 & 0 & H_{\hat{\phi}} \end{bmatrix} q_i$$

$$H_u{}^T = \begin{Bmatrix} H_{u1} \\ H_{u2} \\ H_{u3} \\ H_{u4} \end{Bmatrix} = \begin{Bmatrix} -4.5s^3 + 9s^2 - 5.5s + 1 \\ 13.5s^3 - 22.5s^2 + 9s \\ -13.5s^3 + 18s^2 - 4.5s \\ 4.5s^3 - 4.5s^2 + s \end{Bmatrix}$$

$$H^T = \begin{Bmatrix} H_1 \\ H_2 \\ H_3 \\ H_4 \end{Bmatrix} = \begin{Bmatrix} 2s^3 - 3s^2 + 1 \\ l_i(s^3 - 2s^2 + s) \\ -2s^3 + 3s^2 \\ l_i(s^3 - s^2) \end{Bmatrix}$$

$$H_{\hat{\phi}}{}^T = \begin{Bmatrix} H_{\hat{\phi}1} \\ H_{\hat{\phi}2} \\ H_{\hat{\phi}3} \end{Bmatrix} = \begin{Bmatrix} 2s^2 - 3s + 1 \\ -4s^2 + 4s \\ 2s^2 - s \end{Bmatrix}$$

where $q_i{}^T = \left[u_1, u_2, u_3, u_4, v_1, v_1', v_2, v_2', w_1, w_1', w_2, w_2', \hat{\phi}_1, \hat{\phi}_2, \hat{\phi}_3\right]$ and $s = \dfrac{x_i}{l_i}$

R. Ganguli et al., *Smart Helicopter Rotors*, Advances in Industrial Control,
DOI 10.1007/978-3-319-24768-7

Appendix D
Shape Functions for Time

$$\begin{Bmatrix} H_{t1}(s) \\ H_{t2}(s) \\ H_{t3}(s) \\ H_{t4}(s) \\ H_{t5}(s) \\ H_{t6}(s) \end{Bmatrix} = \begin{Bmatrix} \frac{1}{24}(-625s^5 + 1875s^4 - 2125s^3 + 1125s^2 - 274s + 24) \\ \frac{1}{24}(3125s^5 - 8750s^4 + 8875s^3 - 3850s^2 + 600s) \\ \frac{1}{12}(-3125s^5 + 8125s^4 - 7375s^3 + 2675s^2 - 300s) \\ \frac{1}{12}(3125s^5 - 7500s^4 + 6125s^3 - 1950s^2 + 200s) \\ \frac{1}{24}(-3125s^5 + 6875s^4 - 5125s^3 + 1525s^2 - 150s) \\ \frac{1}{24}(625s^5 - 1250s^4 + 875s^3 - 250s^2 + 24s) \end{Bmatrix}$$

R. Ganguli et al., *Smart Helicopter Rotors*, Advances in Industrial Control,
DOI 10.1007/978-3-319-24768-7

The manufacturer's authorised representative in the EU is Springer Nature Customer Service Centre GmbH, Europaplatz 3, 69115 Heidelberg, Germany. If you have any concerns regarding our products, please contact ProductSafety@springernature.com

Printed and bound by CPI Group (UK) Ltd, Croydon, CR0 4YY
29/07/2026
02180552-0003